Technological Innovations in Dryland Farming Systems

B.C. Sharma, Ph.D.
Division of Agronomy, Faculty of Agriculture,
Sher-e-Kashmir University of Agricultural Sciences and
Technology of Jammu, Chatha, Jammu-180009.

Anil Kumar, Ph.D.
Division of Agronomy, Faculty of Agriculture,
Sher-e-Kashmir University of Agricultural Sciences and
Technology of Jammu, Chatha, Jammu-180009.

Vikas Sharma, Ph.D.
Division of Soil Science and Agricultural Chemistry, Faculty of Agriculture,
Sher-e-Kashmir University of Agricultural Sciences and
Technology of Jammu, Chatha, Jammu-180009.

A.S. Bali, Ph.D.
Division of Agronomy, Faculty of Agriculture,
Sher-e-Kashmir University of Agricultural Sciences and
Technology of Jammu, Chatha,
Jammu-180009.

Published by

Khushnuma Complex Basement
7, Meerabai Marg (Behind Jawahar Bhawan)
Lucknow 226 001 U.P. (INDIA)
Tel. : 91-522-2209542, 2209543, 2209544, 2209545
Fax : 0522-4045308
E-Mail : ibdco@airtelmail.in

First Edition 2010

Price: Rs 1700.00

ISBN 978-81-8189-532-5

Composed & Designed at :
Panacea Computers
3rd Floor, Agrawal Sabha Bhawan
Subhash Mohal, Sadar Cantt. Lucknow-226 002
Phone : 0522-2483312, 9335927082, 9452295008
E-mail : prasgupt@rediffmail.com

Printed at:
Salasar Imaging Systems
C-7/5, Lawrence Road Industrial Area
Delhi - 110 035
Tel. : 011-27185653, 9810064311

Preface

Indian agriculture is predominantly a rainfed agriculture under which both dry farming and dry land agriculture is included. Dry farming was the earlier concept for which amount of rainfall (less then 500 mm annually) remained the deciding factor for more than 50 years. In modern concept, dry land areas are those where the balance of moisture is always on the deficit side. This deficit in moisture is bound to affect the crop production under dry land situation ultimately resulting into total or partial failure of the crops. Dry lands, besides being water deficient, are characterized by high evaporation rates, exceptionally high day temperature during summer, low humidity and high run off and soil erosion. Accordingly the production is either low or extremely uncertain and unstable which are the real problems of dry land in India.

India has about 108 million hectares of rain fed area which constitutes nearly 75% of the total 143 million hectares of arable land. These areas get an annual rainfall between 400 mm to 1000 mm which is unevenly distributed, highly uncertain and erratic. In certain areas the total annual rainfall does not exceed 500mm. About 47 million hectares of area comes under dry lands out of 108 million hectares of total rain fed area. Dry lands contribute 42% of the total food grain production of the country. These areas produce 75% of pulses and more than 90% of sorghum, millet, groundnut and pulses from arid and semi-arid regions. Thus, dry lands and rainfed farming will continue to play a dominant role in agricultural production.

The farmers in these areas are caught in the vicious cycle of poverty. Innovative technologies are the need of the hour for mitigating the miseries of the resource poor farmers. These technologies by and large include an integration of all enterprises be it crop husbandry, alternate land use systems, and other income generating units at the farm level.

Yield potential of irrigated areas has been more or less realized, however, a vast scope lies in enhancing production from the dry land areas. The future escalating food requirements will have to be met through the realization of the maximum yield potential from these areas. For achieving these targets we will have to be innovative in integrating the different agricultural sciences from water conservation to biotechnological interventions. The present book is an effort to integrate the traditional and modern technologies of dryland agriculture for maximizing the profitability, productivity and sustainability of dryland farming areas.

In this endeavour, we sincerely acknowledge the contributors for their valuable inputs covering wide aspects of dryland agriculture. We are grateful to our family members for their cooperation and understanding during the long hours spent in the preparation of the manuscript.

A deep sense of gratitude goes to our Honorable Vice Chancellor, Dr. B. Mishra for creating conducive environment to take up devotional assignment of this kind.

Appreciation is must for International Book Distributing Co, Lucknow in general and Mr. Suneel Gomber in particular for taking personal interest and putting great efforts in giving practical shape to our thoughts and ideas. Suggestions for improvement of this book will be acknowledged with similar enthusiasm.

2009 **Editors**

Contents

CHAPTER 1

Dry Land Farming: Bridging the Gap Between Research and Practice

U. C. Sharma[1] and Vishal Sharma[2]

[1]*Vice President, international Commission on Water Quality, IAHS.*
[2]*Research Associate, Division of Agronomy, CSKHPKV, Palampur, HP.*

Water is one of the most critical inputs in agriculture. Rain or irrigation water applied to plants is not utilized directly by the plants but it is first stored as soil water. The soil is capable of retaining water against the forces of gravity and applied pressure or suctions. Terrestrial plants extend roots into the soil while their canopy grows in the atmosphere. They take water from the soil through their roots and transport it through their stems and leaves to the atmosphere. To remain fully turgid, the plant foliage must lose water fast enough to meet the demand of the ambient atmosphere. When the transpirational loss falls short of the evaporative demand, the plant develops water stress. So, for producing plant food, huge amount of water is required. By 2021 A.D., India will have to produce about 300 million tonnes of food grains, besides other economic crops, to feed about 1420 million population. This target cannot be realized from irrigated areas alone as we have irrigation potential for 178 million hectares only. Therefore, there is need for appropriate technologies for dry land farming to increase and sustain crop yields under such situations. Second 'green revolution' in Indian agriculture can be realized from rainfed/dry land agriculture. This is important to improve the standard of living of farmers inhabiting these areas. It would require concerted efforts on all fronts viz., evolving appropriate technologies for improving productivity in dry land agriculture, assure timely availability of required inputs to the farmers, gearing-up of extension agencies, water conservation measures as well as to make judicious use of available water resources.

Introduction

Droughts are extreme events characterized by persistency in time and large scale effects. It is a creeping hydrological phenomenon due to below average rainfall. Drought occurrence and severity is not strictly due to climatic fluctuations alone but may include a sudden increase in demand for water because of socio-economic growth and development over space and time, which is beyond the water supply capacity. Droughts are usually hazardous and result in disruption of economic, social and institutional set-up depending on its duration, deficit and impacts on users. Quantitative indicators are needed to determine the timing, duration and severity of droughts. The indicators and methods should take into account hydrological characteristics of the region for which they are intended.

About 108 million ha of the total 143 million ha of arable land in India is rainfed (constituting about 75.5%). India has about 47 million hectares of dry lands out of 108 million hectares of total rainfed area. Thus, dry lands and rainfed farming will continue to play a dominant role in agricultural production. Dry lands, besides being water deficient, are characterized by high evaporation rates, exceptionally high day temperature during summer, low humidity and high runoff and soil erosion. As water is the most important factor of crop production, inadequacy and uncertainty of rainfall often cause partial or complete failure of the crops which leads to period of scarcities and famines. Thus the life of both human being and animals in such areas becomes difficult and insecure. In such areas crop production becomes relatively difficult as it mainly depends upon intensity and frequency of rainfall. The crop production, therefore, in such areas is called rainfed farming as there is no facility to give any irrigation, and even protective or life saving irrigation is not possible. These areas get an annual rainfall between 400 mm to 1000 mm which is unevenly distributed, highly uncertain and erratic. In certain areas the total annual rainfall does not exceed 500mm.

The kandi region of Jammu, with a population of over 3.0 million, is characterized by low rainfall (<950 mm, annually), undulating topography, deep water table, low soil organic matter, light soil texture, land infested with sand stones and frequent droughts. Extreme water stress is experienced during summers and winters and even water for drinking becomes scarce. Due to low rainfall, high evapotranspiration and surface and sub-surface flows, the groundwater recharge in the region is poor. Biophysical features, such as fragility, marginality, low accessibility and resource heterogeneity, are constraints in groundwater management in the region. Frequent droughts cause low agricultural production and dependence on outside sources for food grains and other commodities. Whole economy of the region is in peril due to deficit in rainfall as well as lack of awareness about proper conservation, utilization and management of water resources, especially rainwater. The region is affected by droughts, lasting for 1 to 4 years with a return period of 2 to 3 years, on an average. The human use of groundwater has taken precedence over its environmental impacts. Due to sudden increase in water requirement and unthoughtful planning to satisfy the demand, the practice has left the legacy of degraded aquifers, land subsidence and ecological damage (Sharma, 2001, 2003). The consequences are economic losses, environmental degradation, desertification, soil impoverishment and social disorder. There is adverse effect on water supply schemes due to reduced flow in seasonal streams and reduced aquifer recharge. Until now, no systematic study has been undertaken on occurrence of droughts, mitigation, groundwater recharge and their socio-economic implications. Estimation of drought frequencies in the *Kandi* region is important for groundwater assessment and management.

Despite improvements in agriculture, we have yet not been able to evolve an appropriate package of practices for dry land areas. The income of farmers of dry land regions is still very low.

Features of dry land farming

a. Low and uncertain rainfall, varying in time and space.
b. Droughts are common in the dry land areas. The return period of drought occurrence may vary from place to place. Sometimes the rains are of high intensity, resulting in flood events.
c. The dry land areas have generally undulating topography, run-off is more and *in-situ* retention of rainwater is considerably low.
d. Practice of extensive agriculture or prevalence of mono-cropping is common in dry land areas. Sometimes these areas are left fallow for two to three cropping seasons as in the *Kandi* region of Jammu.
e. The choice of crops is limited.
f. Very low crop productivity due to poor soil moisture conditions and low crop response to applied fertilizers.
g. Similarity of crops raised by the farmers.
h. Poor economic conditions.
i. Poor health of farmers, particularly of women folk and children, due to malnutrition and availability of healthy food.

Problems associated with dry land farming

The major problem which the farmers have to face very often is to keep the crop plants alive and to get some economic returns from the crop production. But this single problem is influenced by several factors such as moisture stress during the growing period of crops.

- According to definition, the dry farming areas receive an annual rainfall of 500 mm or even less. The rains are very erratic, uncertain and having temporal and spatial variations. Therefore, the agriculture in these areas has become a sort of gamble with the nature and very often the crops have to face climatic hazards. The farmers also take up farming half-heartedly as they are not sure of economic returns.
- According to characteristics of dry farming, either there will be no rain at all or there will be torrential rain with very high intensity. Thus, in the former case the crops will have to suffer a severe drought and in the latter case they suffer either flood or water logging and they will be spoilt. In case of very

heavy downpour, the excess water gets lost as run-off. The loss of water takes place in several ways namely run-off, evaporation, uptake through weeds etc.

- Non-availability of appropriate technologies for dry land areas. The technologies generated so far have often failed to live up to moisture stress situations, giving rise to famine like situations.
- There is no optimum response of crops to applied inputs due to moisture stress.

Droughts

Drought is a climatic anomaly, characterized by deficient supply of moisture resulting either from sub-normal rainfall, erratic rainfall distribution, higher water needs or a combination of all the three factors. In agriculture, the drought has been categorized into three types viz., meteorological drought, hydrological drought and agricultural drought.

Meteorological drought

It is a situation when there is significant (> 25%) decrease from normal rainfall over an area.

Hydrological drought

Prolonged meteorological drought results in hydrological drought. There is marked decrease in water in streams, low flows and drying up of reservoirs, lakes and rivers.

Agricultural drought

It occurs when soil moisture and rainfall are inadequate to support a healthy crop growth during the growing season. The US Weather Bureau defines drought as a period of dry weather of sufficient length and severity to cause at least partial crop failure. The National Oceanic and atmospheric Administration defines agricultural drought as a combination of temperature and precipitation over a period of several months leading to substantial reduction (< 90%) in yield. According to British Rainfall Organization, a partial drought is a period of more than 28 days with very small rain per day and absolute drought is a period of at least 15 consecutive days during which the rain does not exceed 0.25 mm. The Australian Bureau of Meteorology mentions that drought occurs when rainfall during the interval is less than the minimum water need. In USSR, drought is defined as a period of 10 days with a total rainfall of not exceeding 5 mm (Ramakrishna et al., 2000).

Probability of drought

The probability of drought occurrence in Indian states is given in Table 1. As expected, the probabilities are higher in arid or semi-arid regions.

Table 1. Probability of occurrence of drought in different meteorological sub-divisions

Meteorological sub-division	Frequency of deficient rainfall (75% of normal or less)
Assam	Very rare, once in 15 years
West Bengal, Madhya Pradesh, Konkan, Bihar and Orrisa	Once in 5 years
South interior Karnatka, Eastern Uttar Pradesh and Vidarbha	Once in 4 years
Gujrat, East Rajsthan, Western Uttar Pradesh, Tamil Nadu, Jammu and Kashmir and Telengana	Once in 3 years
West Rajsthan	Once in 2.5 years

Source: Apparao et al., 1991

Variation in yield

Rainfall is the most significant climate factor that affects the crop growth and production from the rainfed areas in the country. Table 2 shows the variation in the yield of major crops grown in India under irrigated and rainfed conditions. Although moisture supply is the basic constraint in farming in the dry areas and little can be done to increase rainfall, there are good opportunities to more effectively use the rainwater (Gupta et al., 2000). Rainwater storage, harvesting and management seems to be the key elements to improve productivity in dry areas.

Table 2. Yield of principal crops under irrigated and rainfed conditions (kg/ha)
Central Water Commission, 1995

Crop	Irrigated	Un-irrigated	% increase over un-irrigated
Rice	1880	1220	54.1
Sorghum	1242	607	104.7
Pearl millet	1170	596	96.2
Maize	2040	1339	52.4
Ragi	1966	996	97.5
Wheat	2068	1100	88.0
Barley	1836	1127	62.9
Gram	830	548	51.3

Groundnut	1244	844	47.3
Sugarcane	70687	43161	63.6
Rape seed and mustard	893	573	55.9
Cotton	440	195	125.7
Jute	1952	1502	29.9

The Kandi region of Jammu

Availability of water

Major factors affecting water availability in the *kandi* region of Jammu include; type, amount and distribution of precipitation, land use, initial soil moisture, soil infiltration and slope. Meteorological factors as amount and duration of rainfall, temperature, *in-situ* retention of rain water, rate of infiltration and amount of run-off as well as aquifer recharge capacity are important indicators of ground water sustainability in the region. About 41.16 km^3 of rainwater is received annually in kandi region (Table 3). Changes in groundwater regime in the region are linked with climate fluctuations as well as environmental changes over time. Groundwater is, by and large, used for drinking as it is not available in sufficient quantity for irrigation. For extracting groundwater in rural areas, the wells are dug and the water is used by the community (Sharma, 1999). Generally, there are 1 to 4 wells in a village of about 100 to 200 houses.

Table 3. Annual rainwater received in Jammu and Kashmir

Region	Rainwater
Kashmir	10.36
Jammu	20.95
Ladakh and other parts	9.85
Total	41.16
Kandi	4.08

The wells are de-silted and cleaned annually for assured water supply throughout the year. During summers, many wells dry-up and the people find it extremely difficult to get drinking water. The women in rural areas have to bring water for drinking from long distances, carrying earthen and metallic pitchers on their heads. This drudgery is continuing since centuries and need immediate redressal to avoid physical torture (Sharma, 2002). In many areas, the rainwater is harvested and collected in big ponds. Common community land is kept as catchment area and the people are not allowed to pollute this area in any way. The defaulters are fined heavily. However, with the introduction of water supplies from the tube-wells in many areas, the rainwater harvesting has received a set-back. These rainwater harvesting structures are, now, in dilapidated condition and the catchments are damaged and polluted. The well water is frequently contaminated and diseases like dysentery, diarrhea, cholera etc. are not uncommon. The water supply in the region is constrained by natural water scarcity, conflicting demands, intensive development, sudden spurt in population growth, dilapidated infrastructure, allocation among users in time and space, aquifer recharge capacity, contamination and land subsidence. There is a need for a policy to regulate water system covering; planning, construction and operation of hydraulic infrastructure to ensure perennial water supplies. A better relationship between functionaries and users would ensure better coordination.

Rainfall variability index

The water levels of aquifers in the *Kandi* region reach their peak from August to October. High intensity rainfall causes huge soil erosion through runoff from the hills and undulating areas, causing silting of stream beds. The rainfall variability index (RVI) was calculated to know about the variability in groundwater recharge over time (Fig 1). The RVI gives the normalized annual departure and is helpful in giving the status of ground water. The RVI varied from –1.40 to 2.62 in the region. No particular trend in rainfall was observed with regard to deficit or surplus years. The deficit years were 1979, 1980, 1982, 1984, 1987, 1989, 1991, 1992, 1993, 1995, 1999, 2000, 2001 and 2002 while during all other years the rainfall was normal. Some rain water goes away as surface runoff and some infiltrates into the soil zone. The soil continues to store more water until it reaches field capacity. At this point, soil begins to drain and recharge can occur. Recharge may not occur at the same location where the rainfall is received. If the streams which collect the water run into aquifers, runoff recharge occurs. Type and amount of distribution of rainfall, infiltration characteristics, initial soil moisture content and topography has influence on the ground water recharge. The temperature variation is an important factor in ground water recharge and so is the sun-shine hours which affect evaporation and transpiration losses. The evaporation losses are more from March to June when sun-shine hours vary from 7.3 to 9.5 hours per day on an average. *In-Situ* retention of rainwater would depend on the nature of vegetation, soil characteristics and land management practices such as soil conservation measures.

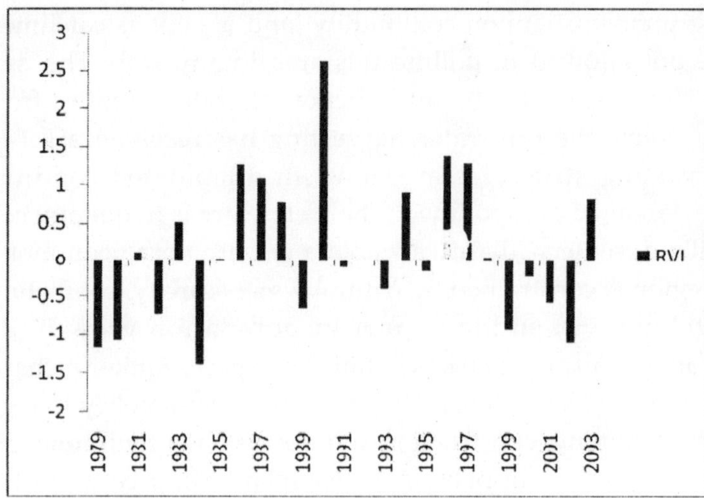

Fig. 1. Rainfall variability index in *kandi* region

Droughts in Kandi region

Depending on the nature of the water deficit, drought can be classified as meteorological (precipitation scarcity) and hydrological (soil moisture, stream flow etc.). Spatial distribution of drought is a relevant characteristics for drought analysis and this demands comprehensive tools for drought studies. Meteorological drought is a creeping hydrological extreme phenomenon characterized by an appreciable below average precipitation, the magnitude of which can be assessed by its duration, deficit and impact on users. It is mainly a demand and supply issue, resulting from rainfall deficiency over a marked period of time. The severity of drought may not necessarily be due to consequences of climatic fluctuations but, may also be due to sudden increase in demand as a result of increase in population, irrigation requirements or other socio-economic reasons.

Maximum temperature during summers (May-September) crosses 40 ^{0}C in June-July on some days, causing huge evaporation from the water and soil surface. However, average highest monthly maximum and minimum temperature was 38.9 and 26.0 ^{0}C, recorded in June. The average evaporation loss was higher in May-June compared to July-August because of high temperature and more sunshine hours during this period. The climatic data helps in modelling aridity index and draw conclusions for better understanding. The evaporation losses were more than rainfall throughout the year, except in July and August. This is the period when groundwater recharge can be expected.

Drought management in Kandi

Socio-economic conditions and land use

Management solutions should aim at the restoration of the more natural, dynamic behaviour of the water system, because this will minimize the volume of water discharged from the area and maximize water conservation. Dynamic surface water control implies that greater water level fluctuations are allowed within the constraints of agricultural activity and safety (Vermeulen et al., 2001). The underlying socio-economic basis for water development in the *kandi* region is characterized by the disparity between urban and rural area needs, socio-cultural rituals with regard to land use and management of rain water. About 11.04 million tonnes of soil and 12.22, 1.46, 9.23, 0.34, 0 .23, 0.85 and 0.66 thousand tonnes of soil, N, P, K, Mn, Zn, Ca and Mg is displaced every year from the region (Table 4). During the last 50 years, attention is being given to improve agriculture in the region to feed fast growing population resulting in higher water demands. Changes in land use over time coupled with their impacts on water holding capacity of soils have affected GWR in the region and drought management.

Policy domain

There is a need for a policy from government side to regulate water system covering; planning, construction and operation of hydraulic infrastructure to ensure water supplies to various users. Inter-related users can be subjected to prioritization scheme in the form of differential pricing and allocation for different user as well as conflict resolution. Such a regulation is normally better if the physical water basin is taken as the basis of water management unit.

Table 4. Annual approximate loss of soil (million tonnes) and nutrients (thousand tonnes) from *Kandi* region.

Parameter	Annual loss	Parameter	Annual loss
Soil loss	11.04	Mn	0.34
N	12.22	Zn	0.23
P	1.46	Ca	0.85
K	9.23	Mg	0.66

Government of India policy on water

Comprehensive and dynamic water policy has remained a debatable issue and two national water polices have been introduced by the government of India (Government of India, 1987, 2002). An outline of major features, intra-alia, of water management policies is given in Table 5. Improper policy and management can induce extreme weather events, degradation and pollution of ground water and change in hydrological cycle. A cordial relationship between the people and policy makers is necessary to honour the legal provisions embodied in government policies on water resources management and their implementation in letter and spirit has to be ensured to avoid excessive groundwater extraction and harmful aquifer depletion. The institutional framework for the management of water resources would include the organization that establish the rules on the use of water resources, the legislative bodies and agencies with regulatory or political functions and responsibilities. Though the rules say that ground water exploitation should not exceed recharge, over-exploitation is going on unabated. A better relationship between functionaries and users would avoid conflictual situations and ensure better coordination. The coordination may be difficult as a regional area is influenced by the climate, physiography, geology, hydrology, soils, plants, and people with invisible elements like cultures, traditions and pattern of governance (Ghosh, 2003).

Table 5. Major features of water policies of 1987 and 2002 in India related to water management (Outline)

1987	2002
1. A periodic assessment of the ground water on scientific basis is stressed, taking into account the quality of water available and economic viability. Exploitation of ground water should not exceed the recharge possibilities. Integrated and coordinated development of surface water and ground water should be envisaged. Over-exploitation should be avoided.	In addition, the state and central governments should effectively prevent over-exploitation of ground water. Ground water recharge projects should be developed and implemented for improving the quality and availability of ground water resources.
2. Water resources planning must be based on hydrological units, drainage basins or sub-basins. Recycling and re-use of water should be an integral part of water resource development.	In addition, non-conventional methods of water resources use such as inter-basin transfers, artificial recharge of ground water and desalinization as well as traditional practices of rain water harvesting, need to be used.
3. Both, surface water and ground water should be regularly monitored for quality. A phased programme should be undertaken for improvement in water quality.	In addition, the effluents should be treated to acceptable levels and standard before discharging them to natural streams. Minimum flow should be maintained in perennial streams for ecology and social considerations.
4. Water conservation consciousness should be promoted through education, regulation, incentives and disincentives.	In addition, resources should be conserved and availability be augmented by maximizing retention, eliminating pollution and minimizing losses.

5. A master plan for flood control and management for each flood prone basin to be prepared.	In addition, strict regulation of settlements and economic activity in the flood plain projects. The flood forecasting activities to be modernized.
6. Water rates should be such as to convey the scarcity value of the resource to the users and to faster the motivation for economy in water use. Water rates for surface and ground water should be rationalized.	The water charges should cover at least the operation and maintenance charge of providing the service initially, and a part of the capital costs subsequently.
7. Soil moisture conservation measures, water harvesting practices, minimization of evaporation losses, development of groundwater potential and transfer of water from surplus areas, if feasible.	In addition, relief works undertaken for providing employment to drought stricken communities, drought prone area development.
8. Participation of farmers and voluntary agencies in various aspects of water management.	Water resources management for diverse uses should incorporate a participatory approach.

Raising productivity in dryland Farming

Interdisciplinary approaches

To boost the crop production under dry farming, there is need to efficiently manage the soil and water resources in the respective areas as dry land farming gets more complex and intractable when droughts occur frequently. An efficient soil and water conservation system will play a vital role in boosting the crop yield in dry farming. Interdisciplinary approaches are aimed at soil and moisture conservation through regulation of run-off, collection of surplus rain water checking evaporation and seepage losses of water. Contouring is practiced on the lands with 3-5% slope. This system consists of constructing earthen bunds and the distance between the two bunds ranges from 30-50 m depending on the degree of slope. This is carried out with an objective to provide a check to the flow of run-off water, which then gets accumulated in the bunded area and is absorbed by the soil. The entire land is laid out into ridges and furrows across the slope. The ridges and furrows are opened before onset of monsoon so that the flow of water may be reduced and erosion may be controlled to the minimum. During rainy season, crops like maize, jowar, bajra, etc. may be grown in the furrows and legumes like soybean, arhar, urd, mung, cowpea, etc. may be grown on the ridges. After the monsoon is over the land is again levelled. This way the furrows are used to accumulate maximum water which will supply moisture for winter season crops.

Cropping systems and patterns

Cropping system refers to an arrangement in which various crops are grown in the same field. The cropping systems followed in dry lands differ from those followed under normal conditions. Only those crops can be grown under dry land conditions which require less water to complete their life cycle or which can stand or yield under drought conditions. This can include both drought resistant and drought tolerant plants. In addition, plants can be grown only where some water is available to sustain the growth of plants. Mixed cropping is also followed to minimize the effect of unpredictability of rain. Mixed cropping may have low yield potential but it works as a buffer against failure under possible unfavourable conditions. Mixed cropping may be defined as sowing of two or more crops simultaneously by mixing their seeds in a definite ratio on the same piece of land. Cropping pattern is defined as sequence of growing crops in a particular field at a particular period. The most common cropping systems, patterns and mixed cropping examples for dry land farming are mentioned hereafter and could be adopted as per suitability.

Maize/bajra + moong – fallow

Maize/bajra + moong/urd – wheat + gram/ linseed

Moong/urd-wheat/barley

Til-wheat+gram

Fallow-wheat/barley

Til+guar/moth/mung

Sorghum-safflower/mustard

Sorghum/mung/urd/cowpea-gram/wheat

Bajra-gram+linseed

Bajra+urd/mung/soyabean-wheat/barley+gram/mustard

Maize-gram/safflower

Conservation of rain-water

Water conservation basically aims at matching demand and supply The strategies for water conservation may be either demand, supply or management oriented. The strategy worth pursuing in the *Kandi* region of Jammu is to reserve a part of the groundwater to meet the drinking water demand in the rural areas and this need not be exploited for irrigation. There is urgent need to harvest the rainwater and store it in big tanks and concrete structures. This water can be used for drinking after proper treatment as well for life saving irrigation to crops at critical stages of growth. This will also avoid over-dependence on groundwater and reduce runoff and sediment load. There is substantial depletion of groundwater due to excessive withdrawals and natural geological formations. The rainwater harvesting is an ecological need in the *Kandi* area or for those areas where availability is poor.

Recommendations for dry farming areas

a. Bunding across the slope and levelling of the land should be done before the commencement of monsoon season.

b. Deep summer ploughings should be followed by surface tillage.

c. Applications of organic manures 20-25 days before sowing and incorporated well in the soil.

d. Placement of fertilizers at a depth of 7.5 to 10 cm in the soil and sowing of seeds about 3 cm above in the furrow.

e. Soil application of BHC (10%) dust @ 25-30 kg/ha for termites and Thimet 20 G @ 15 kg/ha for white grub should be done. These chemicals must be mixed with soil properly while ploughing or at the time of sowing.

f. Sowing of crops and varieties requiring less water for optimum productivity.

g. Proper crop rotation, preferably having one legume crop.

h. Planking the soil after ploughing to conserve soil moisture.

i. Mixed and inter-cropping to ensure at least one crop under adverse conditions.

j. Ensuring proper plant population.

k. Proper weed management.

l. Mulching.

m. More in-situ retention of rainwater by having proper vegetation cover.

n. Rainwater harvesting.

o. Judicious and purposeful utilization of preserved moisture water depends upon soil type, plant type and other factors. The amount of available water to the plants depends upon the depth of plant roots, their proliferation and density. In case of limited moisture condition, the yield directly depends upon the rooting depth. The rooting depth can be desirably increased by mechanical manipulation of the soil. If the planting is very densed and all the plants have same kind of rooting then there will be a tough competition among roots for moisture and scarce moisture condition will result in the wilting of plants. Therefore, utilization of preserved moisture is an art in dry farming. The water collected in ponds or brooks may be used to give protective or life saving irrigation. The widely spaced crops can be intercropped with oilseeds or pulses for increasing the productivity of the land per unit area and per unit time.

p. Harvesting the crop at proper physiological stage to sow the succeeding crop in time.

Conclusions

Drought is a random, low profile natural phenomenon that evolves slowly and requires adequate risk management mechanism for mitigation of its severe effect. Mitigation actions also need to be focused on social causes of drought. While the rainfall is a natural physical event and its amount and intensity cannot be controlled, judicious management and use of rainwater can effectively be done. The land use can be manipulated by having proper cropping patterns and soil and water conservation measures for higher *in-situ* rainwater retention and reduction in runoff to increase groundwater recharge. The future strategies must entail rainwater harvesting as an important aspect of water management in the *kandi* region of Jammu Province. The harvested water can be used for drinking after treatment and also as life-saving irrigation during off-season at the critical sages of plant growth. A well structured institutional framework and efficient capacity building is necessary for efficient drought management in the region. Estimation of drought frequencies is important for efficient water management, reservoir design and management and crop planning. Even after utilizing all the available water resources, about 50% of our culturable area will still depend on rains. Therefore, our agricultural scientists, policy formulators and farmers should appropriately realize the magnitude of role that rainfed agriculture or dryland farming can play. They should thoroughly examine the problems of dry land agriculture from different view points and evolve appropriate technologies, crop varieties, etc. for these areas to better the economic position of the farmers. Dry farming areas, therefore, need a much closer attention. Farmers should utilize well in time whatever improved technology and varieties suitable for dry farming are available. They should be extra careful about the utilization of available rain water, selection of crops and protection of crops from different harmful physiological or biological agencies.

References

Apparao, G. (1991) Droughts and southwest monsoon. Training Course in 'Monsoon Meteorology' 3rd WMO Asian / African Monsoon Workshop, Pune, India.

Central Water Commission (1995) Water and related statistics, 343p, Central Water Commission, New Delhi.

Ghosh, S. (2003) Urban Water Management – Asian Cities. In: Hydrology and Water Resources in Asia – Pacific Region (ed. by K. Takara and T. Kojima), 400-402. Proc. First Int. Conf. of APHW, Kyoto, Japan.

Government of India (1987) National Water Policy. Ministry of Water Resources, Government of India, New Delhi, India.

Government of India (2002) National Water Policy. Ministry of Water Resources, Government of India, New Delhi, India.

Gupta, S.K., Minhas, P.S., Sondhi, S.K., Tyagi, N.K. & Yadav, J.S.P. (2000) Water Resources Management. In: Natural Resources Management for Agricultural Production in India. (J.S.P.Yadav and G.B.Singh eds.), pp 137-244, New Delhi, India.

Ramakrishna, Y.S., Rao, G.G.S.N, Kesava, A.V.R and Vijaya kumar, P. (2000) Weather resources management. In: Natural Resources Management for Agricultural Production in India. (J.S.P.Yadav and G.B. Singh eds.), pp 245-370, New Delhi, India.

Sharma, U. C. (1999) Loss of N through leaching and runoff from two potato-based land use systems on different soils. In: Impact of Land Use Change on Nutrient Loads from Diffuse Sources (ed. by L. Heathwaite), 27-32. IAHS Publ. 257, IAHS Press, Wallingford, UK.

Sharma, U. C. (2001) Effects on farming system type on in- situ groundwater recharge and quality in northeast India. In: Impact of Human Activity on Groundwater Dynamics (ed. by H. Gehrels, N.E. Peters, E. Hoehn, K. Jensen, C. Liebundgut, J. Griffioen, B. Webb & W.J. Zaadnoordijk), 167-169., IAHS Publ. 269, IAHS Press, Wallingford, UK.

Sharma, U. C. (2002). Managing the fragile hydrological ecosystem of the northeast hilly region of India for resources conservation and improved productivity. In: Regional Hydrology: Bridging the Gap between Research and Practice (ed. by H.A. Van Lanen & S. Demuth) 357-364, IAHS Publ. 274, IAHS Press, Wallingford, UK. IAHS Press, Wallingford, UK.

Sharma, U. C. (2003). Impact of population growth and climatic change in the quantity and quality of water resources in the northeast of India. In: Water Resource Systems-Hydrological Risks, Management and Development (ed. By G. Bloschl, S. Franks, M. Kumagai, K. Musiake & D. Rosbjerg , 349-397, IAHS Publ.281, IAHS Press, Wallingford, U.K.

Vermeulen, P., Gehrels, H. Stroet, C. T. & Kremers, T . (2001) Modelling the impact of surface water management on water conservation and water quality. In: Impact of human activity on ground water dynamics (ed. By H. Gehrels, N.E. Peters, E: Hoehn, K. Jensen, C. Leibundgut, J. Griffioen, B. Webb & W.J, Zaadnoordijk), 173-186 IAHS Publ. No. 269, IAHS Press, Wallingford, UK.

CHAPTER

2 Integrated Plant Nutrient Supply System for Productivity Enhancement in Rainfed Agro-ecosystems

Anil Kumar[1], B. C. Sharma[1] and Vikas Sharma[2]

[1]*Division of Agronomy, Faculty of Agriculture, Sher-e-Kashmir University of Agricultural Sciences and Technology of Jammu, Chatha, Jammu-180009.*

[2]*Division of Soil Science and Agricultural Chemistry, Faculty of Agriculture, Sher-e-Kashmir University of Agricultural Sciences and Technology of Jammu,Chatha, Jammu-180009.*

Agriculture will be heavily burdened to feed a world population projected to exceed eight billion by the year 2020. The synergy between the application of inorganic fertilizer and the development of nutrient-responsive modern seed varieties was in no small part responsible for the phenomenal growth in crop yields and food supplies in developed countries over the past thirty-five years. The ability of agriculture to provide for food needs to the year 2020 and beyond is becoming increasingly difficult. In developed countries, over-application of inorganic and organic fertilizers has led to environmental damage, while in developing countries, population pressures, land constraints, and the decline of traditional soil management practices have led to a decline in the fertility of the soil. The integrated management of plant nutrient resources through soil conservation practices and the widespread and responsible use of organic and inorganic fertilizers offers the opportunity to sustain agriculture over the long-term and to maintain and enhance soil fertility, while minimizing any environmental degradation. For Integrated Plant Nutrient Management (IPNM) to be successful at the field level, the farmer must have the knowledge, tools, and support necessary to effectively manage his fields, crops, and soil to their maximum potential.

India's food grain needs

In 1951, we were only 361 million people and now, we are beyond 1.2 billion, nearly trebling our population in the second half of twentieth century. A well-planned and concerted effort by agricultural scientists, extension workers, farmers and as well as the price support policies of the government have made 'green revolution' realizable by increasing grain production at a rate faster than the increase in population. This has kept us away from hunger despite some very bad drought years. Hungry people can hardly think of a quality environment. Three major inputs that have made it possible are: seeds of high yielding varieties of crops, fertilizer and increased irrigation facilities. But then it is not all over and although there are good signs of decline in the rate of population and we are approximately 1.13 billion people (estimate for March 10, 2008) and this comprises approximately one-sixth of the world's population. India's annual population increase is greater than the current populations of 147 of the world's nations and by 2020 we shall be 1.33 billion. This will certainly increase our food demand.

The total foodgrain demand by 2020 is estimated at 294 MT (122 MT rice, 103 MT wheat, 41 MT coarse grains and 28 MT pulses). Thus by 2020 we need to produce about 100 MT of additional food grain/yr from the same or even less area (some more area will go to meet the increasing needs for roads, rails, buildings, etc.). We have no choice but to increase the fertilizer application. During 1980–90 there has been 3–4% decrease in fertilizer nitrogen consumption in Europe and USA, while in Asia it has increased by 74.4%. In India the current fertilizer consumption is much below the mark. Only 19 out of 437 districts in India consume more than 200 kg N + P_2O_5 + K_2O/ha, while 176 districts consume 50 kg/ha or less.

Why do we need fertilizers?

In agriculture we convert solar energy into chemical energy, i.e. food, feed or fibre. It is common knowledge that this is achieved with the help of chlorophyll in plants. Keeping at optimum other conditions such as solar radiation, temperature, etc. the more the chlorophyll the higher the biomass and it is likely that more food is produced. However, for doing this plants need adequate supply of 13 elements (N, P, K, Ca, Mg, S, Fe, Mn, Zn, Cu, B, Mo, Cl) from soil, which are known as essential plant nutrients. Fertilizers are the chemicals that supply these essential plant nutrients, mostly N, P and K, which are removed by crop plants in the largest quantities.

In one study at New Delhi, 28 kg N, 4.4 kg P, 41 kg K, 4.9 kg S, 5.9 kg Ca, 4.2 kg Mg, 400 g Fe, 100 g each of Mn and Zn and 30 g Cu were removed from the soil for each tonne of wheat grain produced. Similarly from the International Rice Research Institute, Philippines it is reported that 10–31 kg N, 1–5 kg P and 8–35 kg K were removed from the soil per tonne of rice grain produced. When crop yields of 5–8 t grain ha^{-1} are taken as with the high yielding varieties of crops, it is not possible for most soils to supply the needed amounts of plant nutrients and that is why fertilizers are needed. Such heavy removal of plant nutrients from soil leads to depletion of soil fertility, which shows up in crop yield decline and lowered factor productivity. From the dryland rainfed areas where yields are low and very little fertilizer applied, a negative balance between crop removal and plant nutrient application of 6.37 million tones/yr (N + P_2O_5 + K_2O) in India has been reported. Thus contrary to lowering the rates of fertilizer application,

Indian farmers may have to think of fertilizing for crop production as well as for soil fertility resilience. Thus fertilizer has been and will continue to be the king input in India's achieving self sufficiency in foodgrain production.

The way out

Social scientists are quite right in pointing out regional imbalances and lack of equity among communities due to modernization of agriculture, but fertilizers do not come in the way of achieving either of these social goals. There are other factors responsible such as lack of infrastructure, lack of small scale agro-industry and the need for appropriate government policies, etc. Only a carefully planned sustainable agriculture can provide a sustainable livelihood security for the poor, which should be the foundation for all development programmes. Gandhiji called such an approach to the development as 'Antyodaya' model, which even foreigners are talking about and in its operational sense implies that the priorities in development should be measured by their potential benefit to the poorest sections of the community. In this exercise food of course will come as the first priority and in countries like India it can be produced only by the judicious use of fertilizers.

As regards environmental problems we have to find out ways to overcome these and reduce pollution hazards. A number of techniques are used by agricultural scientists for reducing nitrate leaching losses. These include growing high-yielding crop varieties, deep placement and split application of nitrogen fertilizers, use of ammonium or amide fertilizers and adopting an integrated plant nutrient system (IPNS).

Dudal and Roy(1995) defined IPNS as an approach which adapts plant nutrition to a specific farming system and particular yield targets, the physical resource base, the available plant nutrient sources and the socio-economic background. The sources of plant nutrients may be mineral fertilizers and/or biological nitrogen fixation and/or organic materials depending upon a particular location. The FAO-IFFCO International Seminar on IPNS (1997) recommended that IPNS should be science-based, associating agronomy, ecology and social sciences. It should use a farming system approach and not limit itself to cropping systems only. It should address both increased productivity and profitability and integrate maintenance and rehabilitation of natural resources.

Some new and promising methods include the use of nitrification inhibitors (Prasad and Power, 1995) and slow-release nitrogen fertilizers (Prasad *et al* 1971) including indigenous (Prasad *et al* 1993, 1995) materials such as neem cake or oil-coated materials. Some of these newly developed eco-friendly fertilizers are expensive (Trenkel, 1997) and their use in field crop production may not be economical. The question is 'should the farmers pay for environmental protection'. After all they are producing food for others, without which this country will have to import and incur heavy expenses. To conclude I would like to quote two sentences from Nobel Laureate Borlaug's (Borlaug *et al* 1994) keynote address at the 15th World

Congress of Soil Science: `Indeed for those concerned with trying to preserve pristine environments or protect endangered species, we would submit that human demographic changes are the greatest threat to the planet Earth in the years ahead. Indeed, if this relentless growth in human numbers goes on unabated, *Homo sapiens* will no doubt end up as an endangered species themselves'. Fertilizers cannot check this growth in population but can help the world and its nations meet their increasing food, feed and fibre demands and in sustaining the humanity. Thus in developing countries reeling under population pressure the sustainable agriculture and efficient fertilizer use must go hand-in-hand for a better tomorrow.

Integrated plant nutrition systems (IPNS)

The basic concept underlying IPNS is the maintenance or adjustment of soil fertility/productivity and of optimum plant nutrient supply for sustaining the desired level of crop productivity through optimization of the benefits from all possible sources of plant nutrients including locally available ones in an integrated manner while ensuring environmental quality. In practical term, a system of crop nutrition in which plant nutrient needs are met through a pre-planned integrated use of mineral fertilizers; organic manures/fertilizers (eg. green manures, recyclable wastes, crop residues, FYM etc); and biofertilizers. The appropriate combination of different sources of nutrients varies according to the system of land use and ecological, social and economic conditions at the local level. The term integrated nutrient management (INM) is also used synonymously.

Integrated Plant Nutrient Management (IPNM)

Integrated plant nutrient management (IPNM) is an important component of sustainable agricultural intensification, as well as crop, pest, soil, and water management. IPNM centers on the management of soils in their capacity to be a storehouse of plant nutrients that are essential for vegetative growth. The goal of IPNM is to integrate the use of all natural and man-made sources of plant nutrients, so as to increase crop productivity in an efficient and environmentally benign manner, without diminishing the capacity of the soil to be productive for present and future generations.

IPNM incorporates many technologies including soil conservation, nitrogen fixation, and organic and inorganic fertilizer application. Soil conservation practices prevent unnecessary losses of nutrients from the field through wind and water erosion. Organic fertilizers play an important role in the improvement of soil structure and organic matter content. They are also often a good source of the secondary and micro-nutrients necessary for plant growth, and contribute a modest quantity of the primary nutrients (nitrogen, phosphorus, and potassium) to the soil. Biological nitrogen-fixation by leguminous plants and by cereals, whereby bacteria-nodules on the roots of the plants synthesize nitrogen for the plant, offer the future potential for plants themselves to meet some of their nutrient needs. Inorganic fertilizers are most desirable and effective when application coincides with the major growth spurts of the plant—when the primary nutrients are needed most intensively—and where necessary to make up for secondary and tertiary nutrient deficiencies in the soil. Further, by enhancing crop growth, inorganic fertilizer application has the added benefit of increasing the biomass of crop residues, which can in turn be reincorporated into the soil as a green manure to improve the structure and organic matter content of the soil. Nutrient application from organic and inorganic sources should thus be at the absolute and relative level required for optimal crop growth and yield, taking into account crop needs, soil nutrient balances, agro-climatic considerations, and improved soil characteristics, while minimizing negative externalities.

If used appropriately, the recycling of organic waste from urban to rural areas is a potential, largely-untapped source of nutrients for farm and crop needs. For example, through irrigation, environmentally undesirable wastewater can be utilized to return nutrients and organic matter to the fields and to improve water quality. Organic household and commercial waste can also be collected and composted to form a safe, nutrient-rich, soil structure improving amendment for application on local farms and gardens. Urban waste needs to be treated and its application monitored to be used safely, however. Untreated sewage used for irrigation can put pathogens in contact with fruit and vegetables. Currently, effective utilization of urban waste is hampered by its high water content, bulkiness, its distance from rural areas, and high, labor- intensive handling, storage, transport and application costs. However, given the cost and the lack of availability of inorganic fertilizers in some areas, and the relative abundance and benefit of waste as an organic soil amendment, the effective utilization of urban waste may yet become an important soil and nutrient amendment in different parts of the world where the economies of waste disposal and conversion are attractive.

The choice for sustaining agriculture through 2020 and beyond is not simply one of either inorganic fertilizer, organic fertilizer, or soil conservation. Inorganic and organic fertilizers and soil conservation are not substitutes, but rather complements to each other. It is the synergy created by using the most appropriate mix of these technologies that will help to sustain agriculture. Effective and efficient management of these resources and technologies, by farmers specifically through integrated plant nutrient management practices, will help to make it possible.

Knowledge and technology

Knowledge and technology are key for the farmer to manage soil fertility to benefit present and future generations. First, farmers need to know the condition of their soils. Widespread soil testing needs to be undertaken to gather data on the nutrient cycle and nutrient balances in representative areas. Second, once the condition of the soil is known, the farmers can then select the most appropriate mix of technologies to manage soils and yields in the short- and long-term, while minimizing environmental externalities, and taking into account their particular financial and resource circumstances. Here researchers and extension service providers have a role to play in making farmers aware of the various technology options and their relative cost and effectiveness. To be used effectively, researchers and extension service providers will need to develop the local knowledge necessary to make appropriate recommendations. Some of this knowledge can be obtained through interaction with farmers—often the local technology and agricultural condition experts. Through their interactions with farmers, researchers have the opportunity to learn and evaluate traditional soil management techniques, and have the responsibility to disseminate the knowledge about the most cost effective and nutritionally beneficial practices for a particular farmer's situation in a variety of regions. Third, the farmer needs to have the knowledge and technology to test soils and plant nutrients in order to monitor the condition of her fields and crops, and to intervene when appropriate to improve yields when crops would otherwise suffer from hidden hunger or failure. As soils are dynamic systems and different crops use different nutrients in different quantities, soil testing and regular monitoring will help ensure that an environment conducive for optimal plant growth and crop yield can be established through nutrient application and conservation, and soil reclamation where necessary. Once the base knowledge has been acquired, other techniques can be implemented by farmers to boost efficiency, such as precision farming practices that use the Global Positioning System, in conjunction with soil sampling, the physical characteristics of the field, and monitoring yields, to maximize crop yields, reduce costs and minimize environmental damage by reducing nutrient over-application within a field.

As per FAO the aims of Integrated Plant Nutrition Systems (IPNS) are-

- Promote adequate and balanced nutrient supply and evaluate impacts of imbalanced use;
- Develop technical and managerial innovations to:
    - improve yields and the productivity and efficiency of nutrients mobilized from all sources,
    - improve production capacity of cropping systems,
    - limit plant nutrient losses, with particular attention to environmental impact;
- Develop on-farm plant nutrition strategies in conjunction with soil fertility management practices to:
    - reverse nutrient mining and promote nutrient capital build-up,
    - improve nutrient flows to meet crop's needs.

IPNS for rice and maize based rainfed agro-ecosystems in Hills and Mountainous Agro-ecosystems

The hills and mountains region of the country has an area of 60 million ha and inhabitates more than 51 million people. Partly or fully, it stretches through 13 states viz., J&K, Punjab, U.P., Sikkim, Manipur, Tripura, Asam, Himachal Pradesh, Haryana, West Bengal, Nagaland, Mizoram and Meghalaya. Looking from the global perspectives the hills and mountainous agro-ecosystems also covers parts of the countries like Afganistan, Pakistan, China, Nepal, Bhutan, Bangladesh and Mayanmar. Therefore, any research findings applicable to the hills and mountainous regions of our country is likely to benefit these countries also which fall in the same agro-ecosystem. Coming back to hills and mountainous regions of the country, diagnostic survey conducted revealed that the average number of animals per farm was 5 in H.P., 3 in Jammu and 7 in Almora region. The adoption levels of HYV,s of major crops by sample farmers in H.P., Srinagar, Jammu and Almora regions stood at 62, 38, 76 and 36 per cent, respectively. The cropping intensity in different regions varied from 142% (Srinagar) to 200%(Almora region). In respect of use of N, P and K by the different farmers the survey indicated that they are not using balanced and recommended levels of fertilizers for different crops. It was observed that more than 80% of labour for crop operations constituted for family labour in all the study regions. The yield of crops was lower than that obtained at recommended level and much lower than the national level. The quantity of FYM/ha increased with the farm size in H.P., Almora and Jammu region while in Srinagar it decreased with an increase in farm size. The farmers in all the regions used maximum quantity of FYM in maize. Alternative of FYM such as leaf material, green manure, bedding material that could be made available locally were identified. The use of these alternatives was found to be quite low. Expensive chemicals in H.P. and Jammu regions and inadequate irrigation in Almora region occupied the first rank among short term problems for rice in rice-wheat system while expensive fertilizers, credit problems and inadequate irrigation ranked first as short term problems in maize in maize-wheat system. Also, expensive chemicals, credit problems and inadequate irrigation for wheat in rice-wheat system was the major problem. The findings have clearly shown that there is a predominance of small holdings and farming is of traditional nature. Based on these findings and the general statistics available it may be inferred that the yields of these major crops which together account for over 80 per cent cropped area have remained very low compared to the national averages and far low compared to the developed countries and the irrigated areas of the plains. The reasons for low yields may be summarised as under

- Insufficient and imbalanced use of nutrients, the mean consumption ranging from 17-48 kg $N+P_2O_5+K_2O$ kg/ha.
- Low fertilizer use efficiency particularly during monsoon season when rice and maize are grown.

- high cost of fertilizers and other chemicals
- Low adoption of high yielding varieties.
- Crops grown under rainfed situations as about 84 per cent of the area of the region is rainfed. Acute stress of soil moisture experienced particularly during sowing periods and during critical growing stages of crops.
- undulating topography and crops grown on bench terraces leading to more nutrient and soil losses thus further decreasing the fertilizer use efficiency.
- small holdings of the farmers (80% being small and marginal) make them less enterprising leading to low use of fertilizers.
- lack of technical know-how.

Though 84 per cent of cultivated land is rainfed but the average rainfall of the region remains 1000 mm, most of which is received during maize and rice growing crop season. Therefore, though almost entire maize cultivation in the region is done under rainfed conditions, but moisture availability conditions are invariably as good as that of irrigated situations. Coupled with this, the low night temperatures prevalent in hills provide the most congenial environment for getting highest possible yields of C_4 plants like maize. Similarly, moisture availability situations for rice are also satisfactory. However, if the rainfall distribution data over years are analysed, invariably it is observed that long dry spells are experienced during October, November and December months. This is the period during which the sowing of rabi crops is to take place. Dry spells are also experienced during the months of March/April coinciding with the critical stages of wheat and other *rabi* season crops for moisture requirement. Here also the use of IPNS may help in better storage of soil moisture and the temporary drought condition overcome by the standing crops. At times, onset of monsoons get delayed, thus delaying the sowing of *kharif* season crop too. Yet better rainfall particularly during *kharif* season provide an opportunity to increase the fertilizer consumption in these crops. But the scattered holdings, and small and marginal farmers have low risk bearing capacity which has resulted into low consumption of fertilizer nutrients in the hills. This necessitates the use of organic manures to supplement the chemical fertilizers. With the increase of chemical fertilizer use which is necessary for attaining the higher yields of these crops so as to meet increased food needs, the use of organic manure i.e. IPNS becomes more important. The most commonly used organic manure in the region is Farm Yard Manure which is available in limited quantities only. There is no scope of recycling crop residues as they are fed to cattle. The survey has brought out that the problems of supplementing inadequate availability of FYM should be met by searching for such locally biomass like *Lantana, Eupatorium, Alnus, Leucaena* or such crop residues like mustard straw which otherwise does not have much economic use. There is also little scope for growing *in-situ* green manure crops for want of irrigation water and scarcity or no rains received in May/June and

October/November months during green manure crops are required to be grown. Longer duration which the field crop take for growing in the hills leave very little for growing green manure crops. The socio-economic conditions of the farmers is such that they cannot afford oilcakes. Therefore, the integration of chemical fertilizer nutrients both at lower levels as well as higher levels with FYM, green leaf manure growing over waste lands/ field bunds or gathering from nearby forests and growing some grain legumes /green manure crops *in-situ* as intercrops provide some alternatives. In some parts of the region particularly in Kashmir valley and parts of North East due to low winter temperature it is not possible to grow wheat after maize/rice. Past studies have shown that rapeseed/ mustard can be grown after *kharif* crops in these areas, thus doubling the cropping intensity. IPNS therefore needs to be studied with these crop sequences in these areas.

Rice and maize are most important food crops of the region, together accounting for over 70 to 80 per cent of net sown area. The average yields of these crops are very low, 16-23 q/ha of maize, 16-19 q/ha of wheat and 16-24 q/ha of rice. The low use of chemical fertilizer nutrients (17 – 48 kg $N+P_2O_5+K_2O$/ha) necessitates the use of IPNS. The nutrient use for maximising the yields are still higher. Sharma and Sharma (1993) recorded 11.2 to 12.9 t/ha/year grain yield of rice-wheat system with combined use of FYM@10t/ha/crop season+150 % NPK to both crops with increased plant population of 40-50%. Similarly in maize-wheat system, Rameshwar(1996) recorded 11.4 t/ha/year grain yield with the use of FYM(10t/ ha) +150% NPK to both the crops. In both the systems, FYM could supplement NPK upto 50% of the recommended dose under mid-hill sub-tropical conditions of HP. In another similar study, FYM@ 20t/ha +200%NPK, the productivity maize-wheat could be increased to 16 t/ha/year. Such a higher use of fertilizer nutrients which though may be a necessity in coming decades but will be hazardous without IPNS. Anyhow, the socio-economic condition of these hill farmers is such that such a higher level of fertilizer use is not possible. IPNS, therefore, becomes a necessity to the farmers of the region. FYM is used by the farmers even without telling of scientists but its availability is not adequate. Green manuring is not feasible due to lack of irrigation facilities and less growing season for these crops available in hills. Crop residues cannot be spared for recycling as they are fed to cattle. Hedge and Dwivedi (1993) have emphasized the need of alternate organic sources so as to make IPNS more popular as the most commonly used organic source, FYM, is not available in sufficient quantities. They have also indicated the need for studying the effect of IPNS on soil physical properties with direct impact on moisture retention, root growth, nutrient conservation etc rather than taking observations on yield alone. In order to find IPNS so as to supplement the scarcely available FYM, locally available organic waste material were used at varying soil fertility levels in the cropping systems in NATP project on IPNS. Sharma and Sharma (1994) reported that upto 30 kg N/ha could be substituted by application of Lantana or Eupatorium compost @ 15t/ha. Various grain legumes or green manure crops were also tried as inter crops with maize so as to provide *in-situ* organic manures. On a conservative estimate it is expected that the IPNS technology will add 2.65 million tonnes to the food grain pool of the region, conservation of soil and nutrients not withstanding will save approximately 53,000 tonnes of N through chemical fertilizers, benefit 1020 thousand small and marginal farm families and improve the economic status of the rural people specifically farm women substantially.

Integrated Nutrient Management in Rice-based systems

Intensive rice cropping with short-duration high-yielding varieties along with increased use of mineral fertilisers and improved irrigation facilities have resulted in spectacular increases in crop productivity. This has, however, led to gradual replacement of organic manures as sources of plant nutrients. There has been a sharp increase in the prices of P and K fertilisers following withdrawal of subsidy, which has led to their decreased consumption by the farmers. The low purchasing power of the farming community and the issue of soil health have again renewed interest in organic recycling. Organic sources available for use in rice production include the bulky organic manures like FYM, quick growing leguminous shrubs grown in the cropping sequence, leguminous trees grown in alley formations and using their loppings as mulch materials, forage or food legumes properly inoculated with *Rhizobia* and grown in the sequence, blue green algae and Azolla. Yield potential of both the crops in rice-based cropping systems can be realized by organic manuring of *Kharif* rice with the available sources along with mineral fertilization of both the crops in the acid lateritic soil conditions of eastern India.

Long-term experiments have shown that neither organic sources nor mineral fertilisers alone can achieve sustainability in crop production. Continuous use of FYM is effective in stabilizing rice productivity under low to medium cropping intensity where the nutrient demand is relatively small. Nonetheless, integrated use of organic and mineral fertilisers has been found to be more effective in maintaining higher productivity and stability through correction of deficiencies of secondary and micronutrients in the course of mineralisation on one hand and favourable physical and soil ecological conditions on the other. Organic manuring also improves the physical and microbial conditions of soil and enhance fertiliser use efficiency when applied in conjunction with mineral fertilisers. Thus, all the major sources of plant nutrients such as soil, mineral, organic and biological should be utilised in an efficient and judicious manner for sustainable crop production in rice-rice cropping system.

Green manure of legume shrubs or tree loppings has been known to be beneficial for sustaining rice productivity. Sunnhemp and dhaincha are popular legumes for green manuring in rice and can accumulate up to 100 kg N/ha in 50-55 days. Incorporation of these green manures *in-situ* before transplanting rice supplies about 45-60 kg N/ha, besides providing a significant residual effect to the succeeding crops. Fertilizer use efficiency is improved when a legume crop such as *Sesbania cannabina or Lathyrus sativus* is introduced in rice-rice cropping system. Adding loppings of leguminous trees like *Leucaena leucocephalla and Glyricidia napus* grown in alleys can also meet the crop N requirement substantially. The productivity of rice-rice cropping system can be increased by about 1 t/ha besides a net saving of 30 kg fertilizer N/ha by including a short-duration legume such as cowpea or greengram and incorporating its residues into the soil after harvesting the grains. Similarly, blue green algae culture in the rice field can contribute about 25 kg N/ha to the rice crop. Algae multiply and cover the field like a carpet which when incorporated into the soil, decomposes and releases N for rice crop. Azolla can be grown in tanks or in rice fields and incorporated into the soil after 4-6 weeks.

In two rice based experiments under NATP (ICAR) sponsored project conducted at the Research farm of Division of Agronomy, SKUAST- Jammu starting from June 2001 ; the FYM along with organic residues *Leucaena* biomass and higher rate (150%RFD) of NPK fertilizers proved to be the best combination for augmenting the grain and straw yields of wheat crop. *Leucaena* tree is readily available in this region and its leaves, green twigs were used as organic source and the results were found to be very promising also with respect to their economic viability. These organic residues could be used as a substitute for almost 50% RFD as these organic residues not only increased the yield in I[st] crop but also improved the soil fertility status. The residual effects of IPNS treatments were clearly depicted in the successive crops of 2 wheat-rice cycles in terms of encouraging yield trends due to gradual improvement in soil fertility.

Table 1. Effect of IPNS on productivity (q/ha) of wheat and rice in three wheat-rice cycles.

Treatments	Mean grain yield of wheat after 3 crop-cycles	Mean grain yield of rice after 3 crop-cycles	Total productivity of wheat-rice system after 3 crop-cycles
T_1: Control	21.64	20.40	42.04
T_2: FYM@5/tha+ 50%NPK	28.38	32.90	61.28
T_3: FYM@5t/ha+ 100%NPK	36.02	35.79	71.81
T_4: FYM@5t/ha+ 150%NPK	42.05	34.42	76.47
T_5: Leucaena biomass+ 50%NPK	26.83	33.12	59.95
T_6: Leucaena biomass+ 100%NPK	33.62	38.63	72.25
T_7: Leucaena biomass+ 150%NPK	41.15	36.05	77.20
T_8: Paddy straw + 50%NPK	24.43	28.98	53.41
T_9: Paddy straw +100%NPK	30.93	37.13	68.06
T_{10}: Paddy straw + 150%NPK	34.98	32.98	67.96

T_{11}: FYM@2.5 t/ha +Leucaena biomass + 50% NPK	31.17	30.07	61.24
T_{12}: FYM@2.5 t/ha +Leucaena biomass + 100% NPK	34.26	31.29	65.55
T_{13}: FYM@2.5 t/ha +Leucaena biomass + 150% NPK	43.79	32.25	76.04
T_{14}: FYM@2.5 t/ha +Paddy straw+ 50% NPK	25.38	32.19	57.57
T_{15}: FYM@2.5 t/ha +Paddy straw+ 100% NPK	32.05	36.47	68.52
T_{16}: FYM@2.5 t/ha +Paddy straw+ 150% NPK	37.13	35.44	72.57

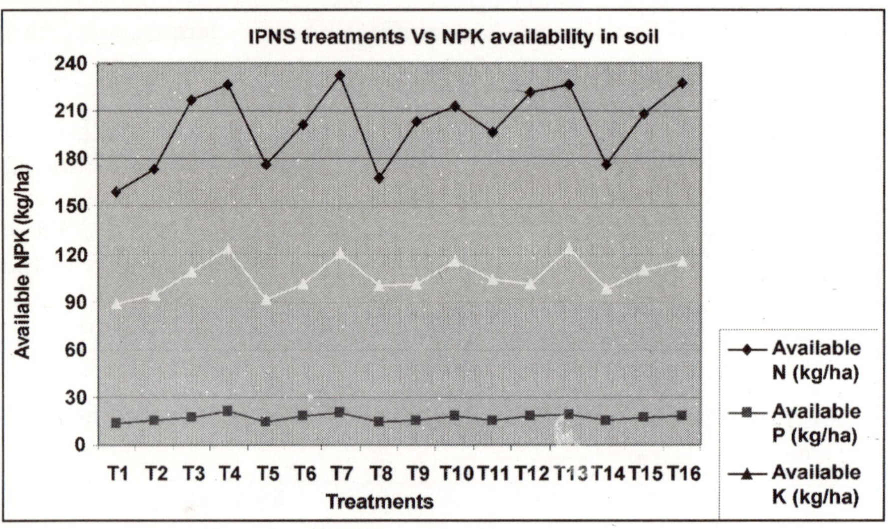

Fig.1. Effect of different treatments on NPK availability in soil after 3-wheat-rice cycles.

In direct seeded rice crop, highest grain and straw yield values (Table 2) of rice were obtained with 100% Rec. NPK as well as when dhaincha was used as green manure crop which proved to be a good substitute for 50% RFD or 5t FYM/ha. Maximum net returns were found to be in treatment T_1: Pure rice +100% RFD which was closely followed by treatment where Dhaincha was applied as green manure along with 50% RFD, both the treatments were economically feasible than all other treatment combinations. The results were well confirmed by the 'On farm' trials where dhaincha was intercropped in transplanted low land rice fields and was incorporated at 30-35DAS. In 'on-farm' trials, treatment T_4: Dhaincha-GM+100% RFD observed to obtain maximum net return values as compared to other treatment combinations. The other farmers were very much interested in adopting IPNS-technology. They were also interested in raising rice crop by direct seeded method along with intercropping with dhaincha locally named as *jantri* provided that the proper equipment for direct drilling could be provided at the farmer's field.

Table 2 Effect of different treatments on productivity of direct seeded rice-wheat system after 2 crops- cycles.

Treatments	Mean grain yield of rice after 2 crop-cycles	Mean grain yield of wheat after 2 crop-cycles	Productivity of rice-wheat system after 2 crop-cycles.
T_1: Pure rice (100%NPK)	42.0	34.93	76.93
T_2: Rice+Dhaincha (50%NPK)	41.15	33.07	74.22
T_3: Rice+Soybean (50%NPK)	38.46	31.42	69.88
T:$_4$ Rice+Sunhemp (50%NPK)	33.11	27.12	60.23
T_5: Rice+Cowpea (50%NPK)	37.54	27.85	65.39
T_6: Rice+5t/haFYM (50%NPK)	41.20	33.62	74.82

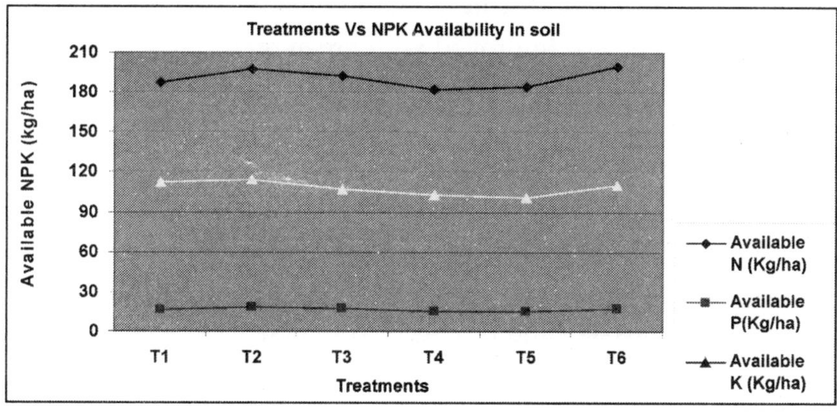

Fig 2: Nutrient availability after 2 crop cycles

Expt. 1.

- The use of higher dose of NPK (150% RFD) in combination with organic residues having narrow C: N ratio (FYM and /or *Leucaena*) applied in wheat crop significantly increased the grain and straw yields.
- Organic residues incorporated in wheat conserved the inorganic fertilizers and thus significant residual effects were observed in the succeeding rice crop.

- In the succeeding rice crop, the yield increased significantly due to residual effects of 100 or 150% RFD in combination with FYM, *Leucaena* and Paddy straw.
- Different organic residues along with inorganic fertilizer application improved soil fertility status but the results were more encouraging where 150% NPK was applied with FYM and *Leucaena*.
- Soil fertility deteriorated in absence of supplementary nutrient application.
- The treatment *Leucaena*+150% RFD recorded maximum values of net returns and B:C ratio in wheat-rice system after 3-crop cycles.

Expt.2.

- The application of FYM@5t/ha and green manuring in rice with Dhaincha+50% RFD registered better grain and straw yields than all other treatments, but were statistically at par with pure rice+ 100%RFD.
- The grain and straw yields of 2nd year rice showed an improvement over the 1st year values. The status of soil fertility improved with FYM application and green manuring (Dhaincha), which saved almost 50% NPK requirement.
- Rice crop could be raised by direct seeding method in lowland conditions and the yield improved if FYM was applied @5t/ha or dhaincha is used as green manure and incorporated in the same field 30-35DAS (*halod*).
- The soil fertility status also improved with green manuring/FYM+50% NPK and also with 100% application of inorganic fertilizers and the values obtained were nearly at par.

The treatment T_1: Pure rice +100% RFD recorded more values of net returns and B:C ratio than all other treatment combinations.

Integrated Nutrient Management in Maize-based systems

Among the maize-based cropping systems (Table 3), the experiment with two organic wastes (wheat straw-Ws and rapeseed straw-Rs), FYM and different doses of NPK (recommended fertilizer dose- N: P_2O_5: K_2O:: 60:40:20 kg/ha), revealed that the treatment FYM @ 5t/ha + 100 and 150 % RFD were the most promising treatment combinations with respect to grain and stover yields for maize crop. In the following wheat crop where the inorganic NPK (recommended fertilizer dose- N: P_2O_5: K_2O:: 60:30:20 kg/ha) dose was reversed from 50, 100 and 150% RFD to 150, 100 and 50% RFD in the respective treatments, again the treatment FYM@5t/ha+150% RFD proved itself better with respect to grain and straw yields of wheat. The available NPK status after four crops was comparatively higher in the treatment FYM@2.5t/ha+Rapeseed straw +50% RFD in maize (150% RFD in wheat-total NPK application=200%RFD), which was at par with treatment FYM@5t/ha+50% RFD in maize (150% RFD in wheat-total NPK application = 200%RFD). The NPK content continued to decline from its initial values after each successive crop in control plots. The maximum net returns were found to be in treatment T_7: Rapeseed straw+50% RFD in maize (150% RFD in wheat), which was economically better than all other treatment combinations.

Table 3. Effect of different IPNS treatments on productivity of maize-wheat system.

Treatments	Mean grain yield of maize after 2 crop-cycles	Mean grain yield of wheat after 2 crop-cycles	Productivity of maize-wheat system after 2 crop-cycles.
T_1: FYM@5t/ha+ 50% NPK	19.75	39.61	59.36
T_2: FYM@5t/ha+ 100%NPK	23.37	35.97	59.34
T_3: FYM@5t/ha+ 150%NPK	23.90	34.46	58.36
T_4: Wheat straw + 50%NPK	18.20	34.88	53.08
T_5: Wheat straw + 100%NPK	18.90	31.80	50.70
T_6: Wheat straw + 150%NPK	20.30	30.15	50.45
T_7: Rape seed straw + 50%NPK	19.35	35.70	55.05
T_8: Rape seed straw + 100%NPK	20.50	33.15	53.65
T_9: Rape seed straw + 150% NPK	21.47	30.75	52.22
T_{10}: FYM@2.5 t/ha +Wheat straw + 50% NPK	17.37	36.71	54.08
T_{11}: FYM@2.5 t/ha +Wheat straw + 100% NPK	19.05	33.87	52.92
T_{12}: FYM@2.5 t/ha +Wheat straw + 150% NPK	20.97	32.76	53.73
T_{13}: FYM@2.5 t/ha +Rapeseed straw + 50% NPK	19.32	37.55	56.87
T_{14}: FYM@2.5 t/ha +Rapeseed straw + 100% NPK	21.10	35.14	56.24
T_{15}: FYM@2.5 t/ha +Rapeseed straw + 150% NPK	21.82	33.04	54.86
T_{16}: Control	14.82	22.83	37.65

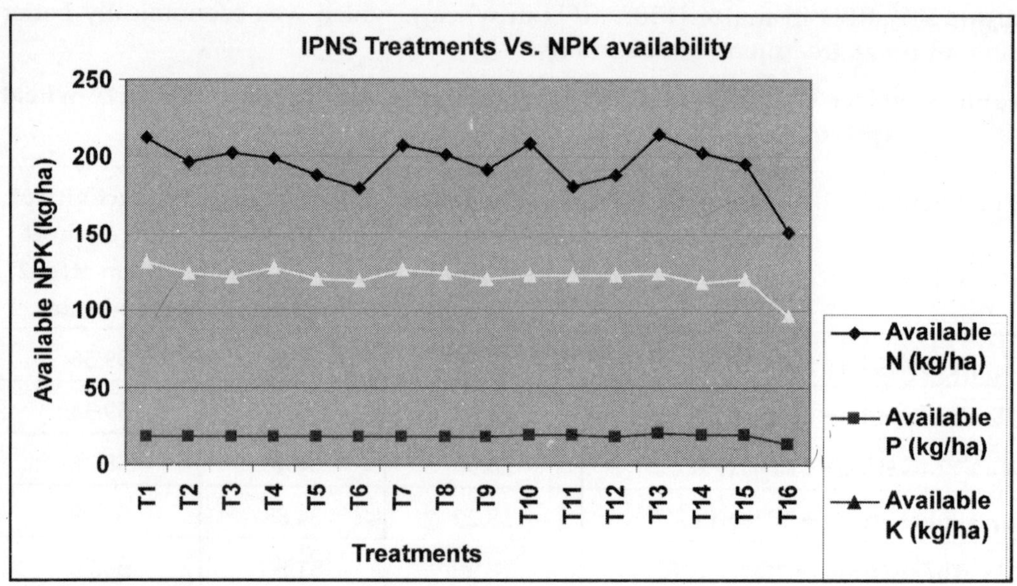

Fig 3. Effect of different treatments on NPK availability in soil after 2 crop-cycles.

In the intercropping maize experiment (Table 4) where maize was intercropped with *In-situ* green manures along with 100% RFD (recommended fertilizer dose N: P_2O_5: K_2O:: 60:40:20 kg/ha), and FYM application, the treatment Pure maize (No FYM) where maize was raised without green manuring, statistically inferior values of grain and stover yields were observed as compared to all other treatment combinations. However, significantly highest values were recorded with treatment T_2: Pure maize+ FYM@5t/ha. In the following wheat, three doses 50,100 and 150% RFD (recommended fertilizer dose- N: P_2O_5: K_2O:: 80:40:25 kg/ha), of inorganic fertilizers were tested in split plot design. In the main plots, the treatment Pure maize+FYM@5t/ha proved superior to all other treatments with respect to grain and straw yields of wheat probably due to residual effect from *Kharif* season. In the sub-plots, 150% RFD responded better in comparison to 100 and 50% RFD. The availability of NPK after two maize-wheat crop cycles increased from its initial values after each successive crop in the treatment Pure maize + FYM@5t/ha whereas, the values obtained with the treatment where green manuring was done with soybean, were also encouraging and the treatment Pure maize (without FYM) observed the least values of NPK availability. The treatment T_6: Maize+soybean-GC obtained maximum net return values due to the higher maize equivalent yield, which was found to be economically more viable.

Table 4. Effect of intercropping and *in-situ* green manuring on productivity of maize- wheat system.

Treatments	Mean grain yield of maize after 2 crop-cycles	Mean maize equivalent yield after 2 crop-cycles	Mean grain yield of wheat after 2 crop-cycles	Total productivity of maize-wheat system after 2 crop-cycles
T_1: Pure maize (without FYM)	15.27	15.27	28.1	43.37
T_2: Pure maize (5t/haFYM)	21.91	21.91	32.05	53.96
T_3: Maize+Cowpea (GM)	20.15	20.15	29.48	49.63
T_4: Maize+Cowpea (Grain crop)	17.41	45.25	28.10	45.51
T_5: Maize+Soybean (GM)	21.0	21.0	31.44	52.44
T_6: Maize+Soybean (Grain crop)	18.10	57.04	29.63	47.73
T_7: Maize+Sunhemp (GM)	19.35	19.35	28.09	47.44

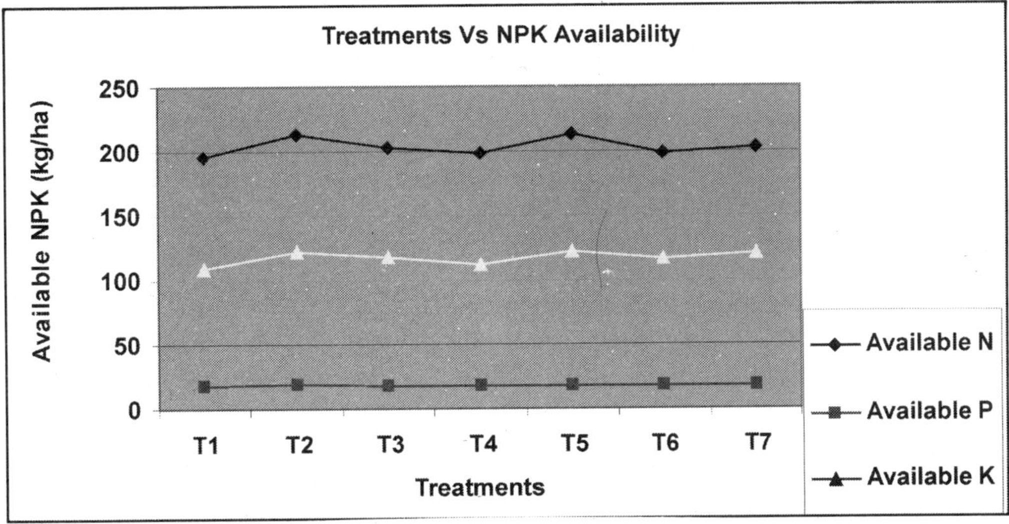

Fig 4. Effect of different treatments on NPK availability in soil after 2nd maize-wheat cycle.

The results were well confirmed by 'on-farm' trials where FYM@5t/ha along with 100% Rec. NPK dose obtained maximum grain and straw yields at all the 5-locations in district Rajouri of J&K state. In 'on-farm' trials, treatment T_2: FYM@5t/ha+100% RFD observed to obtain maximum values of net returns than all other treatments on farmers' field. There was near about 28.44 and 25.48 per cent increase in grain yield of wheat and maize in the treatment where FYM@5t/ha along with 100%RFD was applied over the farmer's practice after 3 crop-cycles.

Expt. 3.

- Among the locally available organic residues FYM proved to be superior to both Rapeseed and wheat straw in respect of grain and straw yield of both maize and wheat crops at the same levels of inorganic fertilizers.
- The treatment FYM@5t/ha + 150% RFD again proved to be superior to all other treatments even when the dose was reversed in the succeeding wheat crop.
- Use of locally available low cost organic residue Rapeseed straw with higher rate of NPK (150% RFD) was also economically viable and obtained maximum net return values and B:C ratio.
- There was a considerable decline in the grain and straw yield along with soil fertility status in the control plot, which was not supplied by any organic/inorganic fertilizers.
- There was an overall improvement in the soil fertility status with FYM@5t/ha/Rapeseed straw + 150% RFD.

Expt. 4.

- The organic residue (FYM) along with 100% RFD excelled itself better in grain and straw yield of maize than all other treatments, but the grain and straw yield values obtained by green manuring with soybean and cowpea were statistically at par.
- The grain yield of maize in the 2nd year showed an increasing trend as compared to the values observed in 1st year.
- The grain crops of soybean and cowpea proved to be economically better option on the basis of maize equivalent yield.
- The fertility status of soil improved significantly where FYM or soybean/cowpea GM) were applied.
- In the succeeding wheat, availability of nutrients increased with increase in NPK-dose from 50 to 150%.
- The treatment T_6: Maize+soybean-GC recorded maximum values of net returns and B:C ratio and treatment T_4: Maize+cowpea also observed almost similar values. This was mainly due to the extra income through intercrops.

It could be easily inferred that FYM could be replaced by locally available cheap organic wastes like *Leucaena* and rapeseed straw. Moreover, *in-situ* green manuring can be equally good when compared with the benefits of GM grown well before sowing of the respective crop besides this when the accompanying legume was grown for grain purpose the cereal equivalent yield assured extra income to the farmer.

Conclusions

In the period since Nobel Chemistry Prize winners Fritz Haber and Karl Bosch developed the process that made the synthetic production of ammonia an economically viable reality, world nitrogen fertilizer production skyrocketed from a mere four thousand metric tons in 1914 to 73 mmt by 1994. The explosion in the use of inorganic fertilizers— particularly nitrogenous fertilizers, was in part a consequence of the adoption of nutrient- responsive, high-yielding modern seed varieties by farmers in the developed world and parts of Asia. Together, this potent combination made a significant contribution to the increase in world food production. In many developed countries however, fertilizer over- application and poor soil management has led to erosion and leaching induced environmental pollution. In many developing countries, by contrast, the decline in traditional soil fertility management practices coupled with insufficient fertilizer application, has resulted in the mining of soil nutrients, a reduction in soil fertility and environmental pollution. With 366 mmt of nutrients projected to be mined from global soils annually, soils will need to be managed more effectively and efficiently for agriculture to be sustained through the year 2020 and beyond.

The widespread and responsible application of inorganic fertilizers will play a critical role in helping to replenish these nutrients. Yet, to help meet the expected nutrient replenishment shortfall and for agriculture to be sustainable over the long-term and recognizing the poor socio-economic conditions in many countries, use of relatively expensive inorganic fertilizer will have to be accompanied by other measures for nutrient requirements to be managed effectively. *First,* soil conservation practices will need to be further adopted to prevent the unnecessary loss of nutrients from the field through leaching and erosion. *Second,* greater use will need to be made of organic fertilizers such as organic manures, cover-crops, and underdeveloped urban waste resources to improve soil structure, rebuild secondary and micro-nutrients, and provide a minimum quantity of the primary nutrients. *Third,* inorganic fertilizers will need to be applied to provide both the primary nutrients at critical plant growth periods and to remedy secondary and tertiary nutrient deficiencies. *Lastly,* genetic engineering offers the potential in the future for the plants themselves to meet some of their nutrient requirements. Together, these nutrient conservation and replenishment methods need to be managed - reflecting the farmer's particular bio-physical and socio-economic situations, in such a way as to provide a cost effective and appropriate level of nutrients to maximize yields and sustain agriculture, without polluting the environment. Furthermore, by sustaining agricultural intensification, effective and efficient nutrient management reduces the need to cultivate and degrade marginal lands.

Governments have both an active and supporting role to play in the adoption and success of IPNM. *First,* because the private sector often does not have a very large incentive to undertake research on technologies that conserve nutrients or more effectively use organic nutrients, governments must provide the financial support. *Second,* for improved soil management, governments a) need to support the establishment of testing and monitoring systems to gather data on the nutrient cycle and nutrient balances in representative areas, b) develop and disseminate the recommendations to farmers and extension service providers, c) encourage closer cooperation and coordination between farmers and researchers and other farmers to exchange, evaluate and disseminate information and technologies, and d) transfer knowledge and technology to farmers to test and monitor soil and plant nutrients. *Third,* at the institutional level, governments have a role to play in the establishment of an efficient market for the productive distribution and use of inorganic and organic fertilizers. In establishing such an environment, governments need to invest in infrastructure, establish property rights, withdraw from the procurement and distribution of goods and services when the private sector can more efficiently undertake these functions, and when necessary encourage fertilizer use through appropriate price incentives.

References

Anonymous. 2004. Annual Progress Report, NATP-IPNS Project, Division of Agronomy, FoA, SKUAST-J, Chatha, Jammu, India.

Borlaug, N. E. and Christopher, R. D., Keynote Address, 15th World Congress of Soil Sci., Mexico, 1994.

Dudal, R., and Roy, R. N., Integrated Plant Nutrient Systems, FAO Fertilizer and Plant Nutrition Bull., 1995, 12, 181–198.

Hedge, D. M. and Dwivedi, B. S. 1993. Fertilizer News. 38 (12) : 49 – 59.

IFFCO, Executive Summary, FAO–IFFCO International Seminar on IPNS, New Delhi, 23–27 November 1997, p. 16.

Kumar, A. and Thakur, N. P. 2004. Annual report, NATP-IPNS (ICAR),Project.

Prasad, R. and Power, J. F., Adv. Agron., 1995, 54, 233–281.

Prasad, R., Rajale, G. B. and Lakhdive, B. A., Adv. Agron., 1971, 23, 337–381.

Prasad, R., Devkumar, S. and Shivey, Y. S., in Neem Research and Development (eds Randhawa, N. S. and Parmar, B. S.), Soc. Pesticide Sci., New Delhi, 1993, pp. 97–108.

Prasad, R., Saxena, V. S. and Devkumar, S., Curr. Sci., 1995, 75, 15.

Rameshwar, K. 1996. Ph. D. Thesis, HPKV, Palampur.

Sharma, G. D. and Sharma, H. L. 1993. Fertilizer News. 38 (11) : 19 – 23.

Sharma, G. D. and Sharma, H. L. 1994. Indian J. Agric. Sci. 64 (3) : 184 – 186.

Trenkel, M. E. 1997. Controlled-Release and Stabilized Fertilizers in Agriculture, International Fertilizer Association, Paris, .P p. 151.

CHAPTER

3 Dryland Farming Systems

Rajesh Singh

Department of Seed Science & Technology, CSKHPKV, Palampur, H.P. 176062.

By 2010 A.D., India will have to produce 300 million tonnes of food grains to feed her 1.5 billion populations (approx.). This target cannot be realized from irrigated areas alone as we have irrigation potential for 178 million hectares only. Therefore, we will have to evolve an appropriate technology for dry land farming. On the other hand, we can say that second 'green revolution' in Indian agriculture can be had in rainfed/dryland agriculture. This is important to improve the standard of living of farmers residing in these areas as well. India has about 108 million hectares of rain fed area which constitutes nearly 75% of the total 143 million hectares of arable land. In such areas crop production becomes relatively difficult as it mainly depends upon intensity and frequency of rainfall. The crop production, therefore, in such areas is called rainfed farming as there is no facility to give any irrigation, and even protective or life saving irrigation is not possible. These areas get an annual rainfall between 400 mm to 1000 mm which is unevenly distributed, highly uncertain and erratic. In certain areas the total annual rainfall does not exceed 500mm. The crop production, depending upon this rain, is technically called dry land farming and areas are known as dry lands.

India has about 47 million hectares of dry lands out of 108 million hectares of total rainfed area. Dry lands contribute 42% of the total food grain production of the country. These areas produce 75% of pulses and more than 90% of sorghum, millet, groundnut and pulses from arid and semi-arid regions. Thus, dry lands and rainfed farming will continue to play a dominant role in agricultural production. Dry lands, besides being water deficient, are characterized by high evaporation rates, exceptionally high day temperature during summer, low humidity and high run off and soil erosion. The soils of such areas are often found to be saline and low in fertility. As water is the most important factor of crop production, inadequacy and uncertainty of rainfall often cause partial or complete failure of the crops which leads to period of scarcities and famines. Thus the life of both human being and cattle in such areas becomes difficult and insecure.

Despite all these improvements in agriculture, we have yet not been able to evolve an appropriate package of practices for our dry land areas. The income of farmers of dry land regions is still very low. For achieving the targets we will have to harness every inch of our cultivable lands, especially dry lands, with utmost care. Dry farming or dry land farming may be defined as: "a practice of growing profitable crops without irrigation in areas which receive an annual rainfall of 500 mm or even less."

Efforts are being made to bring more area under irrigated agriculture and thereby to increase cropping intensity. We continue to stress on intensive agriculture on irrigated land but we can not afford to be complacent with our dry lands. Therefore, improved dry farming is necessary for equity and prosperity. As such we can not achieve stability in food production with unstabilized dry land agriculture. Therefore, we are required to adopt improved technology especially developed for dry land agriculture.

Potential cropping systems

Cropping system refers to an arrangement in which various crops are grown together in the same field. The cropping systems followed in dry lands differ from those followed under normal conditions. Only those crops can be grown under dry land conditions which require less water to complete their life cycle or which can stand or yield under drought conditions. This can include both drought resistant and drought tolerant plants. In addition, plants can be grown only where some water is available to sustain the growth of plants.

A cropping system is the intensification of crops in space and time dimensions. It presents farmers with a range of decisions that need to made such as what combinations of species should be planted and which varieties be selected from within these species, what planting arrangements should be followed and how many plants each components should be sown with and what should be the input level for the component crops?

The length of growing period is the basis that decides the make up of a cropping system. In kharif season the rain fall both in terms of quantum and distribution decides the effective cropping season and it becomes critical in selecting cropping system for a given area (Table 1).

Table 1. Potential cropping systems in relation to rain fall and soil type

Rainfall (mm)	Soil type	Effective growing season weeks	Suggested cropping system
350-600	Alfisols and shallow Vertisols	20	Single rainy season cropping
	Deep Aridisols and Entisols	20	Single cropping during rainy or post-rainy season
	Deep Vertisols	20	Single post-rainy season cropping
	Alfisols. Vertisols and Entisols	20-30	Intercropping
	Entisols Deep vertisols deep Alfisols and Inceptisols	30	Double cropping with monitoring
	Entisols, deep vertisols, deep Alfisols amd inceptisols	>30	Double cropping assured

In India water availability period varies from less than 110 days to more than 250 days in different parts. Depending on soil characters, inter or double cropping can be advantageously practiced under these situations (Table 2). Legumes have special utility in these systems. As a component of understorey crops Legumes establish rapidly, reduce soil erosion, reduce soil temperature and lower weed pressure beside giving yield additions.

Table 2 Efficient cropping systems for different rain dependent regions of India

Water availability period(days)	Inter cropping system		Double Cropping system	
	Base Crop	Inter crop	Rainy season crop	Post rainy season crop
Up to 110	Pearl millet/cluster bean	Cowpea/ Greengram/ blackgram	--	--
	Groundnut/ pearl millet	Pigeonpea	--	--
110-150	Pearl millet/sorghum/ clusterbean	Greengram/ pigeonpea	Cowpea/ blackgram/ soybean	Safflower/ chickpea /mustard/ safflower
			Greengram/ pearl millet	Mustard/ chickpea/ barley
150-175	Pearl millet chickpea	Pigeonpea/moth bean/ barley / linseed/ mustard/ safflower	Greengram	Sorghum
175-200	Pearl millet/ groundnut/sorghum chickpea/ barley	Pegionpea Safflower/ Mustard	Greengram/ Cowpea/ Greengram/ Blackgram / Pearl millet/ Maize/ Rice/ Sesame/ Chickpea	Sorghum/ Safflower/ Barley/mustard/ Chickpea Wheat/wheat+ Chickpea/ Mustard
200-250	Sorghum/ pearlmillet/maize / Soybean Chickpea	Pegionpea Mustard	Sorghum/ groundnut/maize/ Soybean/ Soybean + maize	Chickpea/ safflower/ Wheat/chickpea
>250	Rice/fingermillet / soybean/maize Wheat/barley/ Chickpea	Pigeonpea Chickpea/ mustard	Rice/maize/finger millet/groundnut/ Soybean/ Pearl millet	Wheat/chickpea/ linseed/lentil/ horsegram/ barley/ Wheat

Singh and Das (1995)

Umrani (1995) observes that such a system fulfills the concept of kharif-rabi continuum instead of compartmentalizing kharif and rabi season cropping, which prevails in the hard core of Deccan rabi dry farming tract. Double cropping system are more common in regions having moisture availability period of 175-200 days or more. A wide variety of crops are grown including a short season (70 days) pulse crops (greengram, blackgram, cowpea), cereals (pearl millet, maize, rice) maturing in 95-100 days, and sesame (110 days). Short duration pulses are followed by rabi sorghum, safflower, barley, mustard etc. while cereals /oilseed crops are followed by chickpea mustard or wheat. Sorghum/safflower double cropped system is common in better moisture availability endowed regions (200-250 days). The practice of cultivating soybean in rainy season followed by safflower or wheat in post-rainy seasons is also a common practice in some areas. In areas where moisture availability of 250 days is available, rice is more prevalent although, groundnut and soybean also find a place in double cropping system.

Another important aspect that needs consideration in selection of cropping system for a dry farming situation is the risk of failure. One of the problems in trying to ensure to full use of the potential growing period is that as the required growing period for given system increases, the portability of end season water stress also increases. For instance, on Vertisols with 990 mm rainfall, after a 91 days rainy sorghum crop, there is still quite high probability (73%) of having sufficient water stored in the soil profile for a sequential second crop (assumed to be 100 days) (Willey et al. 1982) (Table 3).

Table 3 Probability of success(%) of different cropping systems at Indore, India calculated from 37 years of rainfall data using a water balance model.

Cropping systems	First crop		Second crop		
	Sufficient moisture for growth	Wet week harvest (>50 mm)	Sufficient moisture for grain (>200 mm)	Insufficient rain for establishment (>200 mm) even though sufficient for growth	Overall probability of success
Sequential system (Ist crop 105 days soghum)	100	24	51	24	27
Sequential system (Ist crop 91-days sorghum)	100	24	73	13	60
Relay system (Ist crop 105-days sorghum)	100	24	73	13	60
Sorghum + pigeonpea (105 days -180 days)	100	24	97	-	97

Willey, et al. (1982)

Following are a few intercropping systems for dryland areas
Moong+Bajra
Guar+Bajra
Til + Guar/moth/mung

Mixed cropping is also followed to minimize the effect of unpredictability of rain. Mixed cropping may have low yield potential but it works as a buffer against failure under possible unfavourable conditions. Mixed cropping may be defined as sowing of two or more crops simultaneously on the same piece of land in separate rows. Examples: Guar + Arhar+ Moong, Bajra +Arhar + Moong and Maize + Urd etc.

Cropping Pattern

Cropping pattern is defined as sequence of growing crops in a particular field at a particular period. The most common cropping pattern for dry land farming are discussed below:

For North Indian conditions

- Sorghum-Safflower/mustard
- Sorghum-Mung/urd/cowpea-Gram/wheat-Gram
- Rice-Gram (for low lying areas)
- Bajra- Gram + Linseed
- Bajra + Urd / Mung / soyabean -Wheat /barley + Gram/mustard
- Maize-Gram/safflower

For central Indian conditions

- Greengram - Rabi sorghum
- Greengram -Safflower

For Bhubaneswar region

- Rice-Horse gram
- Ragi - Redgram
- Groundnut - Niger
- Maize - Niger

Alternative land use system, a land capability based production system which plays important role is stabilizing natural ecosystem has been emphasized in the ecologically sensitive dry land area (Table 4). There are other advantages too.(1) Trees and pastures will be more remunerative and impart stability to dry lands. (2) Perennial vegetable makes use of off season rains and also checks soil erosion and (3) farm plan involving use of fodder, fuel, timber and cash crops meets the energy requirement of rural family.

Table 4. Farming system options for different agro climatic conditions

Annual rainfall (mm)	Soil type	Farming systems	Suitable tree/grass/legume spcies
<500	Shallow (0-30 cm)	Tree farming	*Prosopis cinearia ,P. Jubiflora. Acacia aneura. A. nelotica. A. tortilis, Pitheoellabium dulce*
	Medium (0-45 cm)	Pasture management	*Lasturus sindicus Z(light textured soils) Cenchras setigerus, Sehima rervoium, Stylvanthes scaboh,Clitoria ternatea*
500-750	Shallow (0-30 cm)	Silvi pastoral system	*Acacia nilotica, Celophospermum mopane, Dlbergia sisoo,Cenelirus cilaris C. setigerus. Dicanthium annulatum. Panicum antidotale. Stylosanthes homata Macroptillum antropurpurelum*
	Medium (0-45 cm)	Horti-pastoral system	*Custard apple, Ber, Jamun, Anola, Tamarind, Wood apple, Bact. Cenchrus cilaris, Panicum antidotate, Urochloa mosauficensis. Shylosanthes homata, Macrptillum antropurpureum, Clitoria ternatea.*
>750	Shallow (0-30 cm)	Ley farming silvipastoral system	3 years *Stylosanthes hamata* and 4th year arable crop (sorghum on heavy soil, pearlmillet on lighter soils) silvi-pastoral system as above
	Medium (0-45 cm)	Ley farming hortipastoral system	Ley farming as above. Mango, Sapota, Guava, Anola, *Stylosanthes hamata, Macroptillium autropurpureum.*

Source: Singh and Osman (1995)

Crop production in drylands is a gamble with monsoon. The per capita land, ideal for crop production is decreasing progressively. Therefore, cropping has been extended to marginal lands leading to their fast degradation. Accordingly to National Remote Sensing Agency, culturable waste lands in India occupy 38.8 M ha. Income from drylands could be increased through alternate land use system adopting improved technology. The term alternate land use system is applicable to all classes of land to generate assured income with minimum risk through efficient use of available resources. It aims at a suitable farming system, matching the land capability classes. Eight will recognized land capacity classes have been identified. Need base alternate land use systems are:

Food (Class II & III)	Alley cropping, agri-horticulture, intercropping with nitrogen fixing trees
Fodder (Class IV & V)	Horti-pastoral, silvi-pastoral, ley farming, pasture management
Fuel/timber/fibre(Class VI & VII)	Tree farming, timber cum fibre farming.

Alley Cropping For Arable Lands

Food crops are grown in alleys formed by hedge rows of trees or shrubs in arable lands. It is also known as hedgerow intercropping. Hedgerows are cut at about one meter height at planting and kept pruned during cropping to prevent shading and to reduce competition with food crops. It is recommended for humid tropics, primarily, as an alternative to shifting cultivation. In semiarid regions of India, alley cropping provide fodder during dry period since mulching the crop with hedgerow pruning usually does not contribute to increased crop production. Advantages of this system are:

- Provision of green fodder during lean period of the year,
- Higher total biomass production per unit area than arable crop alone,
- Efficient use of off season precipitation in the absence of a crop,
- Additional employment during off season, and
- It serves as a barrier to surface runoff leading to soil and water conservation.

Based on the objectives, three types of alley systems are recognized.

- Forage alley cropping,
- Forage-cum-mulch system, and
- Forage-cum-pole system.

In all the three systems, crops are grown in alleys and forage obtained from loppings of hedgerows.

Forage-Alley cropping System

In this system both yield of crop and forage assume importance. Only a few tree species are suitable for hedge rows: *Leucaena leucocephala, Colliendra* and *Sesbania*. Pigeon pea or castor crops are suitable for growing in the allies of *Leucaena*. Crop yield decreases with decrease in row width. Increase in cutting height of hedgerow decreases the crop yield due to shading effect. Longer harvest intervals narrow the yield differences between the cutting heights.

Forage-Cum-Mulch System

In this system, hedgerows are used for both forage and mulch. Loppings are used for mulching during the crop season and used as fodder during off season. Substantial reduction in crop yields of sorghum; groundnut, green gram and black gram have been observed at several places.

Forage-Cum-Pole System

Leucaena alleys are established at 5 m intervals along the contours. Hedge rows are established by direct seeding and topped every two months at 1.0 m height during crop season and every four month, during off season A *Leucaena* plant for every 2 m along hedgerows is allowed to grow into a pole. Crop yield is usually reduced due to competition from hedgerows. However, gross returns are higher in all alley cropping system than with sole crop.

Agri-holticulture for Arable Lands

It is one form of agro forestry in which the tree component is fruit tree. It is also called as food-cum-fruit system in which short duration arable crops are raised in the interspaces of fruit trees. Some of the fruit trees that can be considered are guava, pomegranate, custard apple, sapota and mango. Pulses are the important arable crops for this system. However, depending on the requirements, others like sorghum and pearl millet can be grown in the interspaces of fruit trees.

Reasons for this system not being widely adopted are:

- Economic position of farmers may not permit awaiting income after 5 or 6 years,
- Watering of fruit trees, till their establishment is a problem in summer period, and
- Marketing problems for perishable horticultural produce.

HORTI/SILVI -PASTORAL SYSTEM FOR NON-ARABLE LANDS

Class IV and above soil categories are uneconomical for arable crop production and are termed as non-arable lands. Horti-pastoral system is an agro forestry system involving integration of fruit trees with pasture. When fruit tree is replaced by a top-feed tree, it is called as silvi-pastural system.

Guava, custard apple and ber suits well in horti-pastoral system with grass like *Cenchrus ciliaris* (anjan), *C. setigerus* (bird wood), *Pabicum antidotale* (blue panic), *Dicenthicum annualatum* (marvel) and *Chloris gayana* (Rhodes) and legumes like *Stylosanthes hamanta* (hamata), *S. scabra* (stylo) and *Macroptilum atropurpuream* (siratro).

Top-fed trees ideal for silvi-pastural system are:

- *Acacia nilotica* (babul)
- *Acacia senagal* (gum Arabica)
- *Dalbergia sissoo* (shisham)
- *Harwickia binata* (anjan)
- *Leucaena leucocephala* (subabul)
- *Sesbania grandiflora* (agasthi).

Grasses and legumes indicated under horti-pastural system are also suitable for silvi-pastural system.

LEY FARMING FOR NON-ARABLE LANDS

A rotation system which includes a pasture(ley) for grazing and conservation is called alternate husbandry of mixed farming. A farm or part of it, entirely cropped with leys which are reseeded at regular intervals is termed as ley farming. It helps to conserve soil and improve its structure and fertility. For dryland it is a low risk system. A 4 year rotation system involving *Stylosanthes hamata* and sorghum has yield advantage of sorghum besides improvement in soil physical properties.

Veerashankar (1984) observed that soil organic carbon increased(0.56%) with three years of stylosanthes pastures leys compared to castor, sorghum or fallow rotations (Table 5). Similar trend was noticed with available nitrogen. There was a build up of 35 Kg N/ha with three years stylo pasture. At least two years are required to buildup soil nitrogen level.

Table 5. Soil fertility as influenced by duration of stylosanthes leys

Treatment	Before fourth year Sorghum		After fourth year
	OC %	Soluble N (Kg/ha)	Sorghum available N (Kg./ha)
C-S-C-S	0.22	110	82
C-S-Sh-S	0.31	113	97
S-Sh-Sh-S	0.35	141	107
Sh-Sh-Sh-S	0.56	145	106
C-F-F-S	0.15	88	63

C-Castor, S-Sorghum, Sh-Stylosanthes hamata, F. Fallow, source-Veerashankar (1984)

TREE FARMING FOR NON ARABLE LANDS

Trees can flourish abundantly where arable crops are not profitable. Farmers of dry lands are inclined to tree farming because of labour cost, scarcity at peak period of farm operations and frequent crop failure due to drought. A number of multipurpose tree systems (MPTS) have been tested for their suitability and profitability under different situations.

Annual rainfall less than 500 mm	Annual rainfall 500 to 750 mm
Acacia nilotica	Acacia nilotica
Acacia aneura	Acacia ferruginea
Acacia tortilis	Albizia lebbek
Acacia albida	Azadirachata indica
Prosopis cineraria	Casuarina equisetifiloa

Prosopis juliflora	Cassia sturti
Pithecallobium dulce	Dalbergia sissoo
	Leucaena leucocephala
	Taqrmarindua indica

TIMBER-CUM-FIBRE SYSTEM(TIMFIB)

Subabul intercropping with agave appears to be more remunerative at Bijapur of Karnataka. High density plantations (10,000 plants ha^{-1}) and short rotation intensive culture (4-5 years) have gained popularity in Nellore and Praksham districts of Andhara Pradesh. About 50% of rainfed area in these districts is under *Casuarina* and *Leucaena.* Such system will not be viable in marginal land degraded lands due to problems associated with survival of seedlings.

Agro forestry

Agro forestry may be defined as an integrated self-sustained land management system, which involves deliberate introduction/retention of woody components with agricultural crops including pasture/livestock, simultaneously or sequentially on the same unit of land, meeting the ecological and socio-economic needs of people. It is also defined as a collective name of land use systems and technologies where woody perennials are deliberately used from the same land management units as agricultural crops and/or animals in some form of temporal sequence. In agro forestry systems, there is both ecological and economic interaction between different components.

Based on the kind of associated agricultural products, major function of the tree component, spatial arrangements of trees and duration of combination, several systems have been identified. In India, agro forestry systems are classified as:

- Agri-silviculture : Crops + trees
- Silvi – pasture : Trees + Pasture
- Agro-silvi-pasture: crops +trees + pasture/animals
- Horti-pasture: Fruit trees + MPTS + Pasture/animals
- Agri-horti-silviculture: Crops + fruit trees +MPTS

Integrated Intensive Farming Systems for small Farmers

To dispel these concerns and to achieve sustainable food security in India, Swaminathan proposes the development of integrated Intensive Farming System (IIFS), which he believes, will faster an "Evergreen Revolution" in agriculture. Ti fulfill the promise of IIFS, he cites seven important areas where appropriate technology and more efficient (and effective) farm management practices are needed:

- Soil; health care
- Water harvesting and management
- Integrated nutrients and pest management
- Energy conservation and management
- Integrated crop/livestock system
- Information management and network systems

Interestingly, in mobilizing these resources to achieve a more sustainable agriculture and food security, Swaminathan does not call for additional research. He does, however, urge a better organization and dissemination of information and appropriate application of existing technologies at the household, farm and village levels.

References

Singh, R.P. and Dass,.S.K. 1995. Agronomic aspects of plant nutrient management in rain dependent food crop production systems in India. In: Sustainable development of dryland agriculture in India (Ed. R.P.Singh), Scientific publishers Jodhpur, Pp. 117-138.

Singh R.P. and Osman, M. 1995. Alternate land use systems for dry land In: Sustainable development of dryland agriculture in India (Ed. R.P.Singh), Scientific publishers Jodhpur, Pp. 375-398.

Umrani, N.K., 1995. Crop production technology for dryland areas. In : Sustainable development of dryland agriculture in India (Ed. R.P.Singh), Scientific publishers Jodhpur, Pp. 241-270.

Veerashankar, P.,1984. Ley farming . M.Sc. (Agri.) Thesis submitted to APAU, Hyderabad, India

Willey, R.W., Rao, M.R., Reddy, M.S. and Natrajan, M. 1982. Cropping system with sorghum pp. 477-490. In prodceedings of sorghum in the eighties. Hyderbad, India:ICRISAT.

CHAPTER 4 Global Scenario of Maize vis-a-vis Yield Gap Analysis and Strategies for Improving its Productivity in Rainfed Jammu Region- A Case Study

B. C. Sharma, Anil Kumar and Amarjit S. Bali

Division of Agronomy, Faculty of Agriculture, Sher-e-Kashmir University of Agricultural Sciences and Technology of Jammu, Chatha, Jammu-180009.

Introduction

Maize (*Zea mays L.)* or Corn belongs to the tribe tripsaceae (*Maydeae*) of the family gramineae and is one of the most important cereal crops in the world agricultural economy both as food for man and feed for animals. Maize provides nutrients for humans and animals and serves as a raw material for the production of starch, oil and protein, alcoholic beverages, food sweeteners and more recently, fuel. Maize is high yielding, easy to process, readily digested and costs less than other cereals. It has a very high yield potential. There is no other cereal on the earth which has so immense potentiality and that is why it is also known as **"queen of cereals"**. Every part of maize plant has economic value : the grains, leaves, stalk, tassel and cob. These can be used to produce a large variety of food and non food products. Today corn or maize is known in every suitable agricultural region of the globe. A corn crop is maturing somewhere in the world every month of the year. It grows from 58° north latitude in Canada and Russia to 40° south latitude in the southern hemisphere. Fields of corn are growing below sea level in the Caspian plain and at altitude of more than 3658 m in the Peruvian Andes. Corn is grown in regions with less than 25 cm of annual rainfall in the semi-arid plains of Russia and in the regions of more than 100 cm of rainfall on the pacific coast of Columbia. No other crop is distributed over so large area except wheat which occupies larger hectarage.

Global scenario

Maize is an important staple food in many countries of the world and the acreage and production of maize in the world have been increasing continuously. The production of maize increased from about 490 million tons in 1991-92 to more than 690 million tons in 2006-07 with ups and downs in in-between years with highest global production of 712 million tone realized during 2004-05 (Fig.1). The volatility in the production is mainly due to the variations in the yield which is affected by lots of factors besides increase or decrease in the area in different years. Factors like weather during crop growth, pest and disease attack, technological advances and development of new hybrids and varieties affect the yield of corn. Though bit erratic but almost similar trend in global acreage of maize like that of production realized under different years from 1991-92 to 2006-07 was observed. The global acreage increased from 133 million hectare in 1991-92 to more than 146 million hectare in 2006-07. The productivity of maize which was 3.68 million tons per million hectare in 1991-92 has increased to 4.76 million tones per million hectare in 2006-07 and it was not only due to increase in area under maize but also due to adoption of high yielding maize cultivars along with their production technologies by the maize growing farmers of the world. The per cent increase in area in 2006-07 over 1991-92 is only 9.7% whereas 40 per cent increase in production has been realized during 2006-07 over 1991-92.

The US has the largest harvested area of corn and contributes one fifth of the world corn harvested area. China, Brazil, Mexico, India and Indonesia are the other countries which contribute significantly to the world harvested area. These six countries have around 60 per cent of the world corn harvested area and the climatic conditions of these countries during the growth period of corn affect the yield and in turn the supply of corn in the world. From the data presented in the figure 1, it is inferred that USA has the lion's share in total global maize production accounting for 30 per cent of the production. Other major maize producers are China (15%), EU-25 (14%), Brazil (4%) and India (3%). Today, corn is grown on more than 146 million hectare of land and it produces an annual grain yield production of more than 690 million tonnes with an global average productivity of more than 47.60 q/ha (table-1). In total world production, it is out ranked only by wheat and rice and is the principle staple food in many countries, particularly in the tropics and sub tropics. On global scale USA ranks first in area (35022 thousand hectare), production (332092 thousand tonnes) and productivity (94.82 q/ha) with per cent contribution of 30 and 42.31 to the total world area and production, respectively. It is followed by China, Brazil, Mexico and India with per cent contribution of 15.0, 4.0, 3.0 and 3.0 in the global acreage respectively (Fig. 2).

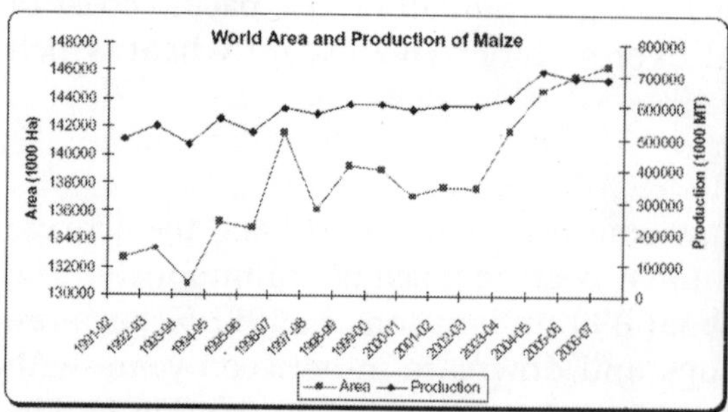

Fig. 1. World area and production of maize

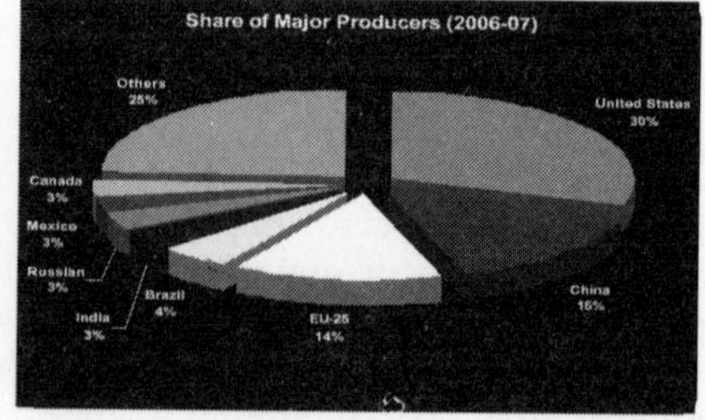

Fig. 2. Share of major producers

Table-1: Productivity of maize in major maize producing countries and the world average(q/ha).

S.No	Country	Productivity	World average
1.	USA	94.82	47.60
2.	Egypt	81.16	
3.	Canada	77.54	
4.	Australia	56.91	
5.	China	54.13	
6.	Brazil	37.30	
7.	Pakistan	32.40	
8.	Mexico	28.84	
9.	Ethopia	27.24	
10.	Russia Federation	25.20	
11	India	18.90	
12	Afghanistan	16.66	

FAO Statistics-May, 2008

National scenario

Maize cultivation in India is practised almost throughout the length and breadth of India on more than 6 million hectare area mainly as rainfed crop during rainy season and only about 25 per cent maize is grown under irrigated conditions. There is wide disparity in the productivity levels, depending on the crop duration, extent and distribution of rainfall and inherent soil fertility. Almost a continuous increase in national acreage, production and productivity of maize has been observed since 1950-51. The increase in acreage, however, remained fluctuating between 5.5 to 6.3 million hectare from 1970-71 till 2006-07. Through and through an increase in production and productivity of maize is realized since 1950-51 to 2006-07 except 1980-81, where a decline was observed. The area under maize increased from 3.4 million hectare in 1950-51 to more than 6 million hectare in 2006-07 and the production and productivity increased from 1.9 million tons to more 14 million tons and 600 to 1890 kgs/ha, respectively. A three fold increase in productivity with only 1.75 times increase in acreage in the country over the years 1950-51 to 2006-07 indicates that the farmers have developed an inclination towards adoption of new high yielding cultivars of maize along with their respective production packages. Under the limiting land resources the farmers have now been realizing that under the constraints of horizontal expansion to increase production, they are left with no option except to adopt the improved technologies for vertical expansion of maize (Fig. 3). Further, as is evident from the data presented in Fig. 4, it can be revealed that for about last 8 years i.e. 1999-02 to 2006-07 there has been almost a negligible increase in acreage under maize in India, however, a mixed trend in production and productivity has been observed during these years with highest production (14 million tons) and productivity (1990 kg/ha) realized during 2003-04.

The major area of the crop is confined to Rajasthan, Madhya Pradesh, Uttar Pradesh, Karnataka, Andhra Pradesh, Gujrat, Himachal Pradesh, Jammu and Kashmir, Bihar and Punjab. On national scale, Andhra pradesh ranks first in per cent share of area (17.0 %) towards national total as well as in productivity (28.25q/ ha). It is followed by Rajasthan (14.0%), Madhya Pradesh (12.0%), Bihar (10%) Uttar Pradesh (9%), Karnataka (8%), Gujrat (6%) and others (24%) in respect of acreage (Fig.5) and with a productivity realization of 21.64, 20.39, 20.04, 19.96, 17.99, 17.66, 17.06, 16.77, 16.62, 16.13 and 15.05 q/ha in Karnataka, Punjab, Maharashtra, West Bengal, Jharkhand, Madhya Pradesh, Gujrat, Tamil Nadu, Bihar, Himachal Pradesh and J&K, respectively and the national average being 18.90 q/ha) (Table 2).

In India, at present about 35 per cent of the maize produced in the country is used for human consumption, 25 per cent each for poultry feed and cattle feed and 15 per cent in food processing (corn flakes, pop corn etc). According to some experts, India may have to produce 20 million tonnes of maize to meet its requirements for human consumption, Piggery, pharma industry and fodder by 2020.

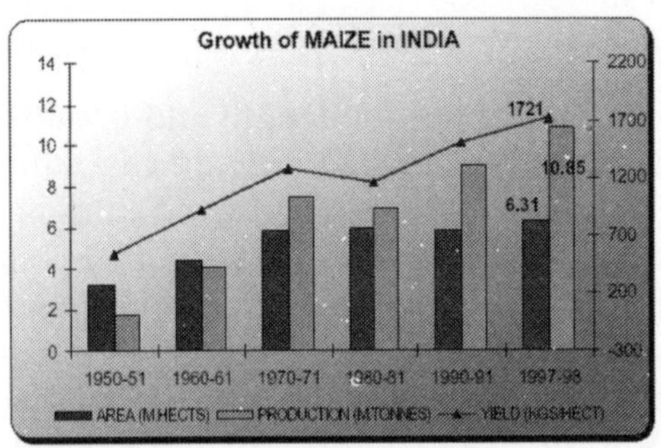

Fig. 3. Growth of maize in India

Fig. 4. Area, production and yield

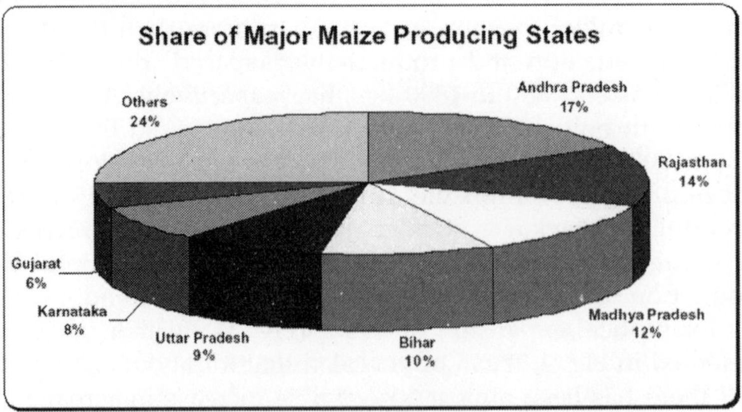

Fig. 5. Share of major maize producing states

Table 2: Productivity of maize in major maize producing states in India and National average (q/ha)

S.No	States	Productivity	National average
1.	Andhra Pradesh	28.25	18.90
2.	Karnataka	21.64	
3.	Punjab	20.39	
4.	Maharashtra	20.04	
5.	West Bengal	19.96	
6.	Jharkhand	17.99	
7.	Madhya Pradesh	17.66	
8.	Gujrat	17.06	
9.	Tamil Nadu	16.77	
10.	Bihar	16.62	
11	HP	16.13	
12	J&K	15.05	

Source: Directorate of economics & statistics, MOA, GOI(2004-05)

State scenario

In Jammu and Kashmir, the crop of maize is cultivated both in hilly as well as plain tracts under rainfed conditions occupying highest area of 323.60 thousand hectares among cereals with average productivity of 15.05/ha . Of the total acreage under maize in Jammu and Kashmir, 101.40 thousand hectares is in Kashmir and 221.90 thousand hectares is in Jammu province. Among the different districts of the state, Udhampur tops in acreage in the state as well as in Jammu Province, where as district Baramulla though at 4th position in the state occupies highest acreage among the maize growing districts of Kashmir province. There is a wide variation in average productivity of both the provinces. It is 10.05 q/ha in Kashmir and is about 17.33 q/ha in Jammu province (Table 4). Further, from the data presented in Table 6, it is clear that there is a continuous increase in area under maize in the state with almost similar trend between production and productivity realized under different years from 1964-65 to 2006-07 and with the highest of 5360 thousands quintals and 17.64 q/ha being during 1995-96, respectively (Table 5). Since there is no increase in the productivity of maize in the state over the years from 1964-65 to 2006-07 which indicates that whatever increase in production has been realized it is only because of bringing of more area under maize and the adoption of improved cultivars and their respective packages has not been well taken off by the farmers.

Table 3: Comparative land use (pattern) in Jammu & Kashmir vis-a-vis national status

Particular	Jammu (Lakh ha)	Kashmir (Lakh ha)	J&K (Lakh ha)	India (m ha)
Net sown area	3.89	3.39	7.48	142.8
Gross sown area	6.68	4.16	10.84	190
Cropping intensity (%)	172	123	148	135
Av. Holding size (ha)	1.02	0.53	0.76	1.41
%Irrigated area	25.7	56.3	41.5	38.5
% Rainfed area	74.3	43.7	58.5	61.5

Source : -Digest of statistics : 2005-06, Directorate of Economics and Statistics, Govt. of J&K & FAO Statistics, May-2008.

Table 4: District wise maize acreage and provincial productivity in J&K during 2006-07.

S.No	District	Area(Thousand ha)	Productivity(q/ha)
1.	Ananatnag	21.94	--
2.	Pulwama	7.31	--
3.	Srinagar	3.96	--
4.	Budgam	13.72	--
5.	Baramulla	28.09	--
6.	Kupwara	26.64	--
	Total Kashmir Province	**101.40**	**10.05**
7.	Jammu	16.93	--
8.	Udhampur	63.64	--
9.	Doda	51.67	--
10.	Kathua	19.52	--
11.	Rajouri	46.55	--
12.	Ponch	23.63	--
	Total Jammu Province	**221.90**	**17.33**
	Total J&K	**323.60**	**15.05**

Digest of statistics 2006-07, Directorate of Economics and Statistics, Govt. of J&K

Table 5: Area, production and productivity of maize in J&K (1964-65 to 2006-07)

S.No	Year	Area (Thousand ha.)	Production (Thousand quintals)	Productivity (q/ha)
1.	1964-65	246.00	3718	15.11
2.	1974-75	265.30	3036	11.37
3.	1985-86	286.98	4939	17.21
4	1995-96	303.87	5360	17.64
5.	1999-00	317.30	4712	15.92
6	2004-05	322.70	4922	15.25
7	2006-07	323.60	4869	15.05

Digest of statistics, J&K Govt, 2006-07

Table 6: Region and statewise maize productivity status in Jammu and Kashmir State (q/ha)

S.No	Year	Jammu	Kashmir	State
1.	1964-65	16.74	12.65	15.11
2.	1974-75	13.55	7.92	11.37
3.	1985-86	20.01	12.22	17.21
4.	1995-96	23.36	8.20	17.64
5.	2005-06	16.36	9.71	14.13
6	2006-07	17.33	10.05	15.05

Digest of statistics, J&K Govt, 2006-07

Global Yield gap

A comparative analysis of the harvested highest and average productivities of maize grains realized at global (9.48 and 4.97 tons/ha), national (2.83 and 2.0 tons/ha) and state level (2.60 and 1.47 tons/ha) reveals that our national and state highest and average productivities are far less to that of worlds highest and average productivities and as such there is quite a wide scope to increase it further (Table 7).

Table 7: Maize productivity scenario at national and international levels (tons/ ha)

S.No	Level	Highest	Average
1	Global	9.48 (USA)	4.97
2	National	2.83 (AP)	1.89
3	State(J&K)	2.60 (Distt Rajouri)	1.47

Sources : FAO, Statistics, May, 2008;Digest of statistics, J&K Govt, 2005-06

Yield gap analysis

Since most of the maize growing areas of Jammu province fall in the hilly tracts of the state, in order to have the true representation of the maize growing areas for yield gap analysis, three districts of Jammu province viz. one plain i.e Jammu and two hilly i.e Udhampur and Rajouri districts were selected to get true representative yield gap picture of the province to draw logical conclusion for the gaps in yield. The three years mean yield data of the field demonstrations on maize conducted by National demonstration scheme in district Udhampur and Krishi Vigyan Kendras in the districts of Rajouri and Jammu under Sher-e-Kashmir University of Agricultural Sciences and Technology and Sher-e-Kashmir University of Agricultural Sciences and Technology-Jammu was analyzed to work out yield differences between field demonstrations and local checks as well as even among the field demonstrations. From the perusal of the data presented in Table 8, it is observed that there also exists a wide gap not only between grain and stover yield of local checks and field demonstrations of maize but even in yields recorded under different maize demonstrations in all the three districts. Gap in grain yield of maize was observed between highest and lowest yielding demonstrations as well as local checks, however the lowest yield recorded under demonstrated maize plot was also even higher than the yields recorded under local checks of maize. In Rajouri district, a positive grain yield gaps of 9.0 and 20.0 q/ha, in Udhampur district 17.72 and 21.0 q/ha and in Jammu 8.81 and 16.34 q/ha were recorded by highest yielding maize demonstration over lowest yielding and local checks of maize, respectively.

A mixed trend in gaps of stover yield was noticed among the highest and lowest yielding maize demonstrations as well as local checks. A grain straw ratio of 1:4, 1:5.25 and 1:7 was observed in maize high and low yielding demonstrations and local checks respectively. Stover yield recorded under lowest yielding maize demonstrations and local checks, however, outyielded stover yields recorded under highest yielding maize demonstrations in district Rajouri and Only under lowest yielding maize demonstrations in the district of Jammu. Stover yield recorded under highest yielding maize demonstration conducted in District Udhampur, however outyielded stover yield of lowest yielding maize demonstration and local checks, whereas stover yield of local check out yielded stover yield of highest yielding maize demonstration in the districts of Jammu. The results of three years mean data of these three districts regarding grain and stover yield of maize presented in Table 9, exhibits that a positive gap of 11.67 and 19.12 q/ha in grain and 1.82 and 9.14 q/ha in stover yield was recorded under highest yielding maize demonstrations over lowest ones and local checks, respectively.

From the above narrated yield results especially in respect of stover yield, it has been worked out that it rarely happens that stover yield of maize crop raised with complete technological package is less as compared to stover yield of crop raised with partial technological package as well as local checks where in equal emphasis was given to the production of grain and stover yields. Moreover, if it has happened in certain situations or specific locations as was in case of Rajouri district (Table -8), even then it is not advisable to sacrifice 9.0 q/ha of maize grains costing about Rs 6000 for just 23 q/ha of stover just costing about Rs 1000/hectare. Moreover, farmers of such areas should be motivated to grow maize as fodder crop in just about 2 kanals of land exclusively by sowing fodder cultivars of maize like African Tall or J-1008 or local cultivar as fodder to compensate for the loss of stover yield and by doing so the farmer will just be sacrificing Rs 3000 as cost of fodder production in 2 kanals of land and 4.5 q of maize grains which otherwise would have been produced from the land being put exclusively under fodder crop and thus this way farmers besides being compensated for the loss of stover, can also save a sum of Rs 3000/hect.

Causes of low yield

Since most of maize dominated pockets of state are rainfed and reel under severe stress of forages for the animals particularly during winter months and maize stover is the only viable forage option for the farmers to feed their animals during these months. The farmers, thus raise crop of maize with their equal emphasis on production of grain as well as stover yield and in doing so they disagree with the crop geometry recommended for high yielding maize cultivars and thus sacrifice both grain and stover yield under the misconception that higher plant population over the recommended shall yield more stover, which otherwise does not happen as is evident from the date presented in Table 9.

(1) Under maize field demonstrations

- Use of excess seed than recommended.
- Sowing by broadcasting of seed without maintaining proper crop geometry.
- Diversion of some of the inputs allotted for field demonstrations, particularly fertilizer to other simultaneously planted crops.
- Casual approach of the farmers.

(2) Under local checks

- Cultivation of local cultivars
- Traditional techniques of cultivation
- Imbalanced nutrition

Table 8: Yield of maize in Jammu province (q/ha)

District	Variety	Yield						Yield gap			
		Grain			Stover			Grain		Stover	
		Demo. Highest	Demo. Lowest	Local check	Demo. Highest	Demo. Lowest	Local check	Demo. Lowest	Local check	Demo. Lowest	Local check
Rajouri	Kanchan-517	48.0	39.00	28.00	192.00	215.00	196.00	9.00	20.00	-23.00	-4.00
Udhampur	Super composite	43.00	25.80	22.00	172.00	141.90	154.00	17.72	21.0	30.10	18.00
Jammu	Kanchan-517	33.67	24.86	17.33	134.68	149.16	121.31	8.81	16.34	-14.48	13.27

Anonymous, 2007

Table 9 : Three years mean yield data of maize(qt/ha) of three districts

S.No	Yield						Yield gap			
	Grain			Stover			Grain		Stover	
	Demons. Highest	Demons. Lowest	Local check	Demons. Highest	Demons. Lowest	Local check	Demons. Lowest	Local check	Demons. Lowest	Local check
1	41.56	29.89	22.44	166.22	164.40	157.08	11.67	19.12	1.82	9.14

Anonymous, 2007

Reasons for Non adoption of improved technology

- Economic status of the farmers.
- Non-availability of inputs in time.
- Less exposure to technological advancements through proper result demonstrations.
- Lack of awareness.
- Some times due to wrong selection of the farmers.

Strategies to reduce yield gaps

In modern agriculture, the higher yield is possible only when there is an appropriate combination of soil, climate, plant and socio economic factors. Specific recommendations for achieving higher yields vis-à-vis and reduce yield gaps between potential and actually realized yields of maize are listed below

(i) Proper land preparation

(ii) Adequate and balanced fertilization

(iii) Use of good quality seed

(iv) Adequate plant density and spacing

(v) Seed treatment

(vi) Appropriate methods of fertilization, especially immobile nutrients such as P&K.

(vii) Top dressing of nitrogen at the right stage of the crop.

(viii)Timely control of insect-pests and weeds

(ix) Use of drought resistant cultivars

(x) Proper crop-rotation

(xi) Maintenance of organic matter

(xii) Harvesting at physiological maturity

In nutshell for minimizing yield gaps, complete adoption of location specific production recommendations are to adhered to in toto.

Conclusions

- Framers should be made aware about the actual loss they are suffering either in kind or cash by not adopting separate package of practices recommended specifically each for grain and fodder crops of maize just for little higher stover yields being realized through the traditional methods of cultivation.
- Farmers must be educated to grow a separate crop of maize for their additional requirement of fodder with exclusive cultivation of fodder variety under its recommended package of cultivation.

- By doing so, the farmers besides producing the additional fodder shall also be able to increase their own as well as the production of grains of Jammu province by three folds over the present.

References

Anonymous, 2008. FAO-statistics, May 2008.

Anonymous, 2006. Digest of Statistics, Govt. of J&K, 2005-06.

Anonymous, 1987. Annual report–National Demonstration (1985-87)

Anonymous, 2007. Annual reports, KVK, Rajouri, 2006-07.

CHAPTER 5

Changing Climate and Contingent Crop Planning for Rainfed Areas

B.C. Sharma, Anil Kumar, and Amarjit S. Bali

Division of Agronomy, Faculty of Agriculture, Sher-e-Kashmir University of Agricultural Sciences and Technology of Jammu, Chatha, Jammu-180009.

Introduction

Rainfed farming is beset with many problems, the paramount being the vagaries of monsoon rains. The crop production in drylands has peaks and troughs depending on the fluctuations in rainfall. The core input in Dryland agriculture, the rain water, is one which is most undependable. The efficient and timely use of rainwater makes all the difference between success and failure of crop production on rainfed lands. There are at least four aberrations in the rainfall behavior.

- The commencement of rains may be quite early or considerably delayed.
- There may be prolonged breaks during the cropping season.
- There may be spatial and / or temporal aberrations.
- The rains may terminate considerably early or continue for longer periods.

Onset of Monsoon

To quantify the aberrations in the onset of monsoon, a long term data on the dates of onset of monsoon are to be studied to work out normal, early or late per cent probabilities of rains for finalizing crop production strategies to be adopted under each situation. For example, if the normal date of onset of monsoon in a particular region is first of July but south-west monsoon set in the region as early as on 18th of June or as late as on 16th of July, analysis of the probability of the date of onset of the south-west monsoon on five days interval reveal that if it rains either on 25th of June or 6th of July, this situation cannot, for the practical purposes, be treated as a serious aberration requiring corrective or remedial measures. If it happens as stated above that is either on 18th of June or 16th of July, these situations, are certainly aberrant and as and when they occur, will require such corrective measures as change of crops or varieties from those being already sown in the region or adoption of other agronomic practices.

Breaks in the monsoon rains

Breaks in the monsoonal rains can be of different durations. Whereas, the breaks of shorter duration like five to seven days may not be of serious concern, the breaks of long duration of two to three weeks or even longer create the situation of plant water stress leading to reduction in agricultural production. The agricultural drought caused due to prolonged breaks in the monsoon rains can be of different magnitude and severity and affect different crops in varying degrees. Many of the crops are drought resistant and, inturn, there yield is not seriously affected by the occurrence of breaks in the monsoon rains, but several crops are quite sensitive to the drought situation and these breaks cause heavy reduction in the yield of the crops.

Another aspect of the drought caused by these breaks is the stage of the crops at which they occur and for the same crop the drought of the same intensity may not be of much consequence at one stage but it may have serious repercussions at some other stage of the growth. It is here that the real application of the meteorological information in terms of the frequency and probability of these breaks and, inturn, the droughts can be made by selecting the crops and varieties in such a way that the water sensitive stage(s) of the crop growth does not coincide with the breaks, most probable and more frequent. Another important factor in relation to breaks in the monsoon rains and the consequent droughts is the physical properties of the soil particularly its water holding capacity. Some of the soils (like deep black soils) have capacity to store as much as 300mm of available water in one meter depth, whereas, other soils (like the desert soils) can store only as little as 100mm or so in one meter depth. Thus, the impact of the drought is more pronounced in the soils having capacity to hold less amount of water than those having storage capacity, 15 days of dry- spell may not effect the crop but in the other soil of low moisture storage capacity with 5 days rainless situation, wilting symptoms may be clearly visible.

Aberrant withdrawl of south-west monsoon

A long term data on the dates of withdrawl of south-west monsoon have to be evaluated for generating quantitative information relating to it for any particular region of the country. For example, if in a particular region the dates of withdrawl of south-west monsoon is the last week of September but it withdrew as early as in the second week of September and as late as if persisted upto 31st of December But the probability analysis of the withdrawn dates of south-west monsoon on fortnightly interval revealed that it is withdrawn from 25th September to 15th October- a situation which cannot be called the aberrant one. But if it gets withdrawn during first fortnight of September or during the month of December, these situations obviously are aberrant and call for the attention of the agricultural planners and the scientists mainly on two fronts.

a) If the monsoon rains do terminate at a date (s) much earlier from the one considered to be normal, what measures are required to be taken to save the crop since by that time the crop had not reached the maturity stage and some soil moisture is still required till its maturity.

b) Whereas, crop and varieties in any given region are selected and grown based on the length of the growing season generally considered to be normal, the persistence of the rains much beyond the normal dates creates an extraordinary situation in the sense that more unusual water becomes available, but action plan may not exist to capitalize this amount of water.

Uneven distribution of monsoon rains in space and time

Such situations are encountered almost every year in one or other part of the country during south-west monsoon period leading to periodical drought and flood situations. As a matter of fact the high variability of rainfall (or more precisely the soil-water) is the single factor which influences the high fluctuations in the crop yields in different parts of the country. Only by having a thorough knowledge of the soil-water pattern, its variability, as judged from the evaluations of long term past climatic records, it is possible to select crops and varieties and formulate cropping systems which will minimize the effects of seasonal droughts and flash floods. Drought leads to moisture stress which, in turn, affects crop production adversely.

The objective of Dryland agriculture is to devise means to stabilize it by evolving contingent crop production strategies for rainfed areas under different weather conditions which are:

- Choice of suitable crops
- Choice of suitable crop varieties
- Alternate crop strategies
- Mid-season corrections
- Crop-life saving measures.

Choice of suitable crops

Often certain crops are grown more for convenience or by convention.This is because of the first priority of a farmer continues to be production of food for his family and feed for his animals. On the other hand, selection of crops suiting to the environment leads to increased and stabilized production.

Choice of suitable crop varieties

An extension of the above philosophy is to be the varieties with the era of hybrids and high yielding varieties of crops, particularly cereals, a replacement of the existing locals with these varieties or hybrids which act as primers for ushering in better crops in drylands.They may not do any better than the traditional varieties in years of poor rainfall, but these are capable of capitalizing on the better seasons.

And rainfalls, being a random phenomenon, good years tend to be a frequent, if not more, as the bad ones in the long run. It is therefore, necessary to capitalize on good rainfall years so that in the long run a farmer is better off if he chooses hybrids and high yielding varieties. Before considering the alternate crop strategies developed to meet weather aberrations, the normal behavior of the south-west monsoon, which is the main source of the rains for our country, need consideration.

Alternate weather crop strategies

Sowing of crops depends on rains in Drylands. But the rains may commence quite early or may be delayed considerably. And the same crop cannot be sown over time as the yields may be reduced because of pests and diseases or shorter growing season.

Mid-season corrections:

If moisture stress is observed during growing season and if it is at a very early stage, i.e. within a week to 10 days of sowing it is better to resow with subsequent rains than allowing inadequate plant stands to persist and yield a poor harvest. On the other hand, if the crop grew for 40-50 days, other means like ratooning or thinning needs to be considered. Thinning of plant population by removing every third row was found to be advantageous to mitigate moisture stress. If the break in the monsoon is very brief, soil mulching was found to be a tool in extending the period of storage of water in the soil profile due to reduction in the evaporation, which in other words, leads to extended periods of moisture availability.

Contingent crop production strategies for Jammu region

The Jammu province of J&K state has a net sown area of 3.89 lakh hectares constituting 52.2% net sown area of the state with a cropping intensity of 174.4% and 25.9% irrigated area. In the region the net area of 2.88 lakh hectares (74.1%) being rainfed is goverened by the monsoons with respect to crop production and its productivity. The province has been divided into three agro-climatic zone viz- 1. Sub-tropical zone which includes irrigated belt of Ranbir, Pratap and Ravi tawi command areas, kandi and unirrigated areas, II-Intermediate zone which consists of areas falling upto 1500 mtrs above mean sea level and III- Temperate zone which includes areas falling above 1500 mtrs above mean sea level. A perusal of the monthly, seasonal, annual as well as mean rainfall data of district Jammu presented in table.1 reveal that a considerable change in amount of rainfall and a shift in its pattern has been observed in past few recent years. This changing rainfall scenario is major setback to the farmers with regards to their usual cultivation pattern. A thorough thought needs to be given to decide upon and recommend to the farmers the contingent crop production strategies under this changing rainfall scenario. Keeping in view the grim situation arising out of prevailing drought like situations either before sowing of the crops or during their lifecycle, an alternative crop plan to meet weather aberrations in different agro climatic zones has been suggested here under:

Table 1: Rainfall of Jammu District (mm)

	Kharif					*Rabi*								
Year	Jun	Jul	Aug	Sept	Season total	Oct	Nov	Dec	Jan	Feb	Mar	Apr	May	Season total
2002	103.0	185.5	280.4	150.1	719.0	29.5	0.0	7.0	19.1	121.9	81.6	7.5	0.5	267.4
2003	80.5	456.9	485.4	139.1	1161.9	2.1	29.4	30.4	84.6	23.6	0.0	40.2	17.5	227.8
2004	136.2	216.8	198.2	38.7	589.9	46.9	4.2	31.0	105.4	138.8	97.4	10.3	10.5	444.5
Normal	83.7	317.8	311.0	122.2	834.7	22.9	12.6	24.6	46.8	52.3	69.9	31.1	24.1	284.3

Zone-1 (Sub-tropical)

Rice: Out of total area of 1.14 lakh hectares under rice about 80% of the area falls under assured irrigation and remaining 20% is cultivated as rainfed rice. In case of normal rainy season,farmers are advised to adopt the recommended package of practices but when rains get delayed, the farmers should adopt the following schedule:

- Use more no. of seedlings per hill
- Transplant the crop at closer spacings
- Add FYM @ 5-10 tonnes at the time of puddling.
- Short duration varieties like IET-1410 may be used.
- Direct seeding with higher seed rate of short duration varieties.
- Adopt efficient weed management practices to minimize crop weed competition.

Long term strategies to mitigate drought strees

Develop water shed management as a holistic approach which should aim at :

- Improvement of water resources
- In- situ soil moisture conservation
- Adoption of water harvesting techniques
- Land use as per its capability
- Improvement in the livestock breeds
- Encourage the plantation of agro-forestry trees viz-*Grewia optiva, Leucaena leucocephala, Albizia lebbeak, Dalbergia sisoo Acacia catechu, Acacia niloticia* and Poplar on the field bunds and farm boundaries for getting fodder and fuel wood etc.
- The installation of tube wells may be encouraged.
- Diversification in agriculture may be encouraged.
- At the tail ends where water scarcity is prevalent, irrigation water supply can be augmented with installation of tube wells.
- Development of drought resistannt varieties may be given priorities.

Zone-II (Intermediate)

Incase of inclement weather conditions with respect to deficit rainfall the following schedule may be adopted

- Instead of high yielding varieties use of local maize cultivars should be encouraged.
- Reduce the fertilizer dose of maize to 45:30:20 kgs NPK/ha.
- Intercropping of maize with pulses (cowpea/moong) may be adopted.
- Short duration pulses like moong may be sown as contingent crop if rainfall is delayed.
- Under very late rainfall conditions, cultivation of fodder (Maize + cowpea) may be adopted.

Zone-III (Temperate)

In case of unusual weather conditions the following recommendations should be adopted:

- Sowing of short duration and less water requiring crops/varieties of local maize and pulses should be preferred over high yielding long duration varieties.
- Split application of fertilizers should be followed when there is likelihood of rains.
- Rice cultivation be taken up in such areas only which have assured irrigation back up.
- Weed control practices should be adopted properly and removed weeds and roughed plants should be spread in- between the crop rows to check the evaporation of the soil.
- Under Doda district conditions, crops like Millets or lesser millets viz: Buck wheat, Amaranthus, Kangni, Salan, Kodo millet, Bajar Bhang and Cheena like crops may be sown in the event of failure of Maize on account of drought.
- Sowing should be done across the contours to conserve moisture.

Mid season corrections

- Encourage use of organic manures.
- Thin out the Plants to have optimum plant population in tune with the available soil moisture.
- Use removed plants and weeds as mulch material.
- Sowing Moong/Fodder can be done as contingent crops.
- Top dressing of N may be done in events of likelihood of rains.

Schedule of work under normal on-set of monsoon followed by long gaps.

1. **Clean cultivation (weed control)** : To check the moisture loss through transpiration of weeds.
2. **Laying of soil mulch:** To check the evaporation from the soil surface use locally available plant material as mulch.
3. **Reduction of plant population (Thinning of crops)** : To check the moisture loss through transpiration of thick sown crops.
4. **Foliar spray of urea (3%)** : During the dry spell spraying of urea (3%) is beneficial.
5. **Inter-row water harvesting**: Ridges and furrows enable the plant roots to derive moisture from the stored moisture within the furrows.
6. **Collection of runoff water** : Collect and utilize harvested water for kharif crop when there is a scarcity of water during any dry spell.
7. Schedule of work under delayed on set of monsoon.
8. **On-set of monsoon as late as second week of July**: Cultivate short duration varieties of normally sown crops.
9. **Dry sowing of seed** : Maize should be sown dry which will germinate just after the rains and thereby save the time in preparation of land.
10. **Minimum tillage**: Sow the seed with the first ploughing to use the available moisture for germination.
11. **On-set of monsoon as late as first week of August**: The cultivation of Pearlmillet, greengram and cowpea for fodder purpose can be taken up.
12. **Mid-season corrections**: When the monsoon is as late as in the third week of August the cultivation of local toria (Crop of 60-65 days) is the only viable option.

References

Anonymous, 2005. Http:\www.contingentcropplan.com\future2829.html

Anonymous, 2006. AICRP on Agrometeorology, SKUAST-J.

Anonymous, 2008. Digest of Statistics (2006-07). Directorate of Economics and statistics, Govt. of Jammu and Kashmir.

Water Resources Management in the Drought Prone Kandi Region

Vikas Sharma[1], K. R. Sharma[1] and Sanjay Arora[2]

[1]Division of Soil Science and Agricultural Chemistry, Faculty of Agriculture, Sher-e-Kashmir University of Agricultural Sciences and Technology of Jammu,Chatha, Jammu-180009.

[2]Central Soil Salinity Research Institute (ICAR), Bharuch, Gujrat

Rainwater is perhaps the most important natural resource in rainfed agriculture. The success and failure of the crop is solely dependent upon this resource who's occurrence is still beyond our control. About 65% of total cultivated area in India is still rainfed and it is estimated that even after exploiting full irrigation potential, about 50% of area will continue to remain rainfed. In India, more than 40% of food grains, 95% of coarse grains and pulses, 75% of oil seeds and 70% of cotton are being produced in rainfed condition (Ramesh & Devasenapathy, 2007). Inadequate soil moisture is the single most important factor that may result in partial or total failure of rainfed crops with the occurrence of mild to severe dry spells during the cropping periods. Mismanagement of this precious resource poses a great tragedy in recent times.

The average production from the rainfed areas has remained largely low as compared to the irrigated areas. A comparison of the recommended doses of nutrients between the irrigated and the rainfed for various crops gives us a clear picture about the limited nutrient use efficiency under dryland conditions. This is invariably due to the moisture stress conditions prevalent in these areas. Occurrence of long dry spells, late onset or early withdrawl of monsoons and skewed pattern of rainfall are important weather phenomenon that adversely impact crop yields under dryland conditions. Deficiency of water at the grain formation stage of kharif crops and at the time of sowing of rabi crop is the single most important factor affecting crop yields. In absence of optimum soil moisture and irrigation facilities; application of fertilizers, good crop seed and other inputs have little impact on crop productivity. Semi-arid regions, however, may receive enough annual rainfall to support crops but it is distributed so unevenly in time and/ or space that rainfed agriculture becomes unviable (Reij et al. 1988). Rockstrom and Falkenmark (2000) noted that due to high rainfall variation in semi-arid regions, a decrease of one standard deviation from the mean annual rainfall often leads to the complete loss of a crop. Pretty and Hine (2001) suggest, there is a 100% yield increase potential in rainfed agriculture in the developing countries, compared to only 10% for irrigated crops. This calls for increased efforts to upgrade rainfed systems globally and, especially in developing countries.

Regional constraints

A prerequisite to solving the water problems is the understanding of its extent and causes.

Climate

Dryland farming implies farming under relatively low annual rainfall (Monteith, 1990). In India, about 15 million ha of dryland area lies in the arid region which receives < 500 mm rainfall; another 15 million ha is in 500-750 mm range, about 42 million ha is in 750 to 1150 mm rainfall zone, with the remaining 25 million ha receiving > 150 mm rainfall per annum (Kanwar, 1999). In spite of areas in Kandi regionreceiving rainfall over 750 mm annually, its skewed distribution leads to moisture stress condition during the major part of the year. Although total rainfall seems to be sufficient for two crops in the area but the unreliable nature of rainfall leads to periodic drought. About 70 to 80 per cent of the annual rainfall occurs from July to September as a result of the South-West monsoon (Fig. 1). The rains are especially erratic in time and space. Most of the rainstorms received in summer season are of short duration and high intensity whereas those received in winter season are of low intensity and inconsistent. Soil moisture in the whole profile depends on the intensity of rainfall more than total rainfall. At high intensities, soil moisture increase only occur in the surface layer, but no significant increases are observed in deeper layers (Ramos & Martínez-Casasnovas, 2006). The Rainfall Variability Index (RVI) varies from -1.68 to +2.02 in monsoon and -1.38 to +3.60 in winter. Such variations in rainfall make agriculture production highly unstable especially in the rabi. All these results in serious problem of soil erosion through rainfall excess in summer monsoon months and soil moisture deficit in the winter months in the region.

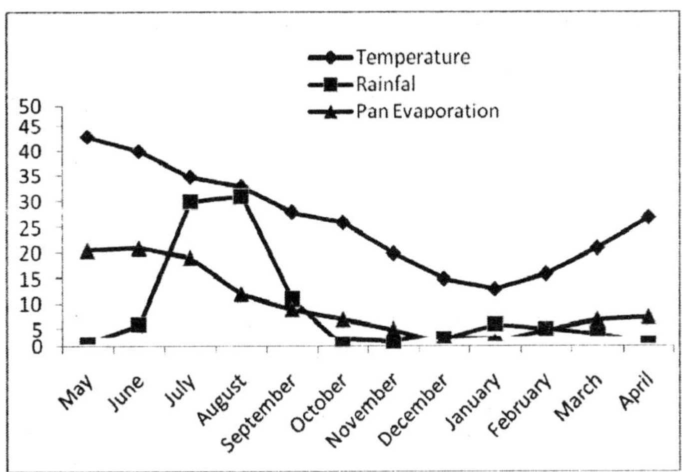

Fig 1. Monthly average rainfall (cm), Pan evaporation (cm) and temperature (°C) in Jammu region (mean of 20 years).

Topography

The lower shivaliks are classified under the physiographic unit Hills ranging from 320 to 720 meters above mean sea level (Rashid and Arora, 2007). The area is mostly dominated by Siwalik, Muree and Subathu group of sediments. The general lithology is sandstone, conglomerate, shale, silt stone and limestone. The weathering of Siwalik rocks has been proceeding at an extraordinary pace. There are gigantic escarpments and dip-slopes, separated by broad longitudinal strike valleys and intersected by deep meandering ravines of the streams (Hussain, 2000). The entire lower shivaliks are drained by a number of major and minor rivers. Hydrological units like rivers and streams flow naturally through their basins. Water can be transported between these basins if such facilities exist, for instance the Ravi-Tawi command system at the lower edge of the Shiwaliks. However, as we move up, it becomes difficult to derive any benefit from the water from the rivers as they flow in hilly terrain and cost involved in lifting the water and construction of canals would be too high compared to the area available for irrigation in the immediate vicinity.

Soils of the region

The geology of the region consists of mainly alluvial detritus derived from upper mountains swept downwards by rivers and streams. The sediment consists of sandstone, siltstone with minor amounts of shales and transported quartzite. Soils of the area are generally coarse in texture, low in organic matter, poor in nutrients and highly erodible (Sharma et al., 2009)(Table 1). The soil surface is infested with stones and water retention capacity is extremely poor. Most of these soils are classified as Ustifluvents, Ustipsamments and Ustochrepts with local haplustalfs and fluvaquents which remain dry for 4-5 months during a year.

Table 1: Physico-chemical properties of some *kandi* soils

Location	OC	OM	Clay	Sand	Texture*
	g kg^{-1}		%		
Hiranagar	4.2	7.24	23.2	28.7	L
Benaid	5.3	9.14	19.7	73.8	SL
Chak Shalla	3.9	6.72	14.1	80.7	SL
Thori	3.5	6.03	10.5	71.3	SL
Palth	5.5	9.48	22.1	39.9	L
Patti	3.7	6.38	12.1	81.5	SL
Raya	4.8	8.28	12.0	56.0	SL
Daboh	3.1	5.34	6.2	76.8	LS
Smailpur	3.3	5.69	13.2	79.2	SL
Dhiansar	3.7	6.38	11.2	79.6	SL

*L=Loam; SL=Sandy loam; LS=Loamy sand (Sharma *et al*, 2009)

Anthropogenic (In)activity

Due to little alternative opportunities available outside the agricultural sector and low land and labour productivity, most of the poverty is concentrated in rainfed regions (Singh 2001). Lack of technical knowledge of some improved rainwater management practices is one of the major constraints (Sharma *et al.*, 2004). The specialized and improved implements were also not available for rainwater management. About 76.6 per cent of the respondents feel that they had lack of skill in handling the practices like contour bunding, staggered trenches etc. About 95 per cent of the respondents had problems due to undulating topography. They had to rely more on human labour rather than machinery because they had sloping and fragmented land holdings. The survey data also reveal that 58 per cent of the respondents even did not know about all the rainwater management practices like bunding, trenching or haloding in maize. Only 31 per cent of the farmers adopt any of the soil and moisture/water conservation measures in the area. Small and fragmented land holdings in the region is another constraint in the adoption of modern mechanical measures of soils and water conservation. A sample survey in the *kandi* area of Jammu showed that 92% of the families have cultivable holdings of less than 2 ha and 80% have less than 1.5 ha (Arora *et al.*, 2006). The tenancy law of transferring the land to all children of the family has made the matter still complicated. Percentage of families having land holdings in different small to large categories of < 1.0 ha, 1 to 5 ha and > 5 ha in the region is hoen in Fig. 2.

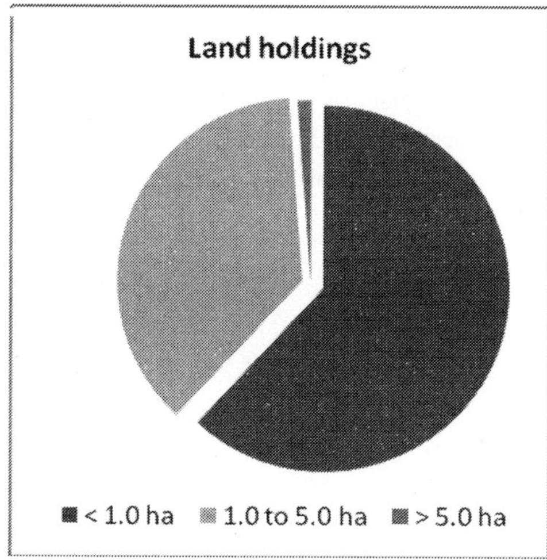

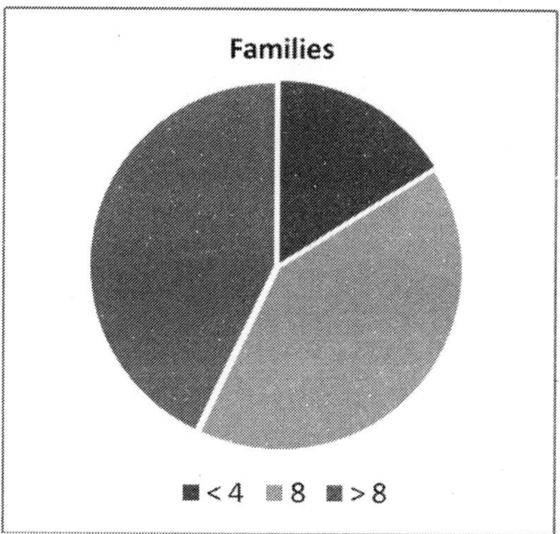

Fig 2. Land holding and family size in *kandi* region of Jammu (Arora *et al.*, 2006)

WATER CONSERVATION

Moisture can be conserved either within the field (*In-situ*) or harvested separately in specialized structures for rainwater harvesting (*ex-situ*). The most efficient and cheapest way of conserving rainfall in the shiwaliks, however, is to hold it in situ. Evaporation losses can also be reduced greatly if rainfall is stored in the soil rather than in structure with a free water surface. Moreover, the water in the soil is readily available to plants, whereas a large investment would be necessary if water is collected elsewhere and brought to the place of application. However collection outside the soil has an advantage that it can be applied at the time of need.

I. *In situ* water conservation

Land treatments

Leveling

The levelling process avoids ponding of water and provides uniformity in soil moisture. Waterlogging-sensitive rainfed crops such as maize, sorghum, pearlmillet and chickpea benefit from this approach. Land levelling with proper grade and provision for disposal of surplus rainwater results in better soil moisture condition and increases the yield of kharif crops like maize, bajre etc. Laser land leveling can reduce the water application rate upto 1509 M^3/ha in comparison with the unleveled fields (Abdullaev et al., 2007). However, laser land leveling is an expensive option for the farmers of this region, conventional methods, therefore, needs to be improved through scientific inputs.

Furrow system

Ridge and furrow in-situ water conservation is yet another promising system for rainfed areas. Maize is cultivated on raised beds (ridges) and furrows collects excess rainwater/ runoff and this moisture is beneficial for rabi crops. Broad beds and furrows is yet another method for conserving soil moisture in the soil. The broad bed and furrow system is laid within the field boundaries. Furrows act as a drainage channel during heavy rainy days.

Bunding

The purpose of graded bunds is to reduce the velocity of runoff water, for *in situ* conservation of rain water and to minimize the soil erosion. It can be laid upto 4-6% slopes. It helps to retain moisture in the field. The graded bund is a small earthen bund with slight grade constructed across the slope for safe disposal of runoff. The graded bunds are recommended up to 10% slope for areas where annual rainfall exceeds 750 mm. However, efficacy of graded bunds gets reduced gradually beyond 4% slope. In these structures, water is moved out of the fields through a graded channel, constructing it on a higher side of the bund at a non-erosive velocity and finally disposed off into a graded waterway. The grades to be provided range between 0.1 to 0.6 per cent which is almost one tenth of the original land slope. In case the length of the bund is between 150 m to 225 m, a uniform grade is provided and if it exceeds upto 400 m, a variable grade is recommended.

Terracing

Bench terraces are flat beds constructed on hills across the slope. In the steep hill slopes, mere reduction of slope length does not affect the intensity of storing of runoff. Bench terracing converts the original sloping ground into level step like fields which reduce the length as well as the degree of slope. The bench terracing is mostly recommended in the slope range of 16-33 per cent and help in slope reduction, reducing soil loss, uniform distribution of soil moisture and ultimately higher productivity. The height of the riser should not be more than one metre and width of bench terrace depends on the degree of slope. Renovation of rainfed bench terraces sloping outwardly has been found cost effective as compared to complete levelling (Juyal et al., 1988). Renovated bench terraces when put under improved management practices resulted in increasing yields over the traditional terraces.

Tillage operations

Tillage is usually practiced by the farmers for loosening the soil. It increases soil porosity and reduces weed incidence thus affecting soil-water-relationships. Tillage may result in initial loss of moisture, however due to formation of dry layer the moisture below is protected. Among the types of tillage, deep tillage has been found to be beneficial for moisture storage. Deep tillage breaks up the hard layer if any in sub surface and opens the soil for better water intake. Deep tillage shows a definite increase in the yields of a variety of crops, along with an increase in soil moisture content. In an on-farm study in a watershed having 1-2 % sloping topography, it was observed that soil moisture storage increased by 9.3, 12.7 and 16.8 per cent with shallow tillage, deep tillage and raised bed sowing of maize respectively over farmers practice at 80 DAS (Hadda et al, 2004).

Off-season tillage helps the rainwater to enter into the soil profile more effectively and in additional help in weed control. Any tillage that is carried out between two crop periods is termed as off-season tillage. This can be practiced with a blade harrow. This tillage aims at keeping the soil open for water entry from off season showers and for weed control.

Integrated Nutrient Management

Sharma and Arora (2008) observed that the application of FYM along with the inorganic fertilizer increase soil moisture in the soil moisture storage both in kharif and rabi seasons (Table 2). Soil without manure dries earlier as compared with the soil treated with manure, indicating that the farmyard manure improves water holding capacity of soil (Hossain and Ishimine, 2007).

Table 2. Effect of manure application on soil moisture storage.

Treatments	Mean % moisture storage (90 cm)	
	Maize	Wheat
NPK (Recommended)	5.35	3.30
NPK + 25% N through FYM	9.11	4.78
NPK + 50% N through FYM	9.64	5.25

(Sharma and Arora, 2008)

Mulching

To obtain maximum storage of moisture under any rainfall condition, the soil must absorb as much water as possible when it rains and losses due to evaporation or transpiration must be kept to a minimum (Xiping *et al.*, 2002). Mulching helps in doing just the same. Mulching is a well known practice for minimizing the evaporation of moisture from soil. Mulching refers to the use of crop residues, organic manures and other litter as well as synthetic material like polyethylene sheet etc., for the purpose of reducing evaporation and thus conserving moisture. Crop residues also play an important role in controlling water erosion. These dissipate the energy of falling raindrops, reduce surface sealing, increase infiltration and decrease runoff velocity (Kukal and Hadda, 1999). Mulches not only help in reducing soil erosion but also helps in conserving soil moisture and improving the yields in rainfed areas. Bhatt et al (2004) reported from a field study that dry matter yield as well grain and straw yield of maize improved significantly in mulched plots over the un-mulched plots in the kandi region of N-W sub-montane region of India. Moisture can be carried over for better wheat germination in mulched plots. Acharya et al. (1998) reported that mulches resulted in 0.06–0.10 m^3 m^{-3} higher moisture in the seed-zone when wheat was sown compared with the conventional farmer practice of soil tillage after maize harvest. The greater the surface area covered by mulch material in a particular mode, the greater is its effectiveness in conserving soil moisture (Arora et al., 2008)(Fig. 3).

Strip cropping

Strip cropping is refered to the system of cultivation where erosion-resistant legumes are grown along with cereals in strips for eg. Maize+mash. It can conserve rain water and reduce the velocity of runoff. In contour strip cropping, erosion resisting versus non-erosion resisting crops are grown across the slope on contours in regular long and narrow strips. On the other hand, field strip cropping is taken up in slopy areas having irregular topography where contour strip cropping is not possible. Strip cropping not only helps in conserving soil and moisture but helps in harnessing the benefits of crop rotation, in sustaining the productivity of the soils. Arora and Hadda (2007) studied the benefits of intercropping and observed that following a combination of maize and mash/cowpea brought the runoff to less than 50% of what was observed in only maize crop (table 3). This difference in runoff is retained in the field and thus, the moisture is available for a longer time. Further strip cropping can be carried out by incurring minimal costs over the actual, and therefore, is suitable for the resource poor farmers of the area. Efficiency of strip cropping may decrease beyond 5% slope.

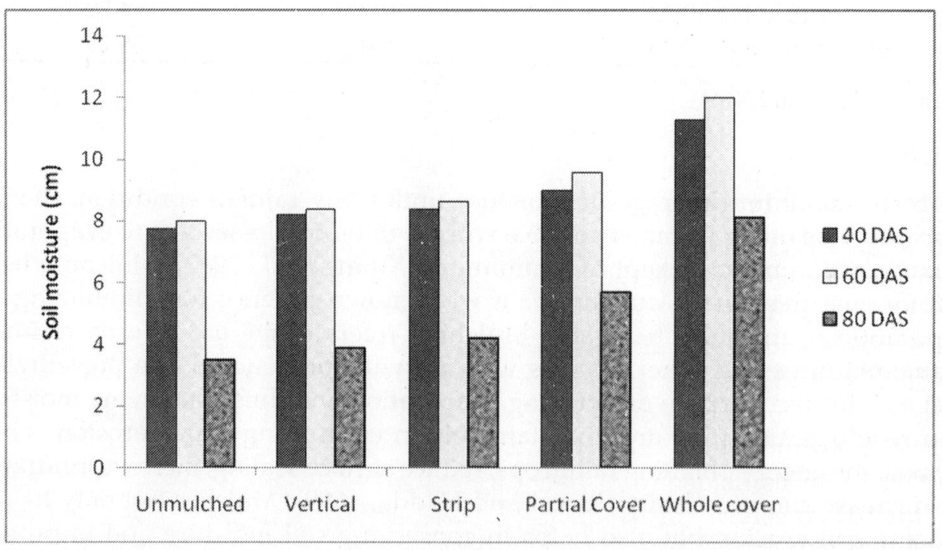

Fig.3.Impact of different types of mulching on soil moisture storage (Arora et al., 2008).

Table 3: Effect of strip/inter cropping on runoff and soil loss

Treatments	Runoff (% of rainfall)	Soil loss (Mg ha^{-1})
Maize sole	36.5	6.45
Mash sole	24.8	5.82
Cowpea sole	18.7	4.63
Maize+ mash	10.6	2.77
Maize + cowpea	14.2	3.34
Cultivated fallow	45.2	8.67

(Arora and Hadda, 2007)

II. Water harvesting (ex-situ)

Surface runoff water harvested in reservoirs can serve as a source of life saving irrigation to the crops and also have many indirect advantages like flood control, recharging of ground water, reclaiming land below the embankment and improvement of ecology etc. Due to limited period of flow and perched water table in the area, a little part of this runoff percolates. Thus, appropriate techniques for tapping this runoff for groundwater recharge needs to be identified which could not only help in maintaining the water table but will also improve moisture status for crop production. Further water harvesting structures can be used for storing water, which can be used during the lean period. There are a large number of indigenous, advanced and combination technologies available for rainwater harvesting.

Farm ponds

Farm ponds are small storage structures used for collecting and storing run-off water. As per the method of construction and their suitability for different topographic conditions farm ponds are classified into 3 categories, viz. Excavated farm ponds suited for flat topography, embankment ponds for hilly and rugged terrains with frequent wide and deep water courses; and excavated-cum-embankment type ponds. Selection of the location of the farm pond is dependent on several factors such as potentiality for yielding sizeable quantity of runoff, rainfall, land topography, soil type and structure, permeability/water-holding capacity, land-use pattern etc. Structurally, the excavated farm ponds could be of 3 types: square, rectangular and circular. All farm ponds must have the provision of removal of excess runoff water by providing 'drop inlet spill-way under normal condition' and 'emergency spill-way' to dispose off overflow of water after heavy rains. Such spill-way should ideally discharge into a grass waterway to avoid excessive erosion.

Rainwater can be utilized by collecting excess runoff in ponds rains in kharif season and then recycled it for supplemental irrigation at critical stages of rabi crop growth. Farm or village ponds are perhaps the oldest way of meeting the water needs of a community in the region. However, it has been observed that over the period of time a substantial number of ponds have become dead due to improper maintenance. The lack of collective will and receding of community feeling, coupled with emerging alternative sources of water for domestic purposes is pushing the remaining ponds towards extinction. There is an urgent need to revive these for ponds as well as construction of new tanks for collecting and storing runoff water. As per the method of construction and their suitability for different topographic conditions farm ponds are classified into 3 categories, viz. Excavated farm ponds suited for flat topography, embankment ponds for hilly and rugged terrains with frequent wide and deep water courses; and excavated-cum-embankment type ponds.

Check dams

Rainfed region of Jammu region has an extensive network of seasonal streams (choes) which receive short duration high intensity storms in summer months leading to massive discharges causing floods. The region is thus, also suitable for construction of numerous check dams. As these will not only help in water storage but also check flash floods that usually occur during high intensity rainfall. Bunds/ dams are manually constructed across the gullies to hold runoff water during the rainy period so as to create flooding in the upstream area temporarily. This stored water can then be used in the lean period. This also helps in controlling soil loss.

Dug-out tanks

Dug-out tanks are necessary in the *Kandi* areas of Jammu, Punjab and Himachal Pradesh because in spite of moisture conservation measures, water stress conditions prevail during long rainless periods. Polythene, brick and cement lining are considered vital because of the texture of the soil. Selection of the location of a dug out tank is dependent on several factors such as potentiality for yielding sizeable quantity of runoff, rainfall, land topography, soil type and structure, permeability/water-holding capacity, land-use pattern etc.

Sunken structure

Sunken structure/ponds are recommended practice in treatment of the upper, middle and lower reaches of drainage lines and accompany the other conservation measures such as contour vegetative hedges, vegetative filter strips, bank erosion stabilization, gully plugging, loose-boulder checks etc. the sunken dug-outs in the lower reaches of drainage-lines can also be linked with recharge wells to provide life-saving irrigation in the adjoining farm lands. These are constructed generally near the foothill and side of drainage lines.

The harvested rainwater in reservoirs can serve as a source of life saving irrigation to the crops and also have many indirect advantages like flood control, recharging of ground water, reclaiming land below the embankment and improvement of ecology etc. Jindal et al (1990) in the submontane Punjab reported that the yield of wheat crop was increased to 55.6 and 96.2 per cent by applying one irrigation after one month of sowing and pre-sowing plus one irrigation after one month of sowing respectively (Table 4). In case of maize, the yield increased to 32.8 per cent with only one irrigation when the crop was under moisture stress (Table 5). The depth of irrigation was 5 cm in each case.

Table 4. Effect of life saving irrigation on wheat crop yields

Year	No irrigation	Pre-sowing irrigation	Irrigation after one month	Pre-sowing irrigation + Irrigation after one month
1984-85	720	1280	1205	--
1985-86	1170	1780	1820	2350
1986-87	1300	1900	1880	2250
Average	1060	1650	1650	2280

(Jindal et al.,1990)

Table 5. Maize yield with one irrigation during stress

Year	Variety	Yield (kg ha^{-1})	
		No irrigation	One irrigation
1986-87	Navjot	23.70	31.50
	Pratap	17.88	21.37

(Jindal et al.,1999)

References

Abdullaev, I., Ul Hassan, M., & Jumaboev, K., 2007. Water saving and economic impacts of land leveling: the case study of cotton production in Tajikistan. Irrigation and Drainage Systems 21, 251-263.

Acharya, C. L., Kapur, O. C., & Dixit, S. P., 1998. Moisture conservation for rainfed wheat production with alternative mulches and conservation tillage in the hills of north-west India. Soil and Tillage Research 46, 153-163.

Arora, S., Hadda, M. S., & Bhat, R., 2008. Tillage and mulching in relation to soil moisture storage and maize yield in foothill region. Journal of the Soil and Water Conservation 7, 51-56.

Arora, S., Vikas Sharma, A. Kohli and V. K. Jalali. 2006. Soil and water conservation for sustainable crop production in *kandi* region of Jammu. *Journal of Soil and Water Conservation.* **5** (2): 77-82.

Deng Xiping, Shan Lun and Inanaga Shinobu. 2002. Assessments on the Water Conservation Practices and Wheat Adaptations to the Semiarid and Eroded Environments. 12th ISCO Conference, Beijing 2002. Pp. 348-360.

Hadda, M.S., Arora, S. and Khera, K.L. 2004. Effect of land, soil and nutrient management practices on moisture conservation and yield of maize in rainfed sub-montane region of Punjab. In : Abstracts, *68th Annual convention, ISSS*, Kanpur.

Hossain, M. A. and Ishimine, Y. (2007) Effects of Farmyard Manure on Growth and Yield of Turmeric (Curcuma longa L,) Cultivated in Dark-Red Soil, Red Soil and Gray Soil in Okinawa, Japan. *Plant Production Science* **10,** 146-150.

Husain, M. 2000. *Systematic Geography of Jammu & Kashmir.* Rawat Publications, Jaipur, India. Pp.250.

Jindal, P.K., Singh, R.P. and Singh, I.B. 1990. Technical feasibility and economic viability of small irrigation dams in the submontane area of Punjab state – a case study. *Proc. Int. Symp. Water erosion, Sedimentation and Resource Conservation, Dehradun,* pp. 507-515.

Juyal, G.P., Katiyar, V.S., Gupta, R.K. and Sewa Ram (1988) Renovation of rainfed terraces in Garhwal Himalayas. *Indian J. Soil Cons.* 13 (1) : 10-15

Kanwar, J. S. 1999. Need for a future outlook and mandate for dryland agriculture in India. In *Fifty years of dryland agricultural research in India.* Central Research Institute for Dryland Agriculture. Hyderabad, India. Pp. 11-19.

Kukal,S.S. and Hadda,M.S. 1999. Assessment of water erosion and its management in submontane Punjab, India- A case study. Pak. J. Soil Sci. 17: 79-82.

Monteith JL (1990) Steps in crop climatology. *In* 'Challenges in dryland agriculture, a global perspective: proceedings of the International Conference on Dryland Farming.' pp. 273-282. Amarillo/Bushland, TX, USA.

Pretty, J.; Hine, R. 2001. Reducing Food Poverty with Sustainable Agriculture: A Summary of New Evidence. Final report of the "safe World" Research Report. University of Essex, UK.

Ramesh, T., & Devasenapathy, P., 2007. Natural Resources Management on Sustainable Productivity of Rainfed Pigeonpea (*Cajanus cajan* L.). Research Journal of Agriculture and Biological Sciences 3, 124-128.

Ramos, M., & Martínez-Casasnovas, J., 2006. Impact of land levelling on soil moisture and runoff variability in vineyards under different rainfall distributions in a Mediterranean climate and its influence on crop productivity. Journal of Hydrology 321, 131-146.

Rashid, G. and Arora, S. 2007. Physiography in relation to sustainable agriculture in Jammu. In: *Natural resource management for sustainable hill agriculture* (eds. Sanjay Arora, S. S. Kukal, and Vikas Sharma). SCSI Jammu Chapter, Jammu. Pp. 93-100.

Reij, C .1988.Impact des techniques de conservation des eaux et du sol sur les rendements agricoles: analyse succincte des donnees disponibles pour le plateau central au Burkina faso. CEDRES/ AGRISK.

Rockstrom, J.; Falkenmark, M. 2000. Semi-arid crop production from a hydrological perspective- Gap between potential and actual yields. Critical Reviews in Plant Sciences 19 (4): 319-346.

Sharma, Vikas, Arora, S., Jalali, V. K. and Kher, D. 2004. Studies on constraints in adoption of improved soil and rainwater conservation practices in *kandi* region of Jammu, *J. Res. SKUAST-J.* **3(2)**: 214-220.

Sharma, Vikas and Sanjay Arora. (2008). Effect of balanced fertilization on moisture storage and yield in maize-wheat system under rainfed conditions. *Proceedings of National Seminar on Policy Interventions for promotion of Balanced Fertilization and Integrated Nutrient Management.* April 10-11, Palampur. Pp. 93-94.

Sharma, Vikas, S.H. Mir and S. Arora, (2009). Assessment of fertility status of erosion prone soils of Jammu Siwaliks. *Journal of Soil & Water Conservation.* 8(1):37-41.

CHAPTER

7 Estimation and Prediction of Runoff from Small Watersheds – A Decision Support System for Resource Conservation

N.K. Gupta[1], Neetu Sharma[2] and Parshotam K. Sharma[2]

[1]*Senior Scientist, Water Management Research Centre, Sher-e-Kashmir University of Agricultural Sciences and Technology of Jammu, Chatha, Jammu-180009*

[2]*Division of Agronomy, Faculty of Agriculture, Sher-e-Kashmir University of Agricultural Sciences and Technology of Jammu, Chatha, Jammu-180009.*

Runoff water is that fraction of precipitation which appears as surface flow in the drainage network of a watershed. Runoff occurs only when the rate of precipitation exceeds the rate of infiltration. Once the infiltration rate is exceeded, water begins to fill the surface depression such as puddles pits and small ponds, this stored water is called depression storage. After filling the depression, water starts flowing over the land as overland flow. The volume of water involved in the build up of head on the ground surface to create runoff is called surface detention and when in the channel it is called channel detention. Before surface runoff can occur, precipitation should satisfy the demands of evapotranspiration, interception, infiltration, surface storage, surface detention and channel detention. The water in the surface storage infiltrates into the soil or evaporates into the atmosphere. The water that infiltrate into the soil but does not reach down to the ground water level moves laterally to join the stream below and is called subsurface flow or interflow. The part of the water that percolates further interflow down may reach the ground water level. Depending upon the hydraulic gradient of stream a fraction of ground water moves and joins the stream below. This is generally called base flow. Thus all the three types of flow contribute to the stream flow, it is the overland flow which reaches first the stream channel, the interflow being slower reaches after a few hours and the base flow being the slowest reaches the stream channel after some days

Thus Direct Runoff = Overland flow + inter flow + base flow.

Estimation of runoff is required for, designing of soil and water conservation structure. For assessing the storage in earthen dams, tanks and ponds etc. estimates of runoff volume or water yield are required.

Factors affecting runoff

The various factors which affect runoff are:

a) Physiography

Size – Both runoff volumes and rates increase as watershed size increases. The size of watershed is an important parameter in determining the peak rate of runoff and peak runoff rate value is needed for design of erosion control structure and channel to carry maximum runoff safely. For the design purpose we must know the rise or fall of stream with respect to time and plotting rise or fall against times gives us hydrograph of flow. The hydrograph is the key tool used in watershed planning; It

- points out the problem
- indicates the type of control needed
- provides design data and
- portrays the effectiveness of the planned or installed control.

By superimposing the rate of precipitation on the watershed hydrograph, it is possible to determine how a watershed will react to various type of storm.

Shape- A watershed may have different shapes. A long narrow watershed will have greater distance to outlet as compared to a wider one in spite of both having the same area. Its time of concentration will be longer its corresponding intensity lower and maximum rate of runoff will also be less. Thus long and narrow watershed are better than square one as time of concentration will be more. Longer the time taken by water to leave the watershed greater the opportunity for the water to infilterate.

Land slope: The speed and extent of runoff depend on slope of the land. The greater the slope, the greater the velocity of flow of the runoff water. Land slope cannot be directly changed but can be shorten by terracing or bunding along the contour thus minimizing the velocity and the resulting runoff. The land slope in percent can be determined from a topographic map by the following formula:

$$S = 100 \times MN/A$$ *(Linsley and Franzini, 1979)*

where M = Total length of all contours within a watershed (m)

N = Contour intervals (m)

A = Area of the watershed (m^2)

The degree of slopes sets limits on land use for annual crops, plantation and even on land reclamation, depending on soil depth, stoniness etc. Hence degree of slope and the length of slope are important factor affecting runoff.

Drainage Density : The drainage density affect runoff pattern in that a high drainage density drains runoff water rapidly, decreases the lag time and increases the peak of hydrograph.

Drainage Density (DD) = Total length of all streams (m)/ Catchment area in (m^2)

Drainage Pattern: Drainage Pattern of an area refers to the design of the stream courses and their tributaries. It is influenced by the slope of the land, lithology and structure. A study of drainage pattern and drainage texture is helpful in the interpretation of geomorphic features. Drainage pattern act as guidelines to locate vulnerable areas requiring different kinds and degree of soil conservation measures.

b) Soils and Geology: The soils and geology of the watershed also determine the amount of water which will percolate and corrective measures which will be used. The soil characteristics also determine the amount of silt which will be washed down into water harvesting structure and the valleys below. Crop residues tilled into the soil and the residual root system from grasses that have been in crop rotations produce a good hydrologic condition, thereby reducing the volume of runoff.

c) Land use: The land in a watershed is used for various purposes such as Crop production, Animal husbandry, Afforestation, Housing, Roads etc. Thus land use will affect rates of runoff, infiltration and erosion.

d) Vegetative cover: The type and quality of vegetative cover on watershed lands influences runoff, infiltration rates, erosion, sediment yield and evapotranspiration and therefore, will determine the hydrological condition (Table 1).

A dense cover of vegetation is most powerful weapon for reducing erosion. Vegetative covers vis-à-vis runoff relationship have to be developed for harnessing and storage of runoff water for different purposes e.g., flood moderation, runoff recycling etc. Vegetation affects runoff in several ways. The foliage provides the sealing of the soil surface from the impact of the rain drops. Some of the raindrops are returned on the surface of its foliage, increasing their chance of being evaporated back to atmosphere. Transpiration of soil moisture from previous rains leaving a greater void in the soil to be filled. Vegetation including its ground litters forms numerous barriers along the path of water flowing over the surface of soil. This lengthens the time of concentration and reduces the peak discharge rate. A good vegetative cover exists in grassland if it covers 75% or more of the ground. A poor vegetative cover exists if it covers < 50% of ground. Grass cover is evaluated on basal area of plot whereas trees and shrubs are evaluated on the basis of canopy cover.

Table 1. Classification of forest & trees crops

S. No.	Vegetative condition	Hydrological condition
01.	Heavily grazed or regular burnt, litter, small trees and brushed are destroyed	Poor
02.	Grazed but not burnt. There may be some litter, but these woods are not protected.	Fair
03.	Protected from grazing, litter and shrubs cover the soil.	Good

e) Precipitation: The amount and nature of precipitation is the most important factor which determines what will happen in a watershed. Rainfall evenly distributed throughout the year has a different impact from sudden, sharp showers or seasonal rainfall. Low intensity (< 3mm/hr) storms over longer spells contribute to ground water storage and produce relatively less runoff. A high intensity (> 8mm/hr) storm or smaller areas covered by it increase the runoff since the losses like infiltration and evaporation are less. If there is a succession of storms, the runoff will increase due to initial wetness of the soil due to antecedent rainfall. Rain during summer season will produce less runoff, while that during winter will produce more. Greater humidity decreases evaporation. The pressure distribution in the atmosphere helps the movement of storms. Snow storage and specially the frozen ground greatly increase the runoff.

f) Time of concentration (T_c): It is the time taken by runoff water to flow from most remotest point (in term of flow) to the outlet of the watershed. When the duration of a storm equals the time of concentration, it is assumed that all parts of the watershed are contributing simultaneously to the discharge at the outlet.

The main variables affecting the time of concentration of the catchment are:

Size: Larger the catchment, longer will be the time of concentratin Peak runoff (if expressed as cumec/km^2) decreases as the catchment area increases due to higher time of concentration.

Topography: Steep rocky catchments with less vegetation will produce more runoff compared to flat tracts with more vegetation. If the vegetation is thick greater is the absorption of water, so less runoff. If the direction of the storm producing rain is down the stream receiving the surface flow, it will produce greater flood discharge than when it is up the stream. If the catchment is located on the orographic side (windward side) of the mountains, it receives greater precipitation and hence gives a greater runoff. If it is on the leeward side, it gets less precipitation and so less runoff. Similarly, catchments located at higher altitude will receive more precipitation and yield greater runoff.

Shape of the catchment: Topography affects the shape of the watershed. A fan shaped catchment produces greater flood intensity since all the tributaries are nearly of the same length and hence the time of concentration is nearly the same and is less, whereas in the fern shaped catchments, the time of concentration is more and the discharge is distributed over a long period. The ratio of the length (L) of the mainstream to the average width (W) of a watershed may be refered to as its shape factor (L/W). An increase in the shape factor value causes a reduction of the peak discharge rate due to the longer time of concentration and vice-versa.

To estimate the time of concentration empirical formula is used:

$$T_c = 0.0195\ K_c^{\ 0.77}$$

Where, $K_c = (L^3/H)^{0.5}$

T_c = Time of concentration in minutes

L = Maximum length of flow in metres

H = Difference in elevation between the most remotest point and the outlet, in metres

Experience has shown that T_c calculated this way is quite short, resulting in an over estimation of the discharge. Ven Te Chow proposed to use:

$$T_c = 0.15\ K_c^{\ 0.64}.$$

g) Conservation Practices: Conservation practices reduces sheet erosion and thereby maintain an open structure of the soil surface. This reduces the volume of runoff but the effect diminishes rapidly with increase in storm magnitude. Contouring & terracing reduces sheet erosion and increases the amount of rainfall withheld from runoff by the small reservoirs. They form gradient terraces increases the distance water must travel and thereby increases the time of concentration. This in turn reduce the peak rate of discharge.

Estimation of Design Runoff rates

The capacity to be provided in a structure that must carry runoff may be called design runoff rates. Temporary structures are designed for a runoff that may be expected to occur once in 10 years. Expensive and permanent structures will be designed for runoff expected only once in 50 or 100 years. There are a number of methods used for estimating runoff from watersheds. The methods commonly used for small watersheds are:

- Rational Method
- Cooks Method
- Hydrologic soil cover complex, number method and
- Table Method/Hydrograph theory

All these methods are based on two assumptions:

1. Rainfall occurs at uniform intensity for a duration at least equal to the time of concentration of the watershed.
2. Rainfall occurs with a uniform intensity over the entire watershed.

C.E. **Ramser** of USA proposed the Rational Method as

Q = CIA/360 where

Q = Peak rate of runoff in cum/sec or m^3/sec for the given frequency of rainfall.

C = Rational runoff co-efficient having values ranging from zero to one depending upon watershed condition. The value of "C" will depend upon, slope, vegetative cover and texture (dimensionless).

I = Intensity in mm/hr for design frequency and for duration equal to time of concentration.

A = Area of watershed in hectares.

Ramser purposed the formula as a result of measurements of rates of runoff and rainfall from small agricultural watersheds. Depending upon the slope, type of soil and vegetative covers, different values of "C" were suggested (Table 2, 3 & 4).

Table-2 Value of C used in Rational method

Vegetative cover and slope (%)	Soil Texture		
	Sandy loam	Clay & Silty loam	Stiff clay
I Wood land/forest			
0-5	0.10	0.30	0.40
5-10	0.25	0.35	0.60
10-30	0.30	0.50	0.60
II Pasture Land			
0-5	0.10	0.30	0.40
5-10	0.16	0.36	0.55
10-30	0.22	0.42	0.60
III Cultivated Land			
0-5	0.30	0.50	0.60
5-10	0.40	0.60	0.70
10-30	0.52	0.72	0.82

Singh *et al*, 1990

Table-3 Runoff coefficient C for various types of catchment

Type of catchment	Value of C
Rocky and Impermeable	0.8-1.0
Slightly permeable, bare	0.6-0.8
Cultivated or covered with vegetation	0.4-0.6
Cultivated absorbent soil	0.3-0.4
Sandy soil	0.2-0.3
Heavy forest	0.1-0.2

Ragunath, (1996)

Table-4 Runoff coefficient C for various types of terrain

Type of Terrain	Value of C
Flat Residential Area	0.4
Moderately steep Residential Area	0.6
Built Up Areas- Impervious	0.8
Rolling Land And Clay Loam Soils	0.5
Hilly areas, forests, clay and loamy soils	0.5
Flat cultivated lands and sandy soils	0.2

Ragunath, (1996)

The choice of the design frequency depends upon the type of structure, whether temporary, semi permanent or permanent. The greater the recurrence intervals the longer will be the life of the structure but the cost will also escalate.

Limitations : This method though very simple have following limitations:

- Since there is hardly ever a rainfall completely satisfying both the assumptions of uniform intensity for atleast the duration of time of concentration and over the entire area of the watershed, the estimation based on this method is rather approximate. Rational method is normally applied to watersheds having areas less than 1300 ha.
- The value of C in this method are based on studies on a broad range of topography, soil and land use conditions. However, the characteristics and condition of small watershed are greatly affected by land use, tillage and cropping pattern/ practices. The effect of such soil conservation and agronomical practices has not been considered separately in this method for the estimation of C value.

In spite of these limitations, this method gives more accurate results & used widely where less costly structure are to be erracted especially in agricultural watershed. The soil conservationists in the field do not have the data on local values of rainfall intensities, duration and rainfall. Figures will serve as a tentative guide with all relevant nomographs and rainfall frequency curves required for runoff estimation by the Rational Method.

Procedure to estimate runoff from small watershed

Step 1: Determine the areas under various land use and soils (A_1, A_2, — — A_n).

Step 2: Determine the value of C from Table 2, 3 and 4 depending upon slope, soil type, land use and nature of surface. In case the watershed comprises more than one land use or soil types, the average value of C is computed for the watershed as given below:

$$C = (A_1C_1 + A_2C_2 + A_3C_3 + \ldots\ldots\ldots A_nC_n)/A$$

Where, A_1, A_2, A_3, ...A_n are the areas in hectares having values of C as C_1, C_2, C_3, – C_n, respectively and A is the total area of the watershed equal to $A_1 + A_2 + \ldots . A_n$ in hectares.

Step 3: Determine the time of concentration (T_c) of the watershed from Fig. 1 by using the maximum length of flow and fall along the line. In estimating fall, sudden drops should not be considered. Alternatively, T_c can be calculated from the following empirical formula (Chow, 1959):

$$T_c = 0.15\ K_c^{\ 0.64}$$

Step 4: Convert the one-hour intensity for design frequency for duration equal to time of concentration by using the formula:

$$I = (KT^{a}) / (t+b)^{n}$$

The values of constants K, a, b and n are given in Table .... or use fig..... and .... Note: K_c is not the same as K and T is recurrence interval in years, t rainfall duration/storm duration in hours.

Step 5: Compute the peak rate of runoff by using the Rational formula.

Q = CIA/360

Example: Estimate the peak rate of runoff expected to occur once in 25 yr from 50 ha watershed located 76°E Longitude and 25°N Latitude with sandy loam soil. The land use comprises 20 ha agricultural land, 15 ha grassland and 15 ha of forest land. The difference in elevation between highest and outlet point is 20m and maximum length of run is 1200 m. The average land slope is 3%.

Solution:

i) The average value of "C" from table 1 is

C = 20x0.3+15x0.1+15x0.1 /20+15+15 = 0.18

ii) Time of concentration = 22.16min.

For longitude of 76° and latitude of 25°, 1 hr. rainfall intensity for 25 year frequency is 120 mm/hr.

From fig (1) rainfall intensity for duration equal to time of conc. 0.4hr is 175 mm/hr.

Q peak = C I A/360 or 0.18 x 175 x 50 /360 = 4.375m³/sec.

Table-5 Intensity-Duration-Return period relationships, India

Zone	Station	K	a	b	N
Northern Zone	Agra	4.911	0.1667	0.25	0.6293
	Allahabad	8.570	0.1692	0.50	1.0190
	Amritsar	14.41	0.1304	1.40	1.2963
	Dehradun	6.00	0.2200	0.50	0.8000
	Jaipur	6.219	0.1026	0.50	1.1172
	Jodhpur	4.098	0.1677	0.50	1.0369
	Lucknow	6.074	0.1813	0.50	1.0331
	New Delhi	5.208	0.1574	0.50	1.1072
	Srinagar	1.503	0.2730	0.25	1.0636
	Northern Zone	5.914	0. 1623	0.50	1.0127

Central Zone	Bagra-tawa	8.5704	0.2214	1.25	0.9331
	Bhopal	6.9296	0.1892	0.50	0.8767
	Indore	6.9280	0.1394	0.50	1.0651
	Jabalpur	11.379	0.1746	1.25	1.1206
	Jagdalpur	4.7065	0.1084	0.25	0.9902
	Nagpur	11.45	0.1560	1.25	1.0324
	Punasa	4.7011	0.2608	0.50	0.8653
	Raipur	4.683	0.1389	0.15	0.9284
	Thikri	6.088	0.1747	1.01)	0.8587
	Central Zone	7.4645	0. 1712	0.75	0.9599
Western Zone	Aurangabad	6.081	0.1459	0.50	1.0923
	Bhuj	3.823	0.1919	0.25	0.9902
	Mahabaleswar	3.483	0.1267	0.00	0.4853
	Nandurbar	4.251	0.2070	0.25	0.7704
	Vengurla	6.863	0.1670	0.75	0.8683
	Veraval	7.787	0.2087	0.50	0.8908
	Western Zone	3.974	0.1647	0.15	0.7327
Eastern Zone	Agartala				
	Dumdum	8.097	0.1177	0.50	0.8191
	Guwahati	5.940	0.1150	0.15	0.9241
	Gaya	7.206	0.1557	0.75	0.9401
	Imphal	7.176	0.1483	0.50	0.9459
	Jamshedpur	4.939	0.1340	0.50	0.9719
	Jharsuguda	6.930	0. 1307	0. 50	0.8737
	North	8.596	0.1392	0.7 5	0.8740
	Lakhimpur	14.070	0.1256	1.25	1.0730
	Sagar Island	16.524	0. 1402	1.50	0.9635
	Shillong	6.728	0.1502	0.75	0.9575
	Eastern Zone	6.933	0.1353	0.50	0.8801
Southern Zone	Bangalore	6.275	0.1262	0.50	1.1280
	Hyderabad	5.250	0.1354	0.50	1.0295
	Kodaikanal	5.914	0.1711	0.50	1.0086
	Madras	6.126	0.1664	0.50	0.8027
	Mangalore	6.744	0.1395	0.50	0.9374
	Tiruchirapalli	7.135	0.1638	0.50	0.9624
	Trivandrum	6.762	0.1536	0.50	0.8158
	Visakhapatnam	6.646	0.1692	0.50	0.9963
	Southern Zone	6.311	0.1523	0.50	0.9465

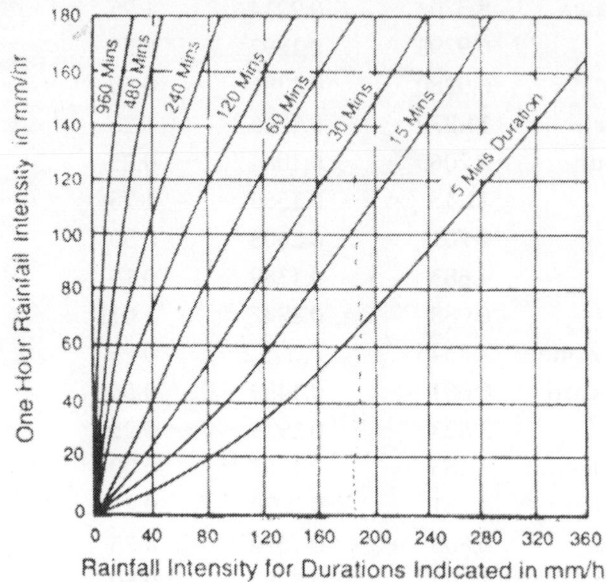

Fig.-1: Relation of one-hour rainfall intensities to intensities at other durations

References

Chow V. T. (1959). Hand Book of applied Hydrology (Edtd.). Mc. Graw Hill Book Company. pp. 427. New York.

Linsley and Franzini. (1979) Water resources engineering. Mc. Graw Hill Book Company. (3rd Ed.) pp. 78. New York.

Ragunath, H.M. (1996) Hydrology Principles, analysis, design. New Age International publisher Pvt. Ltd. pp 112,235, New Delhi, India.

Singh, G., Venkataramanan, C., Sastry, G. and Joshi, B. B. (1990). Manual of Soil and Water Conservation Practices. Oxford and IBH publishing Company Pvt. Ltd. Pp. 20, New Delhi.

CHAPTER

8 Drought Stress Management Under Dryland Farming System

B. K. Sinha[1], A. K. Tiku[1] and Reena[2]

[1]*Division of Biochemistry and Plant Physiology SKUAST-J, FOA, Main Campus Chatha, Jammu J&K State, 180009.*

[2]*Dryland research Sub-station, SKUAST-Jammu, Rakh Dhiansar, Bari Brahaman, J & K*

Agricultural production has increased considerably during the last three decades. This has happened largely due to the development and large scale cultivation of new high-yielding dwarf varieties, increase in area under such varieties and greater application of water and nutrient. Despite the development of impressive irrigation potential, which ensured food security during last three decades, agriculture is still considerably affected by climatic variability. Droughts have been frequent in different parts, throughout its history and are responsible for many famines, rural poverty and migration. Similarly, temperatures, soil salinity, etc during critical stages are known to affect food production due to their effects on various crop growth and yield processes.

The increased demand for food can no longer be met only by higher yields from irrigated areas. Greater efforts are needed today to understand and enhance the contribution of dryland areas to overall agricultural production by developing and applying location-specific technologies. It is this large area that needs to be tapped in future to increase production. In general, the potential productivity of most crops is much higher than the average yield in farmers field (Aggarwal *et al.*, 2000a). The drought stress is mostly accompanied by high temperature stress, salt stress and low temperature stress is associated with drought stress. The contribution due to osmotic stress is a common denominator in water stress, salt stress and low temperature stress.

Abiotic stresses are the primary sources of yield losses. They account for nearly 70%of total reduction in yield in crop plants. It is estimated that only 10% of the world's cultivated area is not subjected to any abiotic stress. As majority of the world's cropped area is under rainfed, amongst abiotic factors, drought accounts for nearly 26% of arable land (Ravikumar and Patil, 2003). Over the past few decades, human activities have resulted in great climate changes viz., increased atmospheric temperatures due to increased levels of CO_2 and other greenhouse gases, etc. Analysis of temperature data of last thirty years indicates a slight rising trend in temperature in North-western India. This may partly be responsible for the observed yield decline in the region (Aggarwal *et al.*, 2000b). The mean temperature in India is projected to increase by 0.1 to 0.3^{0}C in kharif and 0.3 to 0.7^{0}C during rabi by 2010 and 0.4 to 2.0^{0}C during kharif and 1.1 to 4.5^{0}C in rabi by 2070. Such a global climatic change can reduce crop duration, increase crop respiration rates and alter photosynthate partitioning to economic parts (Watson *et al.*, 1998). At the same time, there is an increased possibility of climatic extremes such as the timing of onset of monsoons, intensities and frequencies of droughts, etc. Under such a climate change scenario, the onset of summer monsoon over India is projected to be delayed and often uncertain. This will have a direct effect not only on the dryland crops, thereby putting more constraint on the water availability (Aggarwal *et al.*, 2001 and Aggarwal and Mall, 2002).

Intergovernmental Panel on Climate Change (IPCC) suggested that changes in climate could give rise to regional increase in the incidence of extreme high temperature events and droughts. Developing new crop varieties with increased drought tolerance will contribute to meet the future challenges of productivity needs of the country. Drought is actually a meterological event, which implies the absence of rainfall for a period of time, long enough to cause moisture depletion in soil and water deficit with a decrease of water potential in plant tissues (Kramer, 1980). Drought is a co-occurrence of at least seven environmental stresses (Thomas, 1997):

I. Low soil moisture availability, limiting the supply of water to the roots.

II. High evaporative load due to low humidity, high temperature, high insulation and strong winds. The potential loss of water from the leaves exceeds that which can be taken up by roots, even in well-watered soil. 30–50% of loss in yield during drought is due to low humidity.

III. High temperature causing high respiration and damage to metabolic processes and cell structure.

IV. High solar irradiance, leading to photo-inhibition, photo-oxidation and eventually death of leaves.

V. Soil hardness increases as soil dries, thus adversely affecting the root growth, leading to reduced leaf growth and photosynthesis, especially in seedlings.

VI. Unavailability of nutrients, particularly in the upper soil horizons, which dry most rapidly but are most mineral rich.

VII. Accumulation of salts in the top soil and around the roots leading to osmotic and toxic stress.

What is stress?

Stress is usually defined as an external factor that exerts a disadvantageous influence on the plant. In most cases, stress is measured in relation to plant survival, crop yield, biomass accumulation or the primary assimilation processes (CO_2 and mineral uptake), which are related to overall growth. Stresses can be biotic, imposed by other organism or abiotic arising from an excess or deficit in the physical or chemical environment. Biotic stress arises by attack of bacteria, viruses, fungi, invertebrates and even other plants, whereas abiotic stress occurs due to drought, water logging, high or low temperatures, excessive soil salinity, imbalance of mineral nutrients in the soil, heavy metals and too much or too little light. Stresses trigger wide range of plant responses, from altered gene expression and cellular metabolism to changes in growth rates and crop yields.

So far, drought stress (water deficit or low water availability) is the major abiotic problem, widely distributed world wide (Chaves and Oliveira, 2004; Kijne, 2006 and Passioura, 2007). Under drought conditions, water availability in supporting materials such as soil, vermiculite, perlite and peat moss, is also restricted, thereby causing low water use efficiency (WUE) in plant cells (Blum, 2005; Bloch *et al.*, 2006; Costa *et al.*, 2007; Shao, *et al.*, 2008). Severe water stress may cause inhibition of photosynthesis, disturbance of metabolism and finally the death of plant (Jaleel *et al.*, 2008a). Some aspects of drought stresses can be managed by appropriate management practices and by regional development. So, deeper understanding of the mechanism of water stress mitigating strategies is necessary for increasing the productivity of crop in dry land farming.

Physiological implications of drought

Water stress leads to reduced availability of water for vital cellular functions and maintenance of turgor pressure. Dehydration or osmotic stress induces stomatal closure and consequently, a reduction of the biochemical capacity for carbon assimilation and use. One characteristic cellular feature activated by abiotic stresses is the high production of reactive oxygen species (ROS) in the chloroplasts, mitochondria or in peroxisomes, causing irreversible cellular and tissue damages. However, most plants have developed various adaptation and detoxification mechanisms to deal with stress condition. Considerable knowledge has been gained over the last decade on the activation of plant stress signal and transduction pathways as well as physiological and molecular stress responses. Some of the most common responses for abiotic stress tolerance in plants are overproduction of several compatible organic solutes such as sucrose, betains, proline, etc. for osmotic adjustment and protection of subcellular structures, cellular metabolic changes, anatomical and morphological changes in plant tissues and induction of stress responsive gene expression. Among them, activation of genes involved in signal transduction pathways may lead to complex changes in gene expression resulting in plant adaptation to abiotic stresses.

All soil processes important for agriculture are directly affected in one way or other by abiotic stresses. Changes in precipitation patterns and amount and temperature can influence soil water content, runoff and erosion, soil workability, soil temperature, salinization, soil biodiversity, organic carbon content and nitrogen content. Millions of hectares of land otherwise suitable for agriculture are not cultivated or have low productivity due to high level of salinity. Some aspects of abiotic stresses can be managed by appropriate management practices and by regional development.

Responses of crops to drought stress

To survive abiotic stresses, plants have evolved a number of morphological, physiological and metabolic responses.

A. Morphological responses

I. Decreased number of leaf and their area: Development of optimal leaf area is important to photosynthesis and dry matter yield. Water deficit stress mostly reduces leaf growth as in case of soybean (Zhang *et al.*, 2004). As water content of the plant decreased, the cells shrink and the cell wall relax. This decrease in cell volume, results in lower hydrostatic pressure or turgor (Teiz and Zeiger, 1998a). Significant inter-specific differenced between two sympatric populus species were found in total number of leaves, leaf area and biomass under drought stress (Wullschleger *et al.*, 2005). Leaf growth in wheat was more sensitive to water stress than in maize (Saks *et al.*, 1997), *Vigna unguiculata* (Manivannan *et al.*, 2007a) and sunflower (Manivannan *et al.*, 2007b & 2008). In water stressed plants after a substantial leaf area has developed, leaves senescence and eventually fall off. This leaf area adjustment is an important long-term change that improves the plant fitness for a water limited environment. The production of a thicker cuticle also prevents water loss from the epidermis (Teiz and Zeiger, 1998a).

II. Root modifications: Production of ramified root system under drought is important for above ground dry mass and great variations in this regard have been found in different plant species or varieties of a species (Jaleel, *et al.*, 2009a). A prolific root system can support accelerated plant growth during the early crop growth stage and extract water from shallow soil layers (Johnson *et. al.*, 1992, Shao *et. al.*, 2008). Increased root growth during stress depends on allocation of assimilates to the growing root tips. The enhanced water uptake resulting from root growth is less pronounced in reproductive plant than in vegetative plants (Teiz & Zeiger, 1998a). An increased root growth due to water stress was reported in sunflower (Tahir *et. al.*, 2002) and *Catharanthus roseus* (Jaleel *et al.*, 2008b and 2008a). An increase in root and shoot ratio under drought conditions was related to ABA content of roots and shoots (Sharp and Le Noble, 2002; Manivannan *et. al.*, 2007b)

III Changes in stomatal morphology: Water savers conserve water by means of stomatal control or by employing other xerophytic adaptations. Xerophytes characters that restrict the transpiration water loss include low density of stomata, stomata sunken below the epidermis, a narrow band of bulliform cells, a thick cuticle, short and narrow leaves and the higher number of conducting vessels cm^{-2} in root. Reduced evaporating surface through curling of leaves is most common (Dwivedi, 2000).

B. Physiological and Biochemical Responses

I. Stomatal behaviour during water deficit: Stomatal closure can be considered a third line of defense against drought. Guard cells of stomata are exposed to the atmosphere, so can lose water directly by evaporation, causing the stomata to close by hydropassive closure mechanism. Another mechanism is called hydro active closure, where stomata closes when the whole leaf or the roots are dehydrated and it depends on metabolic processes in the guard cells (Teiz and Zeiger, 1998a). Free flow of O_2 and CO_2 is dependent on stomatal opening which in turn is controlled by turgor of both stomatal guard cells and other epidermal cells. These two metabolic processes are affected by dehydration within the zone of cell turgor (Jaleel *et. al.*, 2008c).

Abscisic Acid (ABA) is synthesized continuously at a low rate in mesophyll cells and tends to accumulate in chloroplasts. When the mesophyll becomes mildly dehydrated, two things happen. First, some of the ABA stored in chloroplasts is released to the apoplast, some of which is further to the guard cells (Cornish and Zeevart, 1985). Second, the rate of net ABA synthesis increases after closure has begun and appears to enhance or prolong the initial closing effects of stored ABA (Teiz and Zeiger, 1998a). In cotton (*Gossypium hirsutum*), nitrogen supply affect ABA accumulation or ABA redistribution or both and thus greatly alter stomatal responses to water stress (Radin and Hendrix, 1988).

II Effect on pigment composition and photosynthesis: Photosynthetic pigments are important to plants mainly for harvesting light and production of reducing powers. Both the chlorophyll a and b are prone to soil drying (Farooq *et al.*, 2009). The Mg^{2+} concentration in chloroplasts influences photosynthesis during water stress through its role in coupling electron transport to ATP production. When sunflower (*Helianthus annus*) plants were grown with different Mg^{2+} levels in the nutrient solution, the plants with the lower tissue Mg^{2+} concentrations maintained higher photosynthetic rates as leaves became dehydrated (Rao *et al.*, 1987). Water stress usually affects both stomatal conductance and photosynthetic activity in the leaf. The foliar photosynthetic rate of higher plants decreased as the relative water content and leaf water potential decreases. Both stomatal and non-stomatal limitation was generally accepted to be the main determinant of reduced photosynthesis under drought stress (Farooq *et al.*, 2009).

Carotenoids are a large class of isoprenoid molecules, which are denovo synthesized by all photosynthetic and many non-photosynthetic organisms (Andrew *et al.*, 2008). They are divided into the hydrocarbon carotenes, such as lycopene and β carotene or xanthophylls (Jaleel *et al.*, 2007c). Carotenes form a key part of the plant antioxidant defense system but they are very susceptible to oxidative destruction. β carotene present in the chloroplasts of all green plants is exclusively bound to the core complexes of PS I and PS II (Havaux, 1998). Protection against oxidative damage generated by drought stress at this site is essential for chloroplast functioning. A major protective role of β-carotene in photosynthetic tissue may be through direct quenching of triplet chlorophyll, which prevents the generation of singlet oxygen and protects from oxidative damage (Farooq *et al.*, 2009).

III Effect on osmotic adjustment: Osmotic adjustment is the net increase in solute content per cell that is independent of the volume changes that result from loss of water. Osmotic adjustment is a process by which water potential can be decreased without an accompanying decrease in turgor (Teiz and Zeiger, 1998a). Drought stress induced break down of carbohydrate and proteins, increases the concentration of solutes in the cell sap resulting in reduction in osmotic potential (Reddy and Reddi, 1999).

Most of the adjustments proceed by increases in concentration of a variety of common solutes, including sugars, organic acids (compatible solutes) and ions (mainly K^+). Compatible solutes, also known as compatible osmolytes, are a small group of chemically diverse organic compounds that are highly soluble and do not interfere with cellular metabolism, even at high concentrations (Bray *et al.*, 2000). The accumulation of amino acids may be due to the hydrolysis of protein and also may be occurring in response to the change in osmotic adjustment in their cellular contents (Manivannam *et al.*, 2007). Their accumulation under stress conditions at all the growth stages indicates the possibility of their involvement in osmotic adjustment (Manivannan *et al.*, 2008). The proline content of *Cajanus cajan* (pigeonpea) plants increased under drought with or without paclobutrazole (PBZ) and Abscisic Acid (ABA) treatments (Jaleel *et. al.*, 2009b). Proline accumulated under stress conditions supplies energy for growth and survival and thereby helps the plant to tolerate stress (Jaleel *et al.*, 2008c). Proline accumulation in plants might be a scavenger and acting as an osmolyte. The reduced proline oxidase may be the reason for increasing proline accumulation (Jaleel *et al.*, 2009b).

Glycine betaine (GB) and protein content increased in water stressed pigeonpea plants (Jaleel *et al.*, 2009b). Increased levels of GB were found under abiotic stresses (Jaleel *et al.*, 2007b) and (Jaleel *et al.*, 2007c). GB may maintain the osmoticum, provided that the basal metabolism of the plant can sustain high rate of synthesis of these compounds to facilitate osmotic adjustment for tolerance to water stress (Jaleel *et al.*, 2008c). Monomeric sugars such as glucose and fructose can be released from polymeric forms (starch and fructans respectively) in response to drought stress. Once the stress is removed, there monomers can be repolymerized to facilitate rapid and reversible osmotic adjustment (Bray *et al.*, 2000). Compatible solutes that accumulate in plants may perform an additional function in minimizing the impact of abiotic stresses on plants. For example, GB prevents salt induced inactivation of Rubisco and destabilization of the oxygen-evolving complex of Photosystem II. Sorbitol, Manitol, myo-inositol, and proline scavenge hydroxyl radicals. This antioxidant activity is a protective role for the former compounds in osmotic stress tolerance developed due to drought stress, distinct from their participation in osmotic adjustment (Bray *et al.*, 2000). Osmotic adjustment increases translocation and helps in increasing grain yield. By osmotic adjustment, production of ABA is reduced. However, at low osmotic potential, photosynthetic activity and rate of hypocotyls extension are reduced due to inhibition of enzymatic activity (Reddy and Reddi, 1999).

IV. Acid Metabolism: Drought stress may induce Crassulacean acid metabolism (CAM) in a plant in which stomata open at night and close during the day and are highly drought resistant. The leaf to air vapour pressure difference that drives transpiration is much reduced at night, when both leaf and air are cool. As a result, the water use efficiency (WUE) of CAM plants are among the highest as compared to the C_3 and C_4 plant species. A CAM plant may gain 1g of dry matter for only 125g of water used, i.e. three to five times greater than the ratio for a typical C_3 plant. Some CAM species display facultative CAM, switching to CAM when subjected for water deficits (Teiz and Zeiger, 1998a). This switch in metabolism due to accumulation of the enzyme phosphoenolpyruvate carboxylase, changes in carboxylation and decarboxylation patterns, transport of large quantities of malate into and out of the vacuoles and reversal of the periodicity of stomatal movements is a remarkable adaptation to drought stress.

V. Disturbance in water flow in the system: When soil dries up, its resistance to flow of water increases very sharply, particularly near the permanent wilting point, when the soil water potential reaches -1.5 MPa. The permanent wilting point is the soil water potential value at which plants cannot regain turgor pressure even at night, in the absence of transpiration. At the permanent wilting point, water delivery to roots is too slow to allow overnight rehydration of plants that have wilted during the day (Teiz and Zeiger, 1998a). The decrease in conductivity as the soil dries is due to the movement of air into the soil to replace the water. Several factors may contribute to the increased plant resistance to water flow during drought stress. As plant cells dry, they shrink. When roots shrink, the root surface can move away from the soil particles that hold water and the delicate root hairs may be damaged as they are pulled away. Outer layer of the root cortex covered with suberin (a water-impermeable lipid), increases the resistance to water flow. Cavitation or the breakage of water columns under tension is an important factor that increases resistance to water flow under drought conditions (Teiz and Zeiger, 1998a).

C. Molecular responses

Studies on molecular responses of plants to various types of stresses indicate that different types of stresses provide the cells with different information. The multiplicity of the information generated by different types of stresses makes the response of plant complex and also the stress signaling pathways (Knight, 2000; Xiong *et. al.*, 2002a and Murata and Los, 1997). Various internal and external signals are required for coordinating the expression of genes during development and for enabling the plant to respond to environmental signals. Such internal as well as external signaling agents typically bring about their effects by the sequences of biochemical reaction, called signal transduction pathways, that greatly amplify the original signal and ultimately result in the activation of genes (Teiz and Zeiger, 1998b). Expression of a variety of genes has been demonstrated to be induced by the drought stress. The products of these genes are classified into two groups (Seki *et. al.*, 2004).

(a) Those which directly protect against stresses. These are the proteins that function by protecting cells from dehydration. They include the enzymes responsible for the synthesis of various osmoprotectants like late embryogensis abundant (LEA) proteins, chaperones and detoxification enzymes.

(b) Gene products include transcription factors, protein kinases and enzymes involved in phosphoinositole metabolism.

Second group of gene products regulates gene expression and signal transduction pathways. The signal transduction pathways in plants under environmental stresses have been divided into three major types (Xiong *et al.*, 2002b).

(a) Osmotic/oxidative stress signaling that makes use of mitogen activated protein kinase (MAPK) modules. It involves the generation of ROS scavenging enzymes and antioxidant compounds as well as osmolytes.

(b) Ca^{+2} - dependent signaling that leads to activation of LEA-type genes such as dehydration responsive elements (DRE).

(c) Ca^{+2} dependent salt overly sensitive (SOS) signaling that regulates ion homeostasis.

Out of three signal transduction networks, two possible pathways are discussed here, which may be involved in drought stress (Figure 1).

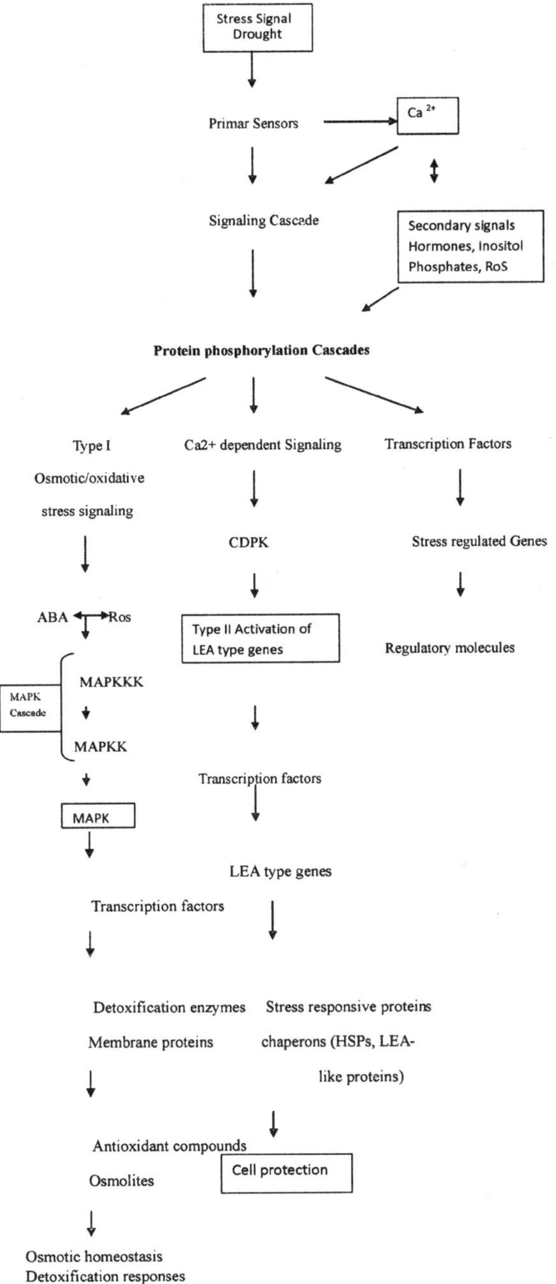

Figure 1: Two possible signal transduction pathway for the drought stress in plants. Pathway starts with signal perception, followed by the generation of second messengers. Second messengers can modulate intracellular Ca^{2+} levels, after initiating a protein phosphorylation cascade that finally target proteins directly involved in cellular protection transcription rolling specific sets of stress regulated genes (Source: Xiong *et. al.*, 2002b & 2003).

(a) Osmotic/oxidative stress signaling: Drought and oxidative stress are accompanied by the formation of Reactive oxygen species (ROS) such as peroxide, hydrogen peroxide hydroxyl radicals, causing extensive cellular damage and inhibition of photosynthesis. Osmotic/oxidative stress signaling is phosphoprotein module used in plants for abiotic stress signaling. MAPK pathways also mediated ROS signaling. The MAP kinase pathways are intracellular signal modules that mediate signal transduction from the cell surface to the nucleus. The MAPK cascade consists of 3 kinases that are activated sequentially by an upstream kinase. The MAP kinase kinase kinase (MAPKKK), upon activation, phosphorylates a MAP kinase kinase (MAPKK) on serine and threonine residues and MAPKK phosphorylates a MAP kinase (MAPK) on conserved tyrosine and threonine residues. The activated MAPK can then either migrate to the nucleus to activate the transcription factor directly, or activate additional signal components to regulate gene expression or target certain signal proteins for degradation (Xiong *et al.*, 2002b).

(b) Ca^{2+} dependent signaling that lead to the activation of LEA- type genes: Ca^{2+} is involved in various intracellular signaling processes. Ca^{2+} concentration in the cytosol is low, and upon stimulation, Ca^{2+} is released from intracellular storage or enters the cell via various Ca^{2+} channels. Drought has been shown to induce transient Ca^{2+} influx into the cell cytoplasm. Calcium dependent protein kinases (CDPKs) are important sensor of Ca^{2+} influx in plants in response to drought stress. The CDPK pathway seems more connected to increasing the expression of LEA proteins for antidesication protection (Serrano *et. al.*, 2003). CDPKs are encoded by multigene family and the expression levels of these genes are spatially and temporally controlled throughout development. Pathway leading to the activation of LEA type genes including the dehydration responsive element (DRE) class of stress responsive genes. The activation of LEA- type genes may actually represent damage repair pathways (Xiong and Xhu, 2002). LEA proteins and chaperons have shown to be involved in protecting macromolecules like enzymes, lipids and mRNAs from dehydration (Yamaguchi *et al.*, 2002).

Crop management strategies under drought stress: Water is the most abundant compound found in nature and is the most limiting factor in dryland agriculture. The objective of its efficient use can be achieved by increasing crop yields and also by reducing the consumptive use of water by various crops (Mishra and Sharma, 2003). Introduction of drought resistant crop for stabilization of crop performance in drought prone environment is needed. The following crop management strategies may be adopted to achieve the maximum crop production under dryland farming systems.

I Reducing water loss by evapotranspiration: In dryland farming, soil moisture is the most limiting factor. It is lost by evaporation from the soil surface and by transpiration from the plant surface. The losses from evapotranspiration can be reduced by use of following.

a. Mulches
b. Antitranspirants
c. Weed control

a. Mulches: Any material that is applied on the soil surface to check evaporation losses is known as mulch. Approximate 60 to 75 percent of the rainfall is lost through evaporation. Various types of mulches can be used in field, such as soil or dust, stubble, straw, plastic and vertical mulch (Reddy and Reddi, 1999). Mulches improve soil water by reduction of evaporation, runoff water and increased infiltration. Mulches obstruct the solar radiation reaching soil surface responsible for formation of vapour molecules. Most of the dry land soils have a high salt content. This salt can be removed from the soil by leaching if precipitation is adequate. However, due to limited precipitation they move only to a limited depth and readily return to the surface as water evaporates from soil surface. If infiltration increases and evaporation decreases, the salts do not accumulate in the surface layers.

b. Antitranspirants: Aproximately 99% of the water absorbed by plants is lost in transpiration. If transpiration is controlled; it may help in maintenance of favorable water balance and improved water use efficiency in dry land farming. Antitranspirant is any material applied to transpiring plant surfaces for reducing water loss from the plant. (Reddy and Reddi, 1999). Based on their mode of action as reported by (Gale and Hagan, 1966 and Davenport *et al.*, 1969) these can be classified into there categories:

I. Film forming compounds
II. Metabolic inhibitors
III. Reflectants

I. Film forming Compounds: Plastic and waxy materials which form a thin film on the leaf surface retard the escape of water due to formation of physical barrier (Reddy and Reddi, 1999). Many compounds such as silicon's, actadecanol, folicate, vapour gard and hexadecanol, etc. have been used as film forming compounds on field crops, horticultural crops and forest plants (Mishra and Sharma, 2003). Forty per cent reduction in transpiration has been noticed by hexadecanol application and its absorption through corn roots (Roberts, 1961). Silicon sprays on foliage penetrates the leaf and may act directly on the cell walls (Haigh and Heinlein, 1973) thereby greatly reducing transpiration losses (Bravado, 1972). Folicate (2.5 to 10%) and vapor gard (1.5 to 5%) reduced transpiration by 43% and 24% respectively in chrysanthemum. Corn yield under moisture stress conditions increased in response to folicate application (Fuehring and Finker, 1983).

II. Metabolic Inhibitors: - Transpiring surfaces of the plant are metabolically active and are affected by various chemical compounds. These chemical compounds may trigger water loss by two ways (Mishra and Sharma, 2003).

1. Chemicals affecting turgor of the guard cells by changing osmotic potential e.g. Abscisic Acid, atrazin, diuron, hydroxy sulfonates, etc.
II. Chemicals which modify the permeability of plasma membrane e.g. phenyl mercuric acetate (PMA) and alkenyl succinic acid (ASA).

The β hydroxyquinoline sulphate prevented the stomatal opening of tobacco leaves. Strawberry roots when immersed in this chemical, it resulted in stomatal closure, thereby reducing water loss and enabling the plants to withstand prolonged drought conditions. (Mishra and Sharma, 2003). PMA reduced transpiration by about 10% (Davenport, 1967) while ASA by 12-33% in barley (Waggoner *et. al.*, 1964).

Aspirin can be successfully used in dryland farming as it is effective in stomatal closure. The aspirin is cheaper and required in very low concentrations and it can be used to save the plants from drought.

III. Reflectants: There are white materials which form a coating on the leaves and increase the leaf reflectance (albedo). By reflecting the radiation, they reduce leaf temperatures and vapour pressure gradient from leaf to atmosphere and thus reduces transpiration (Reddy and Reddi, 1999). A reflectant, kaolin 2%, has been noticed to decrease transpiration by 10-20% under wet regime and 10 to 25% under dry regime in potted wheat. The effect of kaolin on different growth parameters and yield of wheat as reported by (Mishra, 1991) are presented in Table 1.

Table1. Effect of different anti-transpirants on growth, yield, yield attributes and plant water relations of wheat. (Mean Values of 1985-86 and 1986-87 (Source: Mishra, 1991).

Treatments	Tillers per meter row	Height (cm)	Weight of coregent	Length of ear (cm)	Spike lets per ear	Grains per ear	Test weight	Grains yield (kg/ha)	Relative water content %
Control	64.4	61.3	12.16	5.49	10.37	22.70	42.66	94.8	62.00
Kaolin 5%	70.9	64.5	15.17	6.45	12.06	27.64	47.01	127.6	64.03
Suncap 0.1%	68.3	63.1	14.03	6.09	11.29	26.83	45.31	1144	62.9
Silicon 1%	65.2	61.7	12.20	5.54	10.42	23.00	42.93	986	62.4
C.D at 5%	2.61	0.75	0.26	0.19	0.42	0.54	0.84	78	

Note: - Kaolin, Suncap and silicon were sprayed @ 60g, 1.2ml and 12g / 1.2 liters each per plot of 15m^2 at 6 and 10 weeks after sowing.

In dryland wheat (Dash *et. al.*, 1990), and bajra (Bishnoi 1975), application of kaolin 6% greatly reduced transpiration and evapotranspiration rates. Antitranspirants have therefore a great role to play, under dry farming conditions. They have other advantages like soil moisture conservation and improvement in plant performance as a result of increased plant water potential.

c. **Weed control: -** Effective weed control eliminates the competition of weeds with crops under limited soil moisture condition. Transpiration rate of weeds is more as compared to crops. So, proper weed control in dryland agriculture leads to increasing availability of soil moisture to crops and is thus a very useful measure to reduce transpiration loss.

II Use of transgenics: There is a great need of urgency in improving the performance of crops against moisture stress. It is imperative to look for tools that ensure higher productivity levels under water deficit environments. Plant genetic engineering techniques could be effectively utilized to exploit some of the untapped tolerance mechanism to improve the harvestable crop yield under drought. These techniques involve specific genetic manipulations leading to over expression or induction of alien / native genes (Ravi kumar and Patil, 2003). The genes that have proven effective in providing stress tolerance using transgenic approach belong to both structural and regulatory gene categories. The structural genes are the ones that primarily govern synthesis of enzymes involved in stress tolerance related biochemical reactions or pathways. Regulatory genes are the ones that govern expression of structural genes at hierarchically upstream positions such as genes that control expression of transcription factors, signal transduction, components or receptor related proteins (Grover *et. al.,* 2003). A summarized report on a biotic stress tolerant transgenics produced so far, are shown in Table 2. The production of abiotic stress tolerant transgenics in India is a relatively recent development. Being a large country, India has diverse climatic and soil types, varied agriculture patterns and poor infrastructure in farming sector.

Table 2. Selective reports on production of a biotic stress-tolerant transgenic crop (Source: Grover *et. al.,* 2003)

Gene	Protein	Source	Cellular role (s)	Trans-host	Promoter used	Comments	Reference
A. Regulatory genes							
Transcription factor genes							
at-hsfl	Heat shock transcriptional factor 1	*Arabidopsis thaliana*	Transcription factor	*A. thaliana*	CaMV 35S	Transformants exhibited thermotolerance and constitutive expression of the hsp genes at normal temperature.	Lee *et..al.,* 1995.
drebl A	DRE-binding protein	*A. thaliana*	Transcription factor	*A. thaliana*	rd 29 promoter	Transformants showed enhanced expression of various stress-induced genes and showed tolerance to freezing and dehydration. The dwarfed phenotype seen with the CaMV 35S promoter was not seen here.	Kasuga *et. al.,* 1999.

dreb1 and dreb2	DRE-binding protein	*A. thaliana*	Transcription factor	*A. thaliana*	CaMV 35S	Transformants revealed freezing and dehydration tolerance but caused dwarfed phenotypes in transgenic plants.	Liu *et. al.*, 1998.
Signal transduction component genes							
At-dhf2	Cell cycle regulated phosphoprotein	*A. thaliana*	Protein kinase	*A. thaliana*	CaMV 35S	Transformants showed striking tolerance to heat, salt, cold and osmotic stress upon overexpression.	Lee *et. al.*, 1999.
Osedpk7	Calcium-dependent protein kinase	*Oryza. sativa*	Protein kinase	*O. sativa*	CaMV 35S	Overexpression showed induction of some stress responsive genes in response to salinity/drought but not cold	Saijo *et. al.*, 2000.
Detoxification component genes							
apx3	Ascorbate peroxidase	*A. thaliana*	Putative peroxisomal membrane bound ascorbate peroxidase	*Nicotiana tabacum*	Dual CaMV 35S promoter with a terminator	Transformed plants showed increased protection against oxidative stress especially in the peroxisomes but not in chloroplasts.	Wang and Allen., 1999.
hvapx1	Ascorbate peroxidase	*Hordeum vulgare*	Peroxisomal ascorbate peroxidase involved in thermotolerance	*A. thaliana*	CaMV 35S	Transformants were significantly more tolerant to heat stress compared to wild type.	Shi *et. al.*, 2001.
sat	Serine acetyl transferase	*Escherichia coli*	Glutathione biosythesis	*N. tabacum*	Artificial chimeric octopinemannopine promoter with chloroplastic transit peptide	Transformants showed several fold higher SAT activity resulting in resistance to oxidative	Blaszczyk *et. al.*, 1999.
mn-sod	Mn-Superoxide dismutase	*Nicotiana plumbaginifolia*	Dismutation of reactive oxygen inter mediates in mitochondria	*N. tabacum*	CaMV 35S	Transgenic plants overexpressing mitochondrial Mn-SOD in chloroplasts showed enhanced resistance to MV dependent light induced oxidative stress.	Slooten *et. al.*, 1995.
mn-sod	Mn-Superoxide dismutase	*N. plumbaginifolia*	Dismutation of reactive oxygen inter mediates in mitochondria	*Medicago sativa*	CaMV 35S with a chloroplastic and mitochondrial transit peptide	Transformants showed reduced injury from water-deficit stress and increased winter survival.	McKersie *et. al.*, 1996.

mn-sod	Mn-Superoxide dismutase	-	Dismutation of reactive oxygen inter mediates in mitochondria	*Medicago. sativa*	CaMV 35S with a chloroplastic and mitochondrial transit peptide	Transformants showed significantly greater survival in field under water stress and in winter.	McKersie *et. al.*, 1999.
msalr	NADPH-dependent Aldose/aldehyde reductase	*M. sativa*	Detoxification	*N. tabacum*	CaMV 35S	Transformants could resist a period of water deficiency and exhibited improved recovery after rehydration.	Oberschall *et. al.*, 2000.
Fatty acid metabolism genes							
fad7	Omega -3 fatty acid desaturase	*A. thaliana*	Causes reduction of trienoic fatty acids and hexadecatrienoic acid	*N. tabacum*	CaMV 35S	Transformants showing silencing of the gene were able to tolerate higher temperature better.	Murakami *et. al.*, 2000.
Heat shock and other genes							
hsp 17.6A	Heat shock protein 17.6A	*A. thaliana*	Molecular chaperone (in vitro)	*A thaliana*	CaMV 35S	Transformants were tolerant to osmotic stress but not heat stress.	Sun *et. al.*, 2001.
hsp 17.7	Heat shock protein 17.7	*Daucus carota*	Heat shock protein	*D. carata*	CaMV 35S	Transformants expressed the hsp 17.7 gene in the absence of heat shock and showed increased thermotolerance.	Malik *et. al.*, 1999.
hsp 101	Heat shock protein 101	*A. thaliana*	Heat shock protein	*A thaliana*	CaMV 35S	Transformants constitutively expressing hsp 101 tolerated sudden shifts to extreme temperature better than the controls.	Queitsch *et. al.*, 2000.
hsp 101	Heat shock protein 101	*A. thaliana*	Heat shock protein	*O. sativa*	Ubil	Transformants expressing hsp 101 showed enhanced tolerance to high temperature.	Katiyar-Agarwal *et. al.*, 2003
bet B	Betaine aldehyde dehydrogenase	*E. coli*	Glycincbetaine biosynthesis	*N. tabacum*	CaMV 35S	Tansformed plants showed better growth in osmotic stress conditions.	Holmstrom *et. al.*, 1994.
hval	Lea protein	*H.vulgare*	Unknown	*O. sativa*	Rice actin promoter	Transformants were more tolerant to water deficit and salt stress.	Xu *et. al.*, 1996.
imtl	Myo-inositol-o-methyl transferase	*Mesembryanthemum crystallinum*	D-Ononitol biosynthesis	*N. tabacum*	CaMV 35S	Transformants were better adapted to water and salt stress.	Sheveleva *et. al.*, 1997. Vernon *et. al.*, 1993.

mtlD	Mannitol-1 phosphate dehydrogenase	*E. coli*	Mannitol metabolism	*N. tabacum*	CaMV 35S with rbc S 3A gene transit peptide	Tresformants were more tolerant to oxidative stress.	Shen *et .al.*, 1997.
p5cs	O[1] - pyrroline 5-carboxylate synthase	*Vigna aconitifolia*	Proline biosynthesis	*N. tabacum*	CaMV 35S	Transformants accumulated 2-fold more proline than the wild type plants and were more tolerant to water stress.	Kishor *et. al.*, 1995. Hong *et. al.*, 2000.
p5cs	O[1] - pyrroline 5-carboxylate synthase	*V. aconitifolia*	Proline biosynthesis	*O. sativa*	AIPC(ABA - induced promoter complex stress inducoble promoter	Transformed rice plants showed tolerance to salt and water stress.	Zhu *et. al.*, 1998.
Sac B	Leavan sucrase	*Agyneta subtilis*	Fructan biosynthesis	*N. tabacum*	CaMV 35S	Transformants were more tolerant to freezing and PEG-mediated water stress than the wild type.	Pilon-Smits *et. al.*, 1995.
tpsl	Trehalose 6-phosphate synthase	*A.thaliana*	Trehalose biosynthesis (osmolyte accumulation)	*N. tabacum*	CaMV 35S	Transformants were more tolerant to drought and salinity.	Holmstrom *et. al.*, 1996.
tpsl	Trehalose 6-phosphate synthase	*Saccharomyces cerevisiae*	Trehalose biosynthesis (osmolyte accumulation)	*N. tabacum*	CaMV 35S	Transformants exhibited trehalose accumulation and improved drought tolerance.	Romero *et. al.*, 1997.
Atnced 3	Arabidopsis thaliana 9 –cis-epoxy carotenoid dioxygenase	*A. thaliana*	ABA biosynthesis	*A. thaliana*	CaMV 35S	Transformants showed an increase in endogenous ABA levels and enhanced level of transcription of drought and ABA- inducible genes. They also showed a reduced transcription rate in leaves and an improverment in drought tolerance.	Luchi *et. al.*, 2001.
bip	Binding protein	*Glycine max*	Molecular chaperone involved in unfolded protein response (UPR)	*N. tabacum*	CaMV 35S	Transformants were more tolerant to water stress.	Alvim *et. al.*, 2001.

III Use of plant growth regulators: Any substance that influences plant growth and development can be broadly defined as a plant growth regulator (PGR). Therefore, PGRs generally include any compound that promotes or inhibits plant growth and development. These compounds may be synthetic chemicals or the natural products of plant cells. PGRSs produced inside plant cells in small quantities are referred to as phytohormones or plant hormones. The five major classes of naturally occurring plant hormones are auxins, cytokinins, gibberellins, abscisic acid and ethylene. The first three (auxins, cytokinins and gibberellins) are typically considered growth promoter while latter two (abscisic acid and ethylene) are considered growth inhibitors (Bingru, 2007). Other than these, novel PGRs ethephon, benzladanine, uniconazole, brassinolide and triazole are being used for improving tolerance against water strerss. The mechanism of action of plant hormone is very complex and it affects various biochemical, physiological and molecular changes. A brief account of PGRs in relation to drought stress is given below.

1. **Auxins:** Auxins influences several events during plant growth and development, including cell division, elongation, differentiation, apical dominance, tropism, senescence, abscission and flowering. Iodole acetic acid (IAA) is one of the naturally occurring PGRs and it has been successfully used to improve many cash crops suffering from drought stress. Water stress significantly reduced plant height, photosynthetic rate, transpiration rate, stomatal conductance, water use efficiency, relative water content, dry biomass and grain yield in barley (Ashraf *et. al.*, 2006),soybean (Gadallah, 2000) and in aromatic grass, *Cymbopogon martini* and *Cymbopon winterianus* (Farooqi *et.al.*, 2005). Exogenous application of IAA alleviated these adverse effects. Under water deficit conditions, IAA increased soluble protein content, betain and protein content that acts as a free radical scavenger and limits the cytoplasmic acidification, increases qualitative and quantitative mRNA and new molecular weight proteins that play an important role in the readjustment of plant cell osmotic potential and protecting cytoplasmic enzymes (Shehata, 2005). Soaking wheat seeds in IAA increases their germination under saline water irrigation (Sharma and Sharma, 2003).
2. **Gibberellins:** Exogenously applied GA_3 has been reported (Sharma and Sharma, 2003) to increase seed germination in wheat, barley, cotton, and rye. GA_3 influences plant metabolism by:
   a. Increase in synthesis of alpha amylase.
   b. Increase in synthesis of phospholipids and lecithins.
   c. Stimulate the formation of beta amylase, protease, nuclease, RuBP carboxylase.
   d. Inhibit formation of peroxidase and invertase.
   e. Breaking dormancy.

The GA 1/3 content in developing grains indirectly promoted starch accumulation through improving IAA content in wheat (Chun Yan *et. al.*, 2007). The deleterious effects of salinity were counteracted by exogenous application of GA_3 @ 10ppm whereas GA_3 in combination with IAA did not show significant effect (Choubisa and Sharma, 2006).

3. **Cytokinins:** Exogenous application of cytokinins (CKs) has been shown to mimic symptoms of salt stress in plant. Cytokinins are often considered

abscisic acid antagonist and auxin antagonists / synergists in various processes of plants. Cytokinins could increase salt tolerance in wheat plants by interacting with other plant hormones especially auxins and ABA (Iqbal *et. al.*, 2006). Drought generally causes a reduction in cytokinins level in leaves. CKs also react with ABA to control stomatal opening (Sharma and Sharma, 2003).

4. **Brassinosteroids:** Exogenous application of brassinosteroids @ 0.1 mg/liter on soybean (*Glycine max*) significantly increased water potential, photochemical efficiency, photosynthetic pigment, nodules biomass, development of roots and yield under water deficit condition (Zhang *et. al.*, 2004). Brassinolides have been evaluated for use in increasing crop yield and stress tolerance. They were reported to increase ear weight and grain weight in wheat and grain yield in maize (Takematsu *et al.*, 1983; Yokota and Takahashi, 1986).
5. **Abscisic Acid:** Abscisic acid is considered to be a stress hormone and it regulates responses common to many environment stresses (Boussiba *et. al.*, 1975). ABA accumulates in leaves of many species during dehydration (Bunce, 1990). The stomatal closure in maize and sorghum proceeded with increase in ABA level confirming its role in stomatal closure. It has been found to increase water use efficiency (WUE) in wheat and barley seedlings (Jones and Mansfield, 1970). Exogenous application of ABA @ 50 mg/litre significantly increased water potential, photochemical efficiency, photosynthetic rate and grain yield in soybean under water stress (Zhang *et. al.*, 2004).
6. **Triazoles:** Triazole is a new group of growth retardants that has enhanced the potential of chemical growth regulation in agriculture and horticulture (Kaufman and Song, 1988). Triazole group of PGRs (paclobutrazol and uniconazole) are potent gibberellin biosynthesis inhibitors and shoot growth retardants. It induces a variety of other responses in plants. Davis *et. al.* (1988), Gao *et. al.* (1988) and Fletcher and Hofstra (1988) reviewed the biological activities of triazole compound in relation to plant stress tolerance. Plants treated with triazole compounds are better able to withstand several adverse environmental conditions including drought, atmospheric pollutants and extreme temperature. Foliar spray of uniconazole @ 100 mg /liter on soybean under water deficit condition significantly increased water potential, photochemical efficiency, photosynthetic pigment, root biomass and grain yield (Zhang *et. al.*, 2004). The water use in triazole treated sunflower was reduced due to reduced leaf area rather than stomatal diffusive resistance.Treated plants were better able to withstand drought. Soil applied Triadime fon (TDF) increased seed yield of soybean and pea when soil moisture was maintained at half field capacity (Fletcher and Nath. 1984).
7. **Ethephon:** Ethephon (2-chloroethyl phosphoric acid) have been primarily used as antilogging agents in corn grown under optimum conditions (Cox and Andrade, 1988; Gaska and Oplinger, 1988, Norberg *et. al.*, 1988). It can

also be used to reduce water loss by reducing LAI, resulting in extended water availability for critical stages, thereby increasing grain yield under drought stress (Shanahan and Nielsen, 1987, Kasele *et. al.*, 1994, D'Andria *et. al.*, 1997). Foliar application of ethephon @ 0.84 kg a.i./ha on maize (*Zea mays* L.) significantly reduced leaf area index, leaf area index duration (LAID), CGR and total dry matter thus showing a reduction in plant water stress symptoms under water deficit conditions (Shekoofa and Emam, 2008, Kasele, *et. al.*, 1994). Early ethephon application might be advisable for improving grain yield through increased water saving by reducing LAI and plant height under drought stress condition. Ethephon application was most beneficial to yield and yield components at high plant density and under water stress conditions (Shekoofa and Eman, 2008).

References

Aggrawal, P.K., Bandyopadhyay, S.K., Pathak, H., Kalra, N., Chander, S. and Sujith Kumar, S., 2000b. *Outlook Agric.*, 29: 259-268.

Aggrawal, P.K. and Mall, R. K., 2002. *Climate Change*. 52: 331-343.

Aggrawal, P. K., Roetter, R., Kalra, N., Hoanh, C. T., Van Keulen, H. and Laar, H. H., 2001. Land Use Analysis and planning for Sustainable Food Security, with an Illustration for the State of Haryana, IARI, New Delhi and IRRI, Philippines, pp 167.

Aggarawal, P. K., Talukdar, K. K. and Mall, R. K., 2000a. Potential Yields of rice- wheat system in the Indo- Gangetic plain of India, Rice- Wheat Consortium paper series 10. New Delhi, India RWCIGP, CIMMYT, pp.16.

Alvim, F. C., Carolino, S. M. B., Cascardo, J. C. M., Nunes, C. C., Martinez, C. A., Otoni, W. C. and Fontes, E. P. B., 2001. *Plant Physiol.*, 26: 1042–1054.

Andrew, J. S., Moreau, M. Kuntz, G. Pagny, C. Lin, S T and McCarthy, J., 2008. An investigation of carotenoid biosynthesis in Coffea canephora and Coffea Arabica. *J Plant Physiol.*, 165:1087-1106.

Asare- Boamah, N. K. and Fletcher, R. A. 1986. Protection of bean seedlings against heat and chilling injury by tradimefon. *Physiol. Plant.* 67: 353-358.

Ashraf, M. Y., Azhar, N. and Hussain, M. 2006. Indoleacetic acid (IAA) induced changes in growth, relative water contents and gas exchange attributes of barley (*Hordeum vulgare* L.) growth under water stress condition. *Plant growth regulation*. 50(1): 85-90.

Bingru, H. 2007. Plant growth regulators: what and why. GCM P157-160.

Bishnoi, O. P. 1975. 'Seminar on Dryfarming in southern districts of Punjab' PAU, Ludhiana.

Bloch, D., Hoffmann, C. M. and Marlander, B., 2006. Impact of water supply on photosynthesis, water use and carbon isotope discrimination of sugar beet genotypes. *Europ. J. Agron.*, 24: 218-225.

Blum, A., 2005. Drought resistance, water- use efficiency and yield potential are they compatible, dissonant, or mutually exclusive? *Aus. J. Agric. Res.*, 56: 1159-1168.

Boussiba S, Rikin A, Richmond AE .,1975. The role of abscisic acid in cross-adaptation of tobacco plants. *Plant Physiology*. 56: 337-339.

Bravdo, Ben-Ami 1972. Effect of several transpiration suppressants on carbon dioxide and vapour exchange of citrus and grapevine leaves. *Pl. Physiol.* 26:152-156.

Bray, E. A., Serres, J. B. and Wertilnyk, E. 2000. Responses to abiotic stresses. In: Biochemistry and molecular biology of plants. Buchanan, B., Cruissem, W. and Jones, R. Eds. American society of plant physiologists Rockville, Maryland, PP. 1158-1203.

Bunce JA .,1990. Abscisic acid mimics effects of dehydration on area expansion and photosynthetic partitioning in young soybean leaves. *Plant, Cell and Environment* 13: 295-298.

Chaves, M. M. and Oliveira, M. M., 2004. Mechanisms underlying plant resilience to water deficits: prospects for water- saving agriculture. *J. Exp. Bot.*, 55: 2365-2384.

Choubisa, K. K. and Vimal, S. 2006. Interactions of salinity and GA + IAA combinations on certain biochemical parameters in wheat seedling. *National J. of Pl. Improvement.* 8(2): 156-161.

Chunyan, L., chaonian, F. Rong, Z. Ying, Z. Wenshan, G. Xinkai, Z. and Yongxin, p. 2007. relationship of grain starch synthesis with changes of endogenous hormone and sucrose content after anthesis in wheat variety Ningmai. *J.triticeae crops.* 27 (1): 138-142.

Cornish, K., and Zeevart, J. A. D.1985. Movement of abscisic acid into the apoplast in response to water stress in *Xanthium strumarium* L *Pl. Physiol.* 78: 623-626.

Costa,J. M.., Ortuno, F and Chaves, M. M., 2007. Deficit irrigation as a strategy to save water: Physiology and potential application to horticulture. *J. Integra. Plant Biol.*, 49: 1421-1434.

Cox, W.J. and Andrade, H.F. 1988. Growth, yield and yield components of maize as influenced by ethephon. *Crop Sci.* 28: 536-542.

D' Andria, R., Chiaranda, F. Q. Lavini, A. and Mori, M. 1997. Grain yield and water consumption of ethephon-treated corn under different irrigation regimes. *Agron. J.* 89: 104-112.

Dash, D. K., Sen, A. and Mishra, N. M. 1990. Internal power status of dryland wheat as affected by gamma irradiation, evaporation retardants and transpiration suppressants. *Int. Symp. on N.R.M. for S. Agri.* 6-10 Feb.,1990. pp.- 276.

Davenport, D. C. 1967. Effect of chemical antitranspirants on transpiration and growth of grass. *J. Exp. Bot.* 23: 651-654.

Davenport, D. C., Hagan, R. M. and Martin, P. E. 1969. Antitranspirant research and its possible application in hydrology. *Water resources Res.* 5: 735-743.

Davis, T. D., Steffens, G. L. and Sankhla, N. 1988. Triazole plant growth regulators. *Hort. Rev.* 10: 63-105.

Dwivedi, R. S. 2000. Physiology of sugarcane. In: 50 years of Sugarcane Research in India.

Farooq, M., A. Wahid, N. Kobayashi, D. Fujita and S.M.A. Basra, 2009. Plant drought stress: effects, mechanisms and management. *Agron. Sustain. Dev.*, 29: 185–212.

Farooqi, A. H. A., Fatima, S. Khan, A. and Sharma, S. 2005. Ameliorative effect of chlormequat chloride and IAA on drought stressed plants of *Cymbopogon martini* and *C. winterianus. Pl. Growth. Regulation.* 46: 277-284.

Fletcher, R. A. and Hofstra, G. 1988. Triazole as potential plant protectants. In: Beng Plumble edited, Sterol Biosynthesis Inhibitors- Pharmaceutical and Agro- Chemical Aspects, Ellis Harwood Ltd. England, pp. 321-331.

Fletcher, R. A. and Nath, V. 1984. Tradimefon reduces transpiration and increases yield in water stressed plants. *Physiol. Plant.* 62: 422-426.

Fuchring, H. D. and Finker, N. D. 1983. Effect of folicote antitranspirant application on field grain yield of moisture stressed corn. *Agron. J.* 75: 579-582.

Gadallah, M. A. A. 2000. Effects of indole-3-acetic acid and zinc on the growth, osmotic potential and soluble carbon and nitrogen components of soybean plants growing under water deficit. *J. of arid Env.* 44: 451-467.

Gale, J. and Hagan, R. M. 1966. Plant antitranspirants. *Ann. Rev. Pl. Physiol.* 17: 269-282.

Gao, J., Hofstra G, Fletcher RA.,1988. Anatomical changes induced by triazoles in wheat seedlings. *Canadian Journal of Botany.* 66: 1178-1185.

Gaska, J. M. and Oplinger, E. S. 1988. Yield, lodging and growth characteristics in sweet corn as influenced by ethephon timing and rate. *Agron. J.* 80: 722-726.

Grover, A., Aggarwal, P. K. Kapoor, A. Agarwal, S. K. Ararwal, M. and Chandramouli, A. 2003. Addressing abiotic stresses in agriculture through transgenic technology. *Current Sci.* 84: 355-367.

Haigh, W. C. and Heinglein, J. P. 1973. Silicon antitranspirants, A possible mechanism of action. *Pl. Physiol.* 51 (Suppl.):46.

Havaux, M., 1998. Carotenoids as membrane stabilizers in chloroplasts. *Trends Plant Sci.,* 3: 147–151.

Holmstrom, K. O. *et al.,* 1994. *Plant J.,* 6: 749–758.

Holmstrom, K. O., Mantyl, E., Welin, B., Mandal, A. and Palva, E. T., 1996. *Nature,* 379: 683–684.

Hong, Z., Lakkineni, K., Zhang, Z. and Verma, D. P. S., 2000. *ibid,* 122: 1129–1136.

Huang, B. 2007. Plant growth regulators: What and why. Research. GCM pp.157-160

Identified by a Functional Approach in Yeast Monatshefte fur Chemie.134: 1445-64.

Iqbal, M., Ashraf, M. and Jamil, A. 2006. Seed enhancement with cytokinins: changes in growth and grain yield in salt- stressed wheat plants. *Pl. Growth. Regulation.* 50(1): 29-39.

Iuchi, S. *et al.,* 2001. *Plant J.,* 27: 325–333.

Jaleel, C. A., Azooz, M. M. Manivannan, P. and Panneerselvam, R. 2009b. Involvement of paclobutrazol and ABA on Drought- Induced osmoregulation in *Cajanus cajan. Amer. Eura. J. of Botany.* 2(1): 30-36.

Jaleel, C. A., Gopi, R. and Panneerselvam, R. 2007c. Alterations in lipid peroxidation, electrolyte leakage and proline metabolism in *Catharanthus roseus* under treatment with triadimefon, a systemic fungicide, *C.R. Biol.,* 330: 905-912.

Jaleel, C. A., Gopi, R., Gomathinayagam, M. and Panneerselvam, R. 2008c. Calcium chloride effects on metabolism of *Dioscorea rotundata* exposed to sodium chloride-induced salt stress, *Acta. Biol., Cracoviensia Series Bot.,* 50: 63-67.

Jaleel, C. A., Gopi, R., Sankar, B., Manivannan, P., Kishorekumar, A., Sridharan, R. and Panneerselvam, R. 2007b. Alterations in germination, seedling vigour, lipid peroxidation and proline metabolism in *Catharanthus roseus* seedlings under salt stress, *S. Afr. J. Bot.*, 73: 190-195.

Jaleel, C.A., Gopi, R., Sankar, B., Gomathinayagam, M. and Panneerselvam, R. 2008b. Differential responses in water use efficiency in two varieties of *Catharanthus roseus* under drought stress. *Comp. Rend. Biol.*, 331: 42–47.

Jaleel, C. A., Manivannan, P. Wahid, A. Farooq, M. Al-juburi, H. J. Somasundaram, R and Vam, R. P., 2009a. Drought stress in plants: a review on morphological characteristics and pigments composition. *Int. J. Agric. Biol.*, 11: 100-105.

Jaleel, C. A., Manivannan, P., Lakshmanan, G. M. A., Gomathinayagam, M and Panneerselvam, R., 2008a. Alterations in morphological parameters and photosynthetic pigment responses of *Catharanthus roseus* under soil water deficits. *Colloids Surf. B: Biointerfaces.* 61: 298-303.

Jaleel, C.A., Manivannan, P., Kishorekumar, A., Sankar, B., Gopi, R., Somasundaram, R. and Panneerselvam, R. 2007a. Alterations in osmoregulation, antioxidant enzymes and indole alkaloid levels in *Catharanthus roseus* exposed to water deficit. *Colloids Surf. B:Biointerfaces,* 59: 150–157.

Johansen, C., Baldev, B., Brouwer, J. B., Erskine, W., Jermyn, W. A., Juan, L. Li., Malik, B. A., Miah, A. A. and Silim, S. N. 1992. Biotic and abiotic stresses constraining productivity of cool season food legumesin Asia, Africa and Oceania. *In*: Muehlbauer, F.J., W.J. Kaiser (eds.), Expanding the Production and Use of Cool Season Food Legumes, *Kluwer Academic Publishers, Dordrecht, The Netherlands,* pp: 75–194.

Jones, R. J. and Manisfield, T. A. 1970. I annual reviews of plant sciences, C.P.Malik(Ed.). Kalyani publishers, New Delhi.

Kasele, I. N., Nyirenda, F. Shanahan, J. F. Nielsen, D. C. and Andria, R. D. 1994. Ethephon alters corn growth, water use and grain yield under drought stress. *Agron. J.* 86: 283-288.

Kasuga, M., Liu, Q., Miura, S., Yamaguchi-Shinozaki, K. and Shinozaki, K., 1999. *Nature Biotechnol.*, 17: 287–291.

Katiyar-Agarwal, S., Agarwal, M. and Grover, A., *Plant Mol.Biol.*, 2003 (in press).

Kaufman, P.B. and Song, I. I. 1988. Growth retardants as plant hormonal regulators. In: S.S. Purohit edited, *Hormonal Regulation of Plant Growth and Development*, Agro Botanical Publishers, India Bikaner.4: 153-170.

Kijne, J.W., 2006. Abiotic stress and water scarcity: Identifying and resolving conflicts from plant level to global level. *Field Crop Res.*, 97: 3-18.

Kishor, K. P. K., Hong, Z., Miao, G. H., Hu, C. A. A. and Verma, D. P. S., 1995. *Plant Physiol.*, 108: 1387–1394.

Knight, H., 2000. Calcium signaling during abiotic stress in plants. *Int. Rev. Cytol.*, 195, 269–325.

Kramer, P. J., 1980. The role of physiology in crop improvement. In: Linking Research to Crop Production. Staples, R. C. and R. J. Kuhr, Eds., *Plenum press, New York*. pp. 51-62.

Lee, J. H., Hubel, A. and Schoffl, F., 1995. *Plant J.*, 8: 603–612.

Lee, J. H., van Montagu, M. and Verbruggen, N., 1999. *Proc. Natl. Acad. Sci. USA*, 96: 5873–5877.

Lilius, G., Holmberg, N. and Bulow, L., 1996. *Biotechnology*, 14: 177–180.

Liu, Q., Kasuga, M., Sakuma, Y., Abe, S., Miura, H., Yamaguchi- Shinozaki, K. and Shinozaki, K., 1998. *Plant Cell*, 10: 1391–1406.

Malik, M. K., Solvin, J. P., Hwang, C. H. and Zimmerman, J. L., 1999. *ibid*, 20: 89–99.

Manivannan, P., Jaleel, C. A., Kishorekumar, A., Sankar, B., Somasundaram, R., Sridharan, R. and Panneerselvam, R. 2007a. Changes in antioxidant metabolism of *Vigna unguiculata* (L.) Walp. by propiconazole under water deficit stress. *Colloids Surf. B:Biointerfaces*, 57: 69–74.

Manivannan, P., Jaleel, C. A., Sankar, B., Kishorekumar, A., Somasundaram, R., Lakshmanan, G. M. A. and Panneerselvam, R. 2007b. Growth, biochemical modifications and proline metabolism in *Helianthus annuus* L. as induced by drought stress. *Colloids Surf. B: Biointerfaces*, 59: 141–149.

Manivannan, P., Jaleel, C. A., Sankar, B., Somasundaram, R., Murali, P.V., Sridharan, R. and Panneerselvam, R. 2007. Salt Stress Mitigation byCalcium Chloride in *Cajanus cajan*(L.) Wilczek, Acta. Biol. Cracoviensia Series Bot., 49: 105-109.

Manivannan, P., Jaleel, C. A., Somasundaram, R. and Panneerselvam, R. 2008. Osmoregulation and antioxidant metabolism in drought stressed *Helianthus annuus* under triadimefon drenching. *Comp. Rend. Biol.*, 331: 418–425.

McKersie, B. D., Bowley, S. R. and Jones, K. S., 1999. *ibid*, 119: 839–847.

McKersie, B. D., Bowley, S. R., Harjanto, E. and Leprince, O., 1996. *ibid*, 111: 1177–1181.

Mishra, O. R. 1991. Effect of mulching and antitranspirants on yield and yield contributing characters of rainfed wheats. Thesis submitted to the Devi Ahilya Vishwavidyalaya, Indore for the award of Ph.D. degree.

Mishra, V. K. and Sharma, R. A. 2003. Antitranspirants in agriculture, horticulture and forestry In: Adiotic stresses and crop productivity. Maloo, S. R. (Ed.). Geeta Somani agrotech publishing academy Udaipur. 193-210.

Mishra, V. K. and Sharma, R. A., 2003. Antitranspirants in agriculture, horticulture and forestry In : *Abiotic Stresses and Crop Productivity*. Maloo, S. R., (Ed)., *Geeta Somani Agrotech Publishing Academy,Udaipur, India*. pp. 193-210.

Murakami, Y., Tsuyama, M., Kobayashi, Y., Kodama, H. and Iba, K., 2000. *Science*, 287: 476–479.

Murata, N. and Los, D. A., 1997. Membrane fluidity and temperature perception. *Plant Physiol.*, 115, 875–879.

Norberg, O. S., Mason, S. C. and Lowry, S. R. 1988. Ethephon influence on harvestable yield, grain quality and loadging of corn. *Agron. J.*, 80: 786-772.

Oberschall, A. *et al.*, 2000, *Plant J.*, 24: 437–446.

Passioura, J., 2007. The drought environment: physical, biological and agricultural perspectives. *J. Exp. Bot.*, 58: 113-117.

Pilon-Smits, E. A. H., Ebskamp, M. J. M., Paul, M. J., Jeuken, M. J. W., Weisbeek, P. J. and Smeekens, S. C. M., 1995. *Plant Physiol.*, 107: 125–130.

Queitsch, C., Hong, S-W., Vierling, E. and Lindquist, S., 2000. *Plant Cell*, 12: 479–492.

Radin, J. W. and Hendrix, D. L. 1988. The apoplastic pool of abscisic acid in cotton leaves in relation to stomatal closure. *Planta*. 174: 180-186.

Rao, I. M., Sharp, R. E. and Boyer, J. S. 1987. Leaf magnesium alters photosynthetic response to low water potentials in sunflower. *Pl. Physiol.* 84: 1214 - 1219.

Ravikumar, R. L and Patil, B. S., 2003. Strategies of Crop Improvement for Drought Tolerance In : *Abiotic Stresses and Crop Productivity*. Maloo, S. R., (Ed)., *Geeta Somani Agrotech Publishing Academy,Udaipur, India*. pp. 55-119.

Reddy, T. Y. and Reddi, G. H. S. 1999. Dry land agriculture. In: *Principles of Agronomy* Reddy, T.Y. and Reddi, G.H.S (Eds.)., Kalyani Publishers Rajinder Nagar, Ludhiana. pp. 368-407.

Roberts, W. J. 1961. In annual review of Plant Sciences, C. P. Malik(Ed.) Kalyani Publishers, New Delhi.

Romero, C., Belles, J. M., Vaya, J. L., Serrano, R. and Culianez- Macia, F. A., 1997. *Planta,* 201: 293–297.

Sacks, M.M., W.K. Silk and P. Burman, 1997. Effect of water stress on cortical cell division rates within the apical meristem of primary roots of maize. *Plant Physiol.*, 114: 519–527.

Saijo, Y., Hata, S., Kyozuka, J., Shimamoto, K. and Izui, K., 2000. *Plant J.*, 23: 319–327.

Seki, M. *et al.*, 2004. RIKEN *Arabidopsis* full length (RAFL) cDNA and its applications for expression profiling under abiotic stress conditions. *J. Exp. Bot.*, 55, 213–223.

Serrano R, Gaxiola R, Rios G, Forment J,Vicente O, Roc Ros., 2003. Salt Stress Proteins gene from Arabidopsis thaliana in transgenic plants improves stress tolerance.*The Plant Journal* p.35.

Shahi, H. N., Shrivastava, A. K. and Sinha, O. K. Eds. IISR(ICAR) Lucknow pp 73-11.

Shanahan, J. F. and Nielsen, D. C. 1987. Influence of growth retardants (Anti-Gibberllins) on corn vegetative growth, water use and grain yield under different levels of water stress. *Agron. J.*, 79: 103-109.

Shao, H. B., Chu, L. Y., Jaleel, C. A and Zhao, C. X., 2008. Water-deficit stress-induced anatomical changes in higher plants. *Comp. Ren. Biol.*, 331: 215-225.

Sharma, V. and Sharma, N., 2003. Biochemical and molecular basis of abiotic stress tolerance in crop plants In : *Abiotic Stresses and Crop Productivity*. Maloo, S. R., (Ed)., *Geeta Somani Agrotech Publishing Academy,Udaipur, India*. pp. 317-362.

Sharp, R.E. and M.E. LeNoble, 2002. ABA, ethylene and the control of shoot and root growth under water stress. *J. Exp. Bot.*, 53: 33–7.

Shehata, M. M. 2005. Accumulation of osmolytes and stress proteins, alteration of levels of gene expressions and DNA profiles in two wheat cultivars as a response to amelioration of pH stress by IAA. *Biotech.* 4(1): 39-48.

Shekoofa, A. and Emam, Y. 2008. Plant growth regulator (Ethephon) alters Maize (*Zea mays* L.) growth, water use and grain yield under water stress. 7(1): 41-48.

Shen, B., Jensen, R. G. and Bonhert, H. J., 1997. *Plant Physiol.*, 113: 1177–1183.

Sheveleva, E., Chmara, W., Bohnert, H. J. and Jensen, R. G., 1997. *ibid*, 115: 1211–1219.

Shi, W.M., Muramoto, Y., Ueda, A. and Takabe, T. 2001. Cloning of peroxisomal ascorbate peroxidase gene from barley and enhanced thermotolerance by overexpressing in *Arabidopsis thaliana. Gene,* 273: 23–27.

Slooten, L., Capiau, K., van Camp, W., van Montagu, M., Sybesma, C. and Inze, D., 1995. *Plant Physiol.*, 107: 737–750.

Sun, W., Bernard, C., van de Cotte, B., van Montagu, M. and Verbruggen, N., 2001. *Plant J.*, 27: 407–415.

Tahir, M.H.N., M. Imran and M.K. Hussain, 2002. Evaluation of sunflower (*Helianthus annuus* L.) inbred lines for drought tolerance. *Int. J. Agric. Biol.*, 3: 398–400.

Taiz, L. and Zeiger, E. 1998a. Plant Physiology (II Ed). Sinuer Associates Inc., Publishers, sunderland Massachusetts. pp- 725-757.

Taiz, L. and Zeiger, E. 1998b. Plant Physiology (II Ed). Sinuer Associates Inc., Publishers, sunderland Massachusetts. pp 379-407.

Takematsu T, Takenchi Y, Koguchi M., 1983. New Plant growth regulars: Brassinolide analogues, their biological effects and application to agriculture and biomass production. *Chemical Regulation of Plants*. 18: 2-15.

Thomas, H., 1997. Drought resistance in plants. In: Mechanisms of environmental stress resistance in plants. Basra, A. S. and Basra, R. K. Eds., *Harwood Academic Publishers. The Netherlands*. pp. 1-42.

Vernon, D. M., Tarczynski, M. C., Jensen, R. G. and Bohnert, H. J., 1993. *Plant J.*, 4: 199–205.

Waggoner, P. E., Monteith, J. L. and Szeicz, G. 1964. Decreasing transpiration of field plants by chemical closure of stomata. *Nature*, 201: 97-98.

Wang, H. and Allen, R. D., 1999. *Plant Cell Physiol.*, 40: 725–732.

Watson, R. T., Zinyowera, M. C., Moss, R. H. and Dokken, D. J., 1998. Intergovernmental panel on climate change, WMO-UNEP Rep., *Cambridge University Press, UK*, pp. 517.

Wullschleger, S.D., T.M. Yin, S.P. DiFazio, T.J. Tschaplinski, L.E. Gunter, M.F. Davis and G.A. Tuskan, 2005. Phenotypic variation in growth and biomass distribution for two advanced-generation pedigrees of hybrid poplar. *Canadian J. For. Res.*, 35: 1779-1789.

Xiong L, Zhu JK., 2002. Molecular and genetic aspects of plant responses to osmotic stress. *Plant Cell Environ* .25:131-139.

Xiong, L. and Yang, Y., 2003.Disease resistance and abiotic stress tolerance in rice are inversely modulated by an abcisic acid-inducible mitogen- activated protein kinase. *Plant Cell*, 15: 745–759.

Xiong, L., Lee, H., Ishitani, M. and Zhu, J. K., 2002a. Regulation of osmotic stress-responsive gene expression by the LOS6/ABA1 locus in *Arabidopsis*. *J. Biol. Chem.* 277: 8586-8596.

Xiong, L., Schumaker, K. S. and Zhu, J. K., 2002b. Cell signalling during cold, drought and salt stress. *Plant Cell*, S165–S163.

Xu, D., Duan, X., Wang, B., Hong, B., David Ho, T. H. and Wu, R., 1996. *ibid*, 110: 249–257.

Yamaguchi-Shinozaki K, Kasuga M, Liu Q, Nakashima K, Sakuma Y, Abe H, 2002. Biological mechanisms of drought stress response. JIRCAS Working Report 1-8.

Yokota T, Takahashi N.,1986. Chemistry, physiology and agricultural applications of brassinolide and related steroids. In 'Plant growth substances'. (Ed. M Bopp) pp. 129-138. (Springer-Verlag: Berlin/Heidelberg)

Zhang, M. L., Duan, Z., Zhai, J., Li, X., Tian, B., Wang, Z., He and Z. Li., 2004. Effects of plant growth regulators on water deficit-induced yield loss in soybean. *Proceedings of the 4th International Crop Science Congress,* Brisbane, Australia.

Zhang, M., Duan, L. Zhai, Z. Li, J. Tian, X. Wang, B. He, Z. and Li, Z. 2004. Effects of Plant Growth Regulators on Water Deficit-Induced Yield Loss in Soybean. New direction for a diverse planet: Proceeding of the 4th International Crop Science Congress. Brisbane, Australia.

Zhu, B., Su, J., Chang, M., Verma, D. P. S., Fan, Y. L. and Wu, R., 1998. *Plant Sci.,* 199: 41–48.

CHAPTER

9 Crop Diversification Options for Sustaining and Expanding Dryland Agriculture in India

Awnindra K. Singh[1], Premendra Singh[1] and Ashok Sharma[2]

[1]*Dryland Research Sub-station, SKUAST-J, Rakh Dhianasar, Bari Brahmana, J & K.*

[2]*Division of Soil Science and Agricultural Chemistry, Faculty of Agriculture,Sher-e-Kashmir University of Agricultural Sciences and Technology of Jammu, Chatha, Jammu-180009.*

Introduction

India is a home to 18 percent of the world human population and over 15 percent of livestock with just 2 percent land resources of the world. It has nearly 20.2 percent (316.64 km^2), of its geographical area classified as wasteland. Further, there are large areas of degraded forest land, utilized public land and fallow land, which pose a challenge for both agriculture and economic development. The Indian economy is mainly dependent on agriculture, which contributes 30 percent of country capital and 60 percent of employment potential. India made strides in food production during last three decades culminating in self-sufficiency and surplus production. However, feeding the ever-increasing population through the next millennium remains an uphill task. Of the India's total land area of 305 million ha, 152 million ha are under cultivation and out of which, 92 million ha has covers under dryland agriculture with a maximum of 40 percent of the total food grain production of the country. Even after full realization of the irrigation potential of the country, 50 percent of net sown area will continue to be rainfed. This is primarily because due to ignorance this form of cultivation has not been accorded the level of priority it deserves. Hence, an understanding of the rainfall pattern and land characteristics is crucial for optimizing use of available water for dryland areas. Apart from rainfall, the other elements are moisture availability to crops and chemical composition of soil. Since, scientific dryland farming aims at appropriate treatment of land for conservation of moisture, priority needs to be given to proper water and land management .The technology for such farming involves crop rotation and adoption of varieties and practices adjusted to the moisture regimes of an agro-climatically homogeneous area. More specifically, it consists of making the best use of a limited water supply by storing in the soil as much rain water as possible and growing suitable crops by methods that makes the best use of the moisture. In India, context, to sustain an estimated population of one billion plus, the proportion of food grains output from dryland will have to go up to 69 percent from the present level. It is also true that, while the required knowledge and methodology for dryland agriculture is available this has not reached to the farmers at the grassroots level.

The rainfed areas are mostly inhabited by farmers with poor resources and are confronted with several problems like uncertain distributions of monsoons, poor soil fertility and low and unstable crop yield. Consequently, the farm income has remained stagnant over the year. These are mainly due to the lack of widely adaptable and improved crop varieties and a big gap exists between specific utilization of available natural resources and existing production technologies. Hence, the management of agrobiodiversity contributes vitally to agriculture in drylands, as it does in other agroecological systems. The wild relatives of staple food grains often contain genes, which are used for crop improvement for specific agroecosystems. In this context, diversification of crops play a vital role for the integrated as well as efficient utilization of natural resources with cost reduction, which will be the road map for further sustainability in dryland agriculture. This approach aims to promote diversified rural livelihood in dryland agroecosystems, which will be based on sustainable management and exploration of genetic resources and once this process is underway, it will ultimately lead to strengthen the Indian agriculture.

Table 1: Annual average growth rate (at constant prices)

Five Year Plan	Overall GDP growth rate	Agriculture and allied sectors
Seventh Plan (1985-1990)	6.0	3.2
Annual Plan (1990-1992)	3.4	1.3
Eight Plan (1992-1997)	6.7	4.7
Ninth Plan (1997-2002)	5.5	2.1
Tenth Plan (2002-2007)	7.6	2.3
2002-03	3.8	-7.2
2003-04	8.5	10.0
2004-05 (P)	7.5	0.0
2005-06 (Q)	9.0	6.0
2006-07 (A)	9.2	2.7

P; Provisional, Q: Quick estimates, A: Advanced estimates; Note: growth rates prior to 2001 based on 1993-94 prices and from 2000-2001 onwards based on new series at 1999-2000 prices; Source: CSO

Scenario of food demand and resources

The present scenario of rate of growth in Indian population, is that it has already crossed one billion and is expected to reach 1.5 billions by 2025. For ever increasing population the food grain requirement of the country will stand at 245 million tonnes by 2011 and approximately 300 million tonnes by the end of 2025. This will include 114 million tonnes of rice, 80-85 tonnes of wheat, 13 million tonnes maize, 24-26 million tonnes pulses and 9.5 million tones of oilseeds as against the present level of production of various crops i.e. 90.0 million tonnes of rice, 69.4 million tonnes of wheat and 13.4 million tonnes pulses. The County also requires 110.7 million tonnes of vegetables and 70.5 million tonnes of fruits. India should attain annual growth rate of 2.5 percent in food grains, 4.5 percent in pulses, 3.5 percent in oilseeds, 3.8 percent in cotton, 3.5 percent in vegetable and 6 percent in fruits. The green revolution in seventies played a major role in improving production making India self-sufficient in food grains, as the food grain production increased from 50 million tonnes at the time of independence to 208.6 million tonnes in 2005-06. The country has also witnessed increases in production of other commodities including fruits (54.4 million tonnes), vegetables (113.5 million tonnes), spices (5.9 million tonnes), plantation crops (9.8 million tonnes), flowers (0.8 million tonnes) and other (0.5 million tonnes) up to 2007-08. This however, had its own costs in terms of degradation of land and water resources, loss of plant biodiversity, shift of agricultural lands to nonagricultural uses, polluted environment, widening gap between the rich and the poor. Thus, physical access to food has now become the most important cause of hunger-nutrition. Ecological access to food might become the most important concern in the next decade owing to the damage now being done to land, water flora-fauna and atmosphere. The per capita availability of agricultural land in India was 0.46 ha in 1951 and it is estimated to be 0.15 ha in 2000 as against the average 0.6 ha at global level. Number of persons per ha of net cropped area was 3 in 1952 and is estimated to be 6.5 in 2000 and will be 8 persons in 2025. This situation of rapidly declining land to man ratio is likely to worsen further owing to competitive demand for food, fiber, fuel, fodder, timber and development activities such as urbanization and industrialization, mining, road construction and reservoirs etc.

Table 2: International comparison of Yield of selected commodities during 2004 – 05. (*Metric tonnes / hectares*)

Crop - Rice		*Crop - Wheat*		*Crop - Maize*	
Country	Yield	Country	Yield	Country	Yield
Egypt	9.80	U.K.	7.77	U.S.A.	9.15
U.S.A.	7.83	France	7.58	France	7.56
Korea	6.73	China	4.25	Germany	6.69
Japan	6.42	*India*	2.71	China	4.90
India	2.90	Pakistan	2.37	Philippines	2.10

Thailand	2.63	Iran	2.06	*India*	1.18
Myanmar	2.43	Australia	1.64		
Crop - Cotton		*Crop - Oilseeds*			
Country	Yield	Country	Yield		
China	11.10	Germany	4.07		
Brazil	10.96	U.S.A.	2.61		
U.S.A.	9.58	Argentina	2.51		
Uzbekistan	7.98	Brazil	2.48		
Pakistan	7.60	China	2.05		
India	4.64	Nigeria	1.04		
		India	0.86		

Source: Ministry of Agriculture & Cooperation

Table 3: Foodgrains production in India *Metric tonnes*

Crops	Year				
	2001-02	2002-03	2003-04	2004-05	2005-06
Rice	93.3	71.8	88.5	83.1	91.8
Wheat	72.8	65.8	72.2	68.6	69.4
Coarse Cereals	33.4	26.1	37.6	33.5	34.1
Pulses	13.4	11.1	14.9	13.1	13.4
Foodgrains					
Kharif	112.1	87.2	117.0	103.3	109.9
Rabi	100.8	87.6	96.2	95.1	98.7
Total	**212.9**	**174.8**	**213.2**	**198.4**	**208.6**

Source: Ministry of Agriculture & Cooperation

Table 4: Commercial crop production in India (*Metric tones*)

Crops	Year				
	2001-02	2002-03	2003-04	2004-05	2005-06
Groundnut	7.0	4.1	8.1	6.8	8.0
Rapeseed and Mustard	5.1	3.9	6.3	7.6	8.1
Soyabean	6.0	4.7	7.8	6.9	8.3
Other oilseeds	2.6	2.1	3.0	3.1	3.6
Total nine Oilseeds	20.7	14.8	25.2	24.4	28.0
Cotton #	10.0	8.6	13.7	16.4	18.5
Jute & Mesta *	11.7	11.3	11.2	10.3	10.8
Sugarcane	297.2	287.4	233.9	237.1	270.0

Million bales of 170 kgs each, * Million bales of 180 kgs each ; Source: Ministry of Agriculture & Cooperation

Constraints of Production in Rainfed areas

The first green revolution during mid seventies was strenuous mainly on soil rich inorganic carbon through the use of dwarf and fertilizer responsive high yielding varieties of rice, wheat and maize. Most disturbing fact is that if the present scenario of adoption of agricultural technology continues, it will be difficult to catch up with required rate of productivity to meet the demand of food arising from the growing population. Sluggish growth pattern and low yielding agriculture in the rainfed agro-ecosystem depict extremely grim future, making food insecurity most grievous in the face of sharply growing population. Hit by frequent natural disasters such as high-low rainfall, increasing food deficits, low and unstable productivity in agriculture and livestock, the region poses a serious development question to the policy makers. By contrast, the rainfed area of the country was deprived of the benefits of green revolution period, as these areas had limited access to critical inputs such as fertilizer and had controlled water supplies that were crucial to the success of these improved technologies. The rainfed lands suffer from a number of biophysical and socio-economic constraints, which affected the productivity of crops. Access to institutional and infrastructural support, lack of people's participation in the development process and management has been the inhibiting factors for development of dryland agriculture. Yield plateuing in major food crops, shrinkage of land holding, over exploitation of natural resource bases, decline in investment in agriculture, rising inputs cost with declining factor productivity have only reduced competence of Indian agriculture sector.

Identification of viable rainfed technologies

A number of economically viable rainfed technologies have been developed over the years in the country to address the problems of food production in dryland agriculture through dryland agriculture networks. These technologies have been evolved after making the refinement in the farmer's field through Operational Research Project (ORP) sites, IVLP and farm science centers. These include off-season tillage in rainfed alfisols and related soils for better moisture conservation and weed control. By and large, the farmers in the areas of operational research project adopted this practice in few crops (sorghum and castor) and realized the yield advantage by 40 percent over traditional practice in south and central part of the country. The results obtained from dryland areas of eastern Uttar Pradesh has been realized that yield advantage by 41-43 percent in rice, 33 percent in pigeon pea, 68 percent in barley, 63 per cent on wheat, 57-63 percent in pulses and 80 percent in percent by the saving of 20 percent of critical inputs. Conservation tillage through mould board plough and conservation furrows plantation is vital in sustaining rice and pulse based production system.

Table 5: Production of major horticultural crops in India (*Metric tones*)

Crops	Year			
	2002-03	2003-04	2004-05	2005-06
Fruits	49.2	49.8	52.8	54.4
Vegetables	84.8	101.4	108.2	113.5
Spices	3.8	4.0	4.9	5.9
Plantation crops	13.1	9.4	10.4	9.8
Flowers	0.2	0.6	0.7	0.8
Others	0.9	0.3	0.4	0.5
Total	152.0	165.5	177.4	184.9

Source: National Horticultural Board

Diversification options

Diversification is an integral part for the process of structural transformation of an economy at macro level. With the agriculture, some of the sub-sectors like animal husbandry, forestry and fisheries are the key components and placed parallel to crop production system in dryland areas. Within the crop production system, so called super cereals like wheat and rice are progressing faster as compared to coarse cereals like sorghum and pearl millet or minor cereals. Now a day diversification on cropping system of dryland areas through intensification of pulses, oilseeds and light value crops are required to meet the multiple needs of small and marginal farmers and maintain natural resources base especially the soil health to sustain crop productivity. The inclusion of under utilized and more adoptive valuable crop and thier varieties for intensification under various cropping system may be helpful in sustaining the rainfed agro-ecosystem. The factors prompting diversification and the speed with which changes occur vary in different situations. Under diversification the changes are in the nature of shift from one crop to another crop or from one entrepreneur to other entrepreneur. In this context, the diversification for the sustainable crop production under dryland agro-eco regions could suggest the following situations;

(A) Shift from farm to non-farm activities

(B) Shift from less profitable crop to more profitable crop

(C) Shift from less profitable enterprises to more profitable enterprises

(D) Use of diverse resources for complementary activities

Thus it is clear that the concept of crop diversification is relative to crop specialization or crop concentration either in terms of a crop in total or crop area or income of a farm. In its essence, crop diversification is to bring out a desirable change in existing cropping systems or cropping pattern for a more balance and sustainable one.

Table 6: Analysis of crop diversification: macro perspective.

	Anual Compound growth rate of crop area (% per annum)	
Crops	1984 – 85 to 1994 - 95	1994-95 – 2003 -04
Rice	0.53	- 0.05
Wheat	0.91	0.20
Coarse Cereals	- 2.14	- 1.09
Maize	0.49	1.67
Pulses	- 0.21	- 0.59
Oilseeds	4.03	- 1.90
Cotton	0.74	- 0.97
Sugarcane	2.71	0.95

Table 7: Productivity and Profitability of Legume base diversification in Rice - Wheat system in IGP.

Location / Cropping system	Average productivity (kg/ha)		
	Kharif	**Rabi**	**Net returns (Rs./ha)**
Pantnagar			
Rice – wheat	3945	4207	11,218
Rice – Lentil	4299	1506	14,445
Varanasi			
Rice – wheat	3648	3415	8,727
Rice – chickpea	3838	1407	11,770
Kanpur			
Rice – wheat	3614	3560	9,095
Rice – Chickpea	3797	1803	13,815
Blackgram – wheat	1330	3885	11,861
Ludhiana			
Rice – wheat	6111	4372	17,271
Rice – Pea	6481	2064	22,995

(Hedge, 1992)

Effective outcomes of diversification:

- Food & nutrition security
- Income growth
- Poverty alleviation
- Employment generation
- Judicious use of land and water resources
- Sustainable agriculture development, and
- Environmental improvement

Need for Crop Diversification

The small and marginal farming community in Indian sub continent faces instability of farm income. Thus the need for diversification of traditional cropping system arises due to:

1. Increase income on small farm holdings

More than 75% of farm holdings are below 2 ha. To raise the profitability from these small and marginal holdings, it is imperative to diversify the production system through inclusion of high value crops or other farm enterprise to complement the crop production system.

2. Improve competitiveness

The increasing threat of cheap imports consequent to trade liberalization can only be encountered by producing agricultural commodities at a competitive price. A rational diversification is likely to reduce the cost of production and thus help the farmers to exploit.

3. Withstand price fluctuations

Increased dependence on selectively few agricultural crops makes the firm more vulnerable to price feudalism arising due to demand – supply or export – impact equations. A diversified farm with various crops/enterprise can better tolerate the ups and down of prices of various crops and thus ensure economic stability and sustainability to the farm family.

4. Ensure constant incomes - flow

Judicious mix of various crops/enterprise round the years helps the farmer to make the various crops available round the year in the firm and continuous flow of income is received throughout the year.

5. Alleviate hunger and malnutrition

Indian alone houses 200 million malnourished people out of 800 million in the world. Judicious mix of pulses, oilseeds, fruits and vegetables in a farm will diversify the food basket and ensure nutrition security of farm families.

6. Effective management of natural resources

A non – judicious and input intensive production system has accelerated the degradation of natural resource base (soil, water) during post green revolution period as witnessed in North western India. Especially the over exploitations of ground water resources, rising of water table in canal command areas, depletion of soil fertility and deficiency of major and micro nutrients are some of the important issues emerged out of commodity oriented agriculture. Therefore diversification is urgently needed for checking the soil degradation, reduce the over exploitation of ground water, judicious management of canal water through introduction of less water requirement crop, soil fertility builders and inclusion of high value crops in the existing cropping systems.

7. Mitigate ill effects of aberrant weather

Aberrant climatic situations likes erratic rainfall, early withdrawal of monsoon, late onset of monsoon, interfiling drought spell, sudden rise of air temperature etc. are very common to Indian Agriculture. Specialized crops/cropping pattern or farming system with specific demotic requirement may fail under such situation. Therefore crop diversification involving two or more no. of crops or enterprise mix in the farm will help to face the challenges and reduce the risk of crop failure under such weather condition and stabilize the farm productivity under stress conditions.

8. Effective recycling of farm waste

The research based enterprise mix ensures efficient use of farm wastes and by products. For instance, Agriculture and livestock together are complimentary as the by products of either can be used for the other. Such complementarities are not only effective but also eco-friendly.

9. Less dependence on off-farm inputs

Diversification should essentially reduce the dependence of farmer on the inputs purchased from outside the farm. Thus a diversified farm will be more sustainable in economic terms than a farm having single enterprise.

Approaches of Crop Diversification

Since 80's there has been substantial degree of diversification of Indian agrarian economy from favoured region to favoured crop pattern of growth. The next approach of diversification has been from less efficient crop or cropping system to more efficient and profitable crop or cropping system. Now a days need based diversification and demand based diversification are the crop diversification approaches. In the context of global opportunities, value addition diversification and varietal diversification approaches are also graining strength and competitive advantage in the recent times (Rathore and Sharma 2004). The important approaches for crop diversification are:

A. Horizontal Diversification

It is the common and main approach of crop diversification. Also called crop area diversification. It may be of two types.

1. Crop substitution

Replacement of one crop by another crop (In Rice - wheat, rice may be replaced by maize, pigeonpea etc.).

2. Crop Intensification

Addition of more crops to the existing cropping system (example : Rice fallow may be diversified through Rice - greengram).

B. Vertical Diversification

It refers to enhancing economic values of crop produce of different crops on cropping systems by refining and manufacturing value added products i.e. use of guar for extracting guar green, production of malt from barley etc. This is also called crop value diversification. Basically this approach reflects extent and stage of industrialization of crops with practicing enterprises like spices, medicinal and aromatic plants, dryland horticulture, other economic shrubs and livestocks. There are some other approaches of crop diversification are:

Land based approach

In this approach, area identification is the primary focus and all supporting development such as irrigation, infrastructure are secondary e.g. enhance inland productivity/soil productivity, income per unit water.

Water based approach

This approach focuses mainly on the availability or ease of water management in crop production e.g. changes and modifications in crops/cropping systems to advice high productivity per unit area per unit of water.

Varietal based approach

Shift from one variety of crop to other efficient variety of some crop either partially or completely or addition of a member of varieties in a given area is called on varietal diversification i.e. shift from Pusa Basmati -1 variety of rice to PRH – 10, Hybrid Basmati rice variety.

Diversity for Diversification:

The Indian subcontinent is possibly the most diverse region in the world in respect of agro-ecosystem crop species, and possibly in the diversity of pests and diseases that attack these crops. Since the dawn of dryland agriculture, traditional cropping system, crop sequences and crop rotation sustained the production systems for the centuries. With the expansion of irrigation facility, adoption of watershed approaches and in-situ moisture conservation practices and availability of inputs and high yielding varieties for various cropping system, like rice-wheat rotation emerged as the most productive and remunerative. The serious consequence of this system is depletion of ground water reserves, deficiency of micronutrients, and increased vulnerability to diseases and pests. In these circumstances, crop diversification assumes importance of minimizing the risk, conservation of resources, enhancing the income and improving human and soil health. India is endowed with rice diversity of pulse crop and is considered as one of the twelve-

mega gene centers of diversity of crop plants in the world. It is also estimated that India is the homeland of 31 wild species and relatives of various pulses. Most of these by virtue of there own merits assume significance in crop diversification. With the development of extra short duration varieties, the pulses have become an important component of crop diversification. With the short duration varieties of different crops, the four options are readily available for crop diversification. These are inclusion as cash crop, substitution of existing low yielding crops / varieties in prevailing system, intercrop with wide spaced planted crops and introduction of new niches. Besides short duration, other attributes need to be identified so that genotypes find their suitability in different cropping system and new niches.

Issues and strategies for crop diversification in rainfed areas

The country has experienced progressive turn down in production and productivity of food grains from 212.9 million tonnes in 2001-02 to 208.6 million tonnes in 2005-06. With assured supply of cereals, pulses and oilseeds at an affordable price, the main focus of policy makers and planners now is on nutritional security. This can be dominated by monoculture of certain crops as more than 80 percent food comes only from 10 crop species. Thus crop diversification is of paramount importance in mitigating problems of arising due to monoculture. For instance diversifying rice - wheat system with barseem, mustard and sugarcane effectively minimizes *Phalaris minor* infestation whereas inclusion of grain legumes for grain, fodder and green manure improves the fertility and soil health. Most significant examples of crop diversification in past few decades are introduction of rice in Punjab and Haryana, wheat in West Bengal, groundnut in Gujarat, Soybean in M.P. and winter maize in Bihar. Besides cultivation of pulses and oilseeds in Rice fallows of eastern India and that of frenchbean in northern plains are likely to make sizeable differences. There is need to identify promising crops and cropping systems for higher and stable yields and profits under water scare condition.

Crop Diversification under Rainfed condition

In rainfed areas or having limited water availability, different efficient crops were identified based on higher yield compared to traditionally grown crops/cropping system. This is activated either by crop sub situation/replacement or intensification through inter cropping.

1. Substitution

Substitution of cotton by sorghum at Bellary, wheat by chickpea at Varanasi and wheat by taramira at Hissar brought many fold increase in total productivity. As a thumb rule, crops having higher water requirement should not be included in the crop production system under these situation.

2. Intensification

Inter cropping is one of the important ways to increase the productivity and provide income stability under limited soil moisture conditions. Some of the prominent intercropping systems are maize + blackgram at Palampur, Ranching, Banswara; Maize + soybean at Ranchi, maize + cowpea at Karjat, sorghum + soybean at Sehore, sorghum + pigeonpea at Indore, Pigeonpea + green gram at Bichpuri and Hanumangarh, rice + soybean at Kalyan and Jabalpur, wheat + rapeseed at Indore. Net profits from these inter cropping systems were as high as 15 – 20% compared to sole cropping. It was observed that both the component crops should be fertilized at recommended rate to achieve maximum benefit from a diversified intercropping system.

Table 8: Productivity potential and profitability of Maize - legume based Intercropping.

Treatment	Yield (q/ha)				Maize equivalent yield (q/ha)		Net return ('000 Rs./ha)		B:C ratio	
	Maize		Inter crop							
	2001	2002	2001	2002	2001	2002	2001	2002	2001	2002
Maize (Sole)	44.14	45.57	-	-	44.17	45.57	20.89	28.97	2.29	2.97
Maize + Blackgram	43.33	44.87	3.43	3.77	54.10	56.63	24.83	35.17	2.39	3.18
Maize + Green gram	43.28	44.56	4.23	4.75	55.52	59.02	25.40	34.46	2.44	3.33
Maize + Cowpea	42.88	44.01	5.09	5.45	56.40	61.08	26.08	37.96	2.54	3.47
Maize + Syabean	42.21	43.29	5.83	6.33	54.17	57.62	24.55	35.38	2.26	3.07

Table 9: Sugarcane based intercropping systems through pulses.

Treatment	Cane Yield (t/ha)	Intercrop yield (kg/ha)	Sugarcane experiment (t/ha)	Net return over sole sugarcane (Rs./ha)
Sugarcane (S) sole (Co 6304)	81.0	-	81.0	-
Sugarcane + Pea (Rachna)	80.34	626	92.09	3468.00
Sugarcane + Pea (Pusa 10)	81.16	437	88.78	2473.00
Sugarcane + Pea (HFP 4)	76.38	536	85.45	1254.00
Sugarcane + Rajmash (PDR 14)	78.14	721	99.85	11371.00
Sugarcane + Rajmash (VL 63)	81.12	1004	111.13	11371.00
Sugarcane + Chickpea (JG 74)	77.58	1139	96.11	4836.00
Sugarcane + Chickpea (JG 315)	76.58	1190	94.28	4742.00
CD (5%)	NS	NS	0.82	-

Table 10: Rice based intercropping systems with pulses.

Cropping system	Yield (t/ha)			Rice experimental yield (t/ha)
	Rice	Intercrop	Wheat	
Rice + soybean - wheat	2.50	1.01	5.02	10.12
Rice + pigeonpea - wheat	2.28	0.67	4.73	9.40
Rice + cowpea - wheat	2.04	0.36	5.00	8.44
Rice (sole)	3.45	-	4.98	9.17
Soybean (sole)	-	2.52	5.32	10.82
Cowpea (sole)	-	0.69	5.33	7.39
Pigeonpea (sole)	-	1.69	5.15	10.06

(Ram Sewa and Singh, 2002).

Conclusions

Agriculture diversification as it happened in India has played an important role in the growth of agricultural output during the last 3 decades. Though initial incentive for diversification come from technology or demand driven factors, the success and speed of diversification depends on infrastructural factors, availability of finances to buy inputs needed for shift to new production choices, structure of land holdings and of course availability of market for the products.

References

Chalka, M.K. and V. Nepalia (2005) Indian Journal of Agronomy, 50 (2):119-122.

Hedge 1992. Cropping systems research highlights, coordinates report, In : Proceedings of the 20th workshop. Tamil Nadu Agril. Univ., Coimbatore 1-4 June, 1992. PDCSR, Modipuram, Meerut.

Ram Sewa and Singh D. (2002). Alternative cropping systems to rice – wheat, Ann. Rep. 2001- 02, PDCSR, Modipuram, Meerut, 13 – 16 p.

Rathore P.S. and Sharma S.K. 2004. Crop Diversification concepts, needs and approaches. In: Crop Diversification a Paradigm for sustainable agriculture in arid and semi arid areas, winter school 24 Nov. 14th December 2004, Rajasthan Agril. University, Bikaner, Page 1-8.

CHAPTER

10 Prospects of Temperate Horticulture Under Limited Water Conditions

R. M. Sharma, Kiran Kour and M. K. Pandey

Regional Horticulture Research Sub-Station, Sher-e-Kashmir University of Agricultural Sciences and Technology of Jammu, Bhaderwah (Doda)

Temperate fruits are grown in both the hemispheres and those regions lie mostly within latitudes 30° to 50° but are extended into higher latitudes by the moderating influence of nearby water bodies and into lower latitudes by the cooling influence of higher elevations. The general climatic requirements for temperate fruits are as follows:

- Winter temperatures must not be so cold that they kill the plants.
- Winter must be cold enough to give buds adequate chilling to break winter rest.
- The growing season (number of frost free days) must be long enough to mature the crop.
- Temperature and light during the growing season must be adequate for the species in question to develop fruit of good quality.

In India, temperate fruits (apple, pear, peach, plum, apricot, cherry olive, walnut, almond etc.) are grown principally in J & K, HP and Uttarakhand and to some extent in Arunachal Pradesh, Sikkim and Nagaland. The low chilling pears and peaches are being grown successfully in foot hills of north Indian plains. The hilly areas having water scarcity, undulating land, poor soil fertility and depth have least scope of agriculture crop production hence temperate fruit production is the best alternative in these areas because of the following additional reasons.

- Establishment of orchards in rainfed areas serves as an insurance against total or partial crop failure.
- Growing of perennial fruit crops provides additional/supplemental income.
- Fruit crops being deep rooted are better adapted to rainfed situations.
- Temperate fruit crops ensure proper utilization of waste lands.
- Provide nutritious and balanced food to the ever increasing population and help in solving malnutrition problem.
- Provide raw materials for agro-based industries.

The status of temperate fruits in India and their importance are given in Table 1 and 2, respectively.

Table 1: Status of temperate fruits in India

Fruit	Area (ha)	Production (MT)	Productivity (t/ ha)	State wise Production Share(%)			
				J&K	HP	UA	NE State
Apple	241530	1384760	5.73	65.68	27.20	6.47	0.02
Pear*	26390	91360	3.46	38.96	11.90	49.12	----
Peach/ Nectarines	20030	30680	1.81	7.33	18.41	74.25	-----
Plum	26230	26360	1.00	17.71	19.61	62.67	-----
Apricot	15080	15341	1.11	36.63	3.20	60.16	-----
Cherry	2520	7240	2.87	98.89	1.10	---	-----
Almond	26560	10710	0.33	92.15	7.84	----	-----
Walnut	84160	96650	1.14	89.24	1.34	9.40	-----

*Punjab is having highest production (51200 MT) from 2560 ha area but of sub-tropical pear

Water is one of the costly inputs in commercial fruit production. In India, most of the temperate fruit orchards are established in sloppy lands where soils are shallow, erratic and prone to erosion and very little irrigation facilities are available. The usual features of arid ecosystem are extreme of temperature, low and erratic rainfall, low relative humidity, high wind velocity etc.Although winter rains and snow are quite common in temperate regions, the bulk of precipitation takes place during July-September. There is a little rain in April, May and June when drought or semi-drought conditions prevail frequently, causing water stress in deciduous fruits.

Table 2: Importance of temperate fruits

Fruit	Importance
Apple	• Good source of sugars (9-11%) and anthocynins. • Apple enhances dental hygiene and packed with phytochemicals such as quercetin that may help prevent heart disease and cancer • Low in calories, high soluble fibre that helps lowering cholesterol • Exported to Sri Lanka, Nepal, Bangladesh etc. • Can be processed into a variety of products like juice, jam, jelly, dehydrates, cider etc. • Can also be used for cooking purposes
Pear	• Good source of simple sugars, carotenoids anthocyanins (in coloured varieties), ditery fiber, folate and Vit C • Can thrive best in the conditions where other crops are outright failure. • Wide range in cultivars exists in reference to growing conditions • Can be processed to prepare juice, nectar, preserve, candy, leather and perry, and can be dried and canned

Peach/Nectarines	• Good source of beta carotene with useful amount of vitamin C, potassium and anthocynin • Genetically precious • Highly remunerative • Good source of dietary fiber, Wide range in cultivars exists in reference to growing conditions
Plum / Prunes	• Good source of simple sugars, anthocynins, Vit C, fiber, vit.A and Fe and K • Can be processed for canning, drying and wine making
Cherry	• Good source of simple sugars and anthocynins • Low in calorie, almost fat free • High in pectin, a soluble fiber that lowers cholesterol • Processed into canned fruit cocktails and juice making • Used for cooking, salad, confectionary and ice cream making
Apricot	• Good source of carotenes (2612 IU /100g edible portion) and pectin (2.66%) • Rich source of kernel oil (36-60 %) used for body massage, illumination in lamps, cooking and in cosmetic industry • Can be dried and processed into jelly, jam and nectar.
Olive	• Very good source of oil (8-30%) used for cooking, body massage and medicinal purposes. Fruits can also be used for pickle making.
Walnut	• Good source of fat (64%), protein (14.8%), carbohydrate (15.8%) and potassium (450 mg/100g) • Rich source of alpha linolenic acid, having cardio protective effects • Antioxidant (melatonin) induct and regulate sleep in human beings • Good source of foreign exchange as exported to France, Germany, Spain, Portugal, Austria, UK, Kuwait worth of Rs 150 crores from J & K (257 MT in shell +5417 MT kernel) • Can be grown commercially in interior areas
Pecan	• Good source of fat (71%), protein (9.2%) and carbohydrate (14.6 g) and potassium (603 mg/100g) • Good shelling percentage and thin shelled • Can be grown commercially in warm temperate interior areas
Almond	• Good source of fat (54%), protein (19%) and carbohydrate (21%) • Can be grown commercially in interior areas • Can also be used as table purposes
Seabuckthorn	• Good source of vitamin C (769 mg/100g) with good juice yield (65.70%) • Can be used as windbreak, biofence, fuel, fodder and the fruits can be used as fresh, dried for chutney and for making squash, RTs beaverage, jam, chyavanprash, pickle etc. • Leaves can be used as tea. • Good export potential to Germany, Russia, China, Canada, Japan and USA. • More than 200 products are being marketed world over.

Irrigation is quite necessary for the areas receiving precipitation less than 50 cm/ year for successful cultivation of these fruits. Water deficit affects growth and development of temperate fruit trees by following ways

- Growth of the tree by influencing cell division and expansion
- Fruiting by influencing flown bud differentiation
- Fruit quality by decreasing carbohydrate production through altering stomatal aperture and enzyme activities of photosynthesis and respiration.

So keeping in view the availability of water and exploitation of potential of hill state and temperate fruit crops, the following strategies need to be adopted to maintain the pace of fruit production with the growing demand of present days under limited availability conditions.

- Selection of suitable fruit crops
- Varietal suitability
- Rootstock suitability
- Propagation techniques
- Planting techniques
- Orchard soil management practices
- Water management
- Nutrition
- Diseases and pests management

Selection of suitable fruit species

The selection of suitable fruit crop is prerequisite, to be grown under limited water conditions. The order of water requirement of temperate fruit trees is quince > Pear > plum > peach > apple > cherry > walnut > apricot. The fruit crops like kiwifruit, olive and strawberries require assured irrigation hence needs to be discouraged under water scarcity zone, however, they can be grown with the use of micro- irrigation system.

Varietal suitability

The list of drought hardy varieties of temperate fruit species is given in Table 3. Being a highly drought resistant crop, the varietal comparisons of walnut are presented in Table 4-5.

Table 3 : List of fruit specific drought hardy varieties

S. No	Fruit Name	Varieties
1	Apricot	Halman, Rkchey karpo, Tokpopa, Rogan, Safaida, Nari, Australian, Shakarpara, New Castle, Charmagz, Kaisha, St. Ambroise, Moorpark.
2	Seabuckthorn	

	Russian	Chuisksya (heavy fruiting type), Jivko (medicinal use due to high carotenoids), Chuskaya 377-72-31, 579-3-1 (mechanical harvesting suitability), Sibirski Rumyuattes (early ripening with high yield), Krasny Fakel (late ripening, high oil and nutrient contents), Capris (early ripening), Zaruitsa (multipurpose), Podruga (large fruit, suitable for table purpose), Parad (large fruit, good for fresh use, oil production with good shelf life) (Belykh, 1998)
	German	Ashola (early ripening), Dorana (easy to harvest), Frugana (early ripening), Hergo (heavy fruiting), Leikora (ornamental), Poly 1,2,3 and 4 (male clones for pollination) (Albrecht and Schuldt, 1997)
3	Apple	Starkrimson, Mollies Delicious, Morespur Gold, Coop-4. Red Delicious, Granny Smith, York Imperial.
4	Pear	Flemish Beauty, Bartlett, Max Red Bartlett, Conference, Starkrimson, Devoe, LeConte, Patharnakh, Clapps Favorite.
5	Almond	Non Pareil, Thin Shell, Drake, Ne- Plus-Ultra
6	Plum	Red Beaut, Santa Rosa, Burbank, Monfor.
7	Pecan	Stuart, Mahan, Pawnee
8	Persimmon	Hachiya
9	Cherry	Compact Stella
10	Pomegranate	Kandhari, Anardana (wild pomegranate)
11	Peach/ Nectarine	July Elberta, Snow queen
12	Walnut	Tutle-31, Serr, Blackmore, Hartley, Franquette, Chico, Pedro, Payne, Sulaiman, Hamdan, Local Selections

Table 4: Varietial comparison of various indigeneous selections and exotic cultivars of Walnut.

Varieties/ Selection	Leafing out	Lateral Bud fruit-fullness (%)	Dichogamy	Shell Seal	Kernel Colour	Shelling Percentage
Exotic Varieties						
Hartley	Late	60	Protandrous	Medium & Tight	Light	48
Franquette	Late	50	Protogynous	Weak & Tight	Light	51
Amigo	Early	80	Protogynous	Weak & Open	Amber light	53
Payne	Early	80	Protandrous	Weak & Tight	Light	50

Pedro	Mid	80	Protandrous	Medium & Open	Light	47
Chico	Mid	90	Potogynous	Weak & Open	Amber Light	50
Indigenous Selections						
SKAU-W-0003	Early	30	Protandrous	Medium & Open	Amber light	51
SKAU-W-0004	Early	30	Protandrous	Weak & Open	Amber light	51
SKAU-W-0008	Early	50	Protandrous	Medium & Open	Light	52
SKAU-W-0014	Late	70	Protandrous	Weak & Tight	Light	54
SKAU-W-0017	Late	70	Protogynous	Weak & Tight	Light	52
SKAU-W-0022	Late	40	Protogynous	Medium & Tight	Amber light	42

(Sounduri *et al.*, 2005)

Table 5: Quality characters of indigenous walnut selections

S. No	Selections	Nut shape (A)	Shell texture (B)	Shell colour (C)	Shell seal (D)	Shell strength (E)	Shell integrity (F)	Kernal colour (G)
1	SKAU-W-0002	8	7	1	9	7	2	1
2	SKAU-W-0003	1	7	5	5	5	3	2
3	SKAU-W-0004	1	5	1	5	5	3	1
4	SKAU-W-0005	1	5	3	3	5	3	2
5	SKAU-W-0008	1	5	1	3	3	3	2
6	SKAU-W-0009	6	5	7	7	7	3	3
7	SKAU-W-00022	6	7	5	9	7	1	2
8	SKAU-W-00023	5	5	1	5	5	3	1
9	SKAU-W-00024	7	5	5	5	7	3	2
10	SKAU-W-00025	5	7	7	5	5	1	1
11	Sulaiman	1	5	3	3	3	3	2
12	Hamdan	4	3	3	5	5	3	1

(Sounduri and Sharma, 2005)

Legends

- A: 1-Round, 2-Triangular, 3-Broad ovate, 4-ovate, 5-Short trapezoid, 6-Long trapezoid, 7-Broad elliptical, 8-Elliptical, 9-Cordate.
- B: 3-Smooth, 5-Medium, 7-Rough
- C: 3-Light, 5-Medium, 7-Dark
- D: 3-Weak, 5-Intermediate, 7-Strong
- E: 3-Weak, 5-Intermediate, 7-Strong
- F: 1-Incomplete, 2-Intermediae, 3-complete
- G: 1-Extra light, 2-light, 3-Amber

Rootstock suitability

The list drought hardy rootstocks is given in Table 6.

Table 6: Resistant / Tolerant Rootstocks

Apple	Seedlings of crab apple, MM-111, MM 106, M9, KC1, KC-1-48-41
Pear	OH x F 217, Province Quince, Seedling of Kainth, Oregon 211, 249, 260, 261, 264
Peach	Lovell, Halford, Nemaguard, Wild peach seedling, GF 557, GF-677
Plum	Plum seedlings
Apricot	Seedlings of wild apricot, Myroblan 27
Almond	Nemaguard, Lovell, GF-305, GF-677, GF-557, Myrobalan 2082, Marina -2624, Mariana GF-8/1, hybrids of Myrobalan X peach, Myrobalan x almond, Bitter almond, Wild peach.
Walnut	Seedlings
Cherry	Mahaleb, Colt, Seedlings (Paja)
Persimmon	Wild amlook
Pecan	Seedling (Curtis)

Propagation techniques

Application of 120 Kg N, 30 Kg P_2O_5 and 45 Kg K_2O per ha has been found highly effective to attain the buddable/graftable size well in time. Two sprays of BA or GA_3 @ 100 ppm + urea (0.5%) during growth period enhance seedling growth. Tongue grafting is highly effective and widely employed for the propagation of most of the temperate fruits, however, other techniques have also been found equally effective for their multiplication. These new propagation techniques include chip budding for apple, plum, apricot and almond, annular budding for walnut and pecan, patch budding for pecan and veneer grafting for the propagation of walnut and persimmon (Ananda, 2001). Chip budding has been found a successful method for apple multiplication thrice (February, June and September) in a year. It is advisable to use tongue grafting during January-March (depending on the elevation) instead of T-budding if the water scarcity exists during active growth season.

Planting techniques

Application of the principles just pointed out that particular fruits and particular locations are the main deciding factors in the determination of distance of planting for orchard fruits where water supply is most frequently the limiting factor in this connection even though the grower seldom realizes it at the time planting. This is contrary in the way it often works out, to the frequently repeated statement that trees can be planted more closely in a poor than in a good soil. If the soil is poor because it is shallow or of poor water holding capacity unproductiveness will only be increased by closer spacing. In soils that are both fertile and well watered, planting distance should be governed by the size of the plants normally retain; in other words it is largely a matter of fully occupying the space without undue crowding. If they are infertile and well watered, spacing must be relatively wide in order to avoid starvation symptoms or, what is more practicable, planting much as under the preceding circumstances and dealing with the fertility problems through the use of fertilizers. If the moisture is the limiting factor and irrigation is not available, spacing should be wide enough so that water requirement is fully taken care of by the wide ranging roots, even though the area seems to be far from fully utilized by the tops.

Many unirrigated fruit plantations, especially those closely planted bears satisfactory crops of well sized fruits for a period of years. Then as they become older and their roots more fully exploit the area, growth of tops and size of crop and of fruit decline and the planation becomes unprofitable. Profitable productive life can be materially lengthened by wider spacing.

The technique of *in situ* sowing/ planting of hardy types with follow up grafting/ budding with selected scion material eliminates the disadvantages of both poor root system and demand for irrigation when grafted material is transplanted. This technique also helps to the plants to establish better (particularly deep rooted plants like pecan) and further they become avail to tolerate the moisture stress for longer period.

Orchard soil management

Soil management refers to the management of orchard soil in such a way so as to get higher yields of good quality fruits over the years. The chief objective of orchard soil management is to provide the necessary conditions to the plants to grow satisfactorily, utilizing the moisture and nutrient supply in the soil to the best advantage. Soil management is the relationship between the soil and fruit crops to be raised. The choice of soil management system depends on altitude, soil type, kind of fruit variety, rooting depth of trees, topography, climatic conditions particularly temperature and rainfall and economic conditions of the farmers. An ideal soil management system should have the following characteristics:

- Create favourable moisture supply conditions in the soil.
- Supply adequate nutrients and organic matter i.e. balanced supply of nutrients.
- Keep the soil porous and friable i.e. suitable physical conditions in the soil for tree root development and favourable conditions for soil water movement.
- Check soil erosion and water runoff.
- Maintain and improve the status of soil organisms like earthworms, bacteria, mycorrhizae and fungi in the soil by providing favourable environment.

Most of the orchards in the hills are rainfed and the perennial source of water is not available. These orchards should be covered with perennial grasses which prevent soil erosion and also maintain the soil fertility if legumes are grown. When the slope of the land is more than 10%, orchard floor should always be covered with grasses and no clean cultivation in between should be carried out except the basin area around the tree trunk for application of manures and fertilizers. Perennial grasses can also be replaced by nutritional fodders, legumes, such as red clover, lucerne and white clover. On northern aspects, shade tolerant and moisture loving grasses and on the southern aspect, the drought tolerant species such as *Fescue* should be grown.

Sod culture

Quite frequently weeds are allowed to grown during the rainy season in the orchard. Natural gasses (Table 7) are allowed to grow beneath and between the trees. There is a great advantage of sod in sloppy lands as it prevents erosions and lessens the loss of moisture by surface evaporation. Furthermore, cutting grasses and leaving it in place (particularly in basins) is highly effective in conserving moisture in the orchards.

Cover Crops

A cover crop is a non-economic crop grown in the middles between orchard rows with sprayed strips. Cover crops can be annuals, which germinate and die in one season, or perennials that live for more than one year. Both legumes and grasses are available as an annuals or perennials depending species. Additionally, both winter and summer weeds can be allowed to grow and managed like a cover crop.

The appropriate use of cover crops can result in substantial benefits to the cultivated fruit crop itself and/or can cause potential problems. One hopes the benefits out weigh the problems. Since both the benefits and problems can be site and management specific, it is good to review them before planting. The most universal reason for using cover crops is to reduce soil erosion. The established plant roots hold the soil against the forces of moving water. Established cover crops have been shown quite effective in controlling erosion on slopes as well as river bottom soils during flooding.

Table 7: Suitable sods for temperate fruit orchards

S. No.	English common name	Botanical name
Legumes		
1.	Kudju Vine	*Puerania sp.*
2.	Lucerne (alfa alfa)	*Medicago sativa*
3.	White clover	*Trifolium repens*
4.	Red clover	*Trofolium pratenses*
5.	Subterranean clover	*Trifolium subterraneum*

S. No.	English common name	Botanical name
Non-Legumes		
6.	Timothy grass	*Pheum pretance*
7.	Orchard grass	*Dactylis glomerate*
8.	Rye grass	*Lolium sp.*
9.	Toowamba canary grass	*Phalaris aquatic*
10.	Tall fescue grass	*Festuca sp.*
11.	Kentucky blue grass	*Poa pratensis*

Biomass production or the production of plant tissue both above and below ground can be beneficial in both nutrient and soil quality perspective. A good stand of planted annual cover crop can produce 5000 pounds per planted acre of above ground dry matter per season. Add to that the root mass, and the total biomass can be near 7500 pounds/planted acres.

In terms of nutrients, cover crops extract left over nitrogen from the orchard; take up mineralized nitrogen from organic matter and in the case of legumes (such as clovers and vetch) extract nitrogen from the atmosphere. Legumes generally produce twice the nitrogen per pound of dry (3%) matter than grasses (1.5%). Therefore a 5000-pounds/acre dry matter cover crop can produce up to 75 -150 pounds of nitrogen depending on the species mix. In order to utilize all this nitrogen, the cover must be incorporated into the soil usually by disking. In contrast, mowing the cover along with subsequent irrigation will cause some (a lot) of the nitrogen to be lost to the atmosphere. Cover crops which are not incorporated but mowed have not been shown to replace the application of fertilizer to meet the nitrogen requirement.

Cover crop living vegetation and biomass derived from cover crops protect the soil surface from the damaging effects of raindrops and equipment. With the cover crop drying up soil by using water during the winter and covering the surface, orchard access is enhanced. Ruts from equipment are much less of a problem. As the biomass decomposes into primarily polysaccharides (long chain sugar products), they bind soil particles together stabilizing the soil aggregates against the effects of sprinkler and rain droplets and other soil compacting/ crusting forces. Aggregate stabilization along with the channels created by roots enhance the soil's water infiltration characteristics.

By using moisture from the soil during the rainy season, room for more water to infiltrate the soil is made. So when a rainfall event occurs more of the water infiltrates rather than becoming run off which can reduce off-site movement of pesticides. Cover crops can reduce undesirable weed species. In the walnut pest management alliance (WPMA) plots, the cover crop established well and reached maturity at both sites, allowing for reseeding. The number of winter weed species and their dry weight were significantly decreased in the cover crop plot versus the resident vegetation plot. Although the occurrence of spring or summer weeds overall was not significantly different between the plots, certain weeds such as burr buttercup were dramatically reduced where there was a cover crop. Other species such as hairy fleabane were found at low levels only in the unseeded plots and not found in the cover crop blocks.

As with any cultural practice, there are drawbacks to the use of cover crops. The biggest is the use of water. Nearly 300 pounds (36 gal) of water is required to produce one pound of aboveground dry matter. Using this conversion a 5000-pounds/acre cover crop would consume about 6.5 inches of water (180,000 gal). In areas, which receive high rainfall, the cover's use is not usually a determent to the eventual volume of water stored in the root zone. However, in areas of lower winter rainfall, the cover can use a portion of the moisture normally stored in the root zone for later use by the orchard. Hence, more water must be pumped to meet the orchard's seasonal water requirement. Perennial cover crops, which grow year round, compete successfully with the orchard for water. Studies in a mature almond orchard indicate a 10-30% increase in orchard water use when a perennial clover was present when compared to bare soil.

Lower spring temperatures on cover cropped orchards may increase the risk of frost, especially with early varieties. A dense cover crop will reduce the temperature at the surface level in comparison with bare ground. Close mowing of the cover before the frost hazard will reduce the possibility of damage from radiation type frosts.

Cover crop biomass production and orchard sanitation practices used to control navel orange worm (NOW) can be in conflict. Orchard sanitation is the most important element in a NOW control programmme. It requires the mower to be adjusted low enough to shred the nuts before insect emergence. This practice can decrease biomass production and be detrimental to some species.

Another problem that may result from cover cropping is the buildup of pocket gophers and voles. Gophers are particularly attracted to annual and perennial clovers.

Winter annuals that die off in early summer can be allowed to stand and compete (for light) with germinating summer weeds or be mowed. Repeated mowing helps decompose the cover crop residue. Legume residue decomposes very fast where as grasses are slower. Full coverage irrigation systems (flood and sprinkler) speeds decomposition whereas micro-irrigation (drip and micro-sprinklers) leave more residues in the non-wetted areas. This cover crop trash can cause a slower harvest and in wet years, can increase the incidence of mold if walnuts are left on the ground.

When choosing a cover crop species or mix of species, growers should first determine the benefits desired and potential drawbacks mentioned earlier in conjunction with specific orchard conditions, cultural practices, and lastly, costs.

Table 8 contains a number of popular species, growth habits, physiology and common seeding rates which can be used as cover crops.

Table 8: Selected Characteristics of Important Cover Crops

Common name	Growth Habit	Max. Height (inches)	Flowering Period	Maturity Period	Tolerates Close Mowing in Winter	Reliably Self-Reseeding	Seedling Rate (lbs/ac)	Comments
Legumes	**WINTER ANNUALS**							
Burmedic (burclover)	Prostrate to erect	6.15	Feb-Apr	Apr-May	Yes	Yes	15-20	Neutral to alkaline soils
Field pea	Viny	18-30	Mar-May	May-June	No	No	70-120	'Magnus' & Miranda are especially vigorous
Clovers								
Berseem	Erect	18-30	May-June	Jun-Jul	Yes	No	15-20	Needs multiple cutting for best results
Crimson	Erect	12-20	Apr-May	May-June	Yes	Yes	20-25	Fast winter growth
Rose	Semi erect	8-15	Mar-Apr	May-June	Yes	Usually	5-20	'Hykon' is an early and well-adapted variety
Subterranean	Prostrate to semi erect	6-15	Mar-May	Apr-June	Yes	Yes	20-25	Many varieties bury seedhead; most prefer neutral to acid soils; 'Kosla' & 'Clare' tolerate alkalinity
Vetches								
Bell (fava) bean	Erect	36-84	Mar-May	May-June	No	No	120-150	Host for bean aphid
Common	Viny	18-24	Apr-May	May-June	High	Yes	40-80	Winter hardy; has extrafloral nectanes
Hairy	Viny	18-24	Apr-May	May-June	High	Yes	35-50	Very winter hardy; adapted o sandy soils
'Lana; woolypod	Viny	18-24	Mar-May	Apr-June	High	Yes	40-60	Produces some hard seed;
Purple	Viny	18-24	Apr-May	May-June	High	Yes	40-60	Least winter hardy vetch; popular in California
Nonlegumes-Grsses								
Annual Rye grass	Erect	30-60	Apr-May	June-Sep	Yes	Yes	20-35	Rapid growth; high biomass, late maturity may lead to competition with trees and vines
Soft chess ('Blando' brome)	Semi erect	12-30	Mar-Apr	Apr-May	\`Yes	Yes	12-15	Reliable; reseeds well; good for erosion control, grazing

Foxtail fescue ('Zorro')	Erect	12-24	Mar-Apr	Apr	Yes	Yes	8-12	Tolerates poor soils; good for erosion control
Cereals								
Barley	Erect	24-36	Apr-May	May-June	Yes	Yes	80-120	Heat, drought, & salinity tolerant
Cereal rye	Erect	36-72	Apr-May	May-June	Yes	Yes	60-120	;Merced' is drought tolerant; many varieties tolerate waterlogged soils
Oat	Erect	24-60	Apr-May	May-June	Yes	Yes	100-120	Relatively drought intolerant; tolerates wet soils.
Others								
Mustards	Erect	24-27	Mar-May	Apr-June	No	Yes	10-15	Rapid growth; may host brassica crop pathogens
PERENNIALS								
Legumes								
Birdsfoot treefoil	Semi Erect	12-24	June-Sep	Jul-Oct	Yes	No	10-15	Slow establishment
Strawberry Clover	Prostrate	8-12	May-Jun	Jun-Jul	Yes	Yes	10-15	Vigorous; invasive; heat & drought tolerant
White clover	Prostrate	8-12	May-Jul	Jul-Aug	Yes	Yes	10-15	Vigorous; invasive; shade tolerant
Nonlegumes								
Pereninal ryegrass	Semi Erect to erect	8-36	May-Sep	Jun-Oct	Yes	Yes	25-35	Vigorous; competitive
SUMMER ANNUALS **Time of First Flowering (Days After Seed)**								
Legumes								
Cowpea (blackeyed pea)	Erect, Viny	18-36		40-80			35-40	Performs well with minimal irrigation; may attract lygus bugs.
Hemp sesbania	Erect	48-120		60-85			20-25	Drought intolerant; may attract bean aphid
Hyacinth bean (lablab)	Viny	18-36		60-85			20-25	Performs very well with minimal irrigation
Sunnhemp	Erect	48-120		60-85			20-25	Drought tolerant; rapid growth.
Nonlegumes								
Buck wheat	Erect	12-24		25-30			20-30	Drought intolerant; flowers attract beneficial insects, as well as lygus bugs
Sorghum & sudangrass	Erect	36-120		60-80			25-35	Rapid growth; performs well with minimum irrigation.

(Ingles, 1994)

Notes:

Optimum seeding rates may very based on local conditions and planting dates. Listed rates are for monocultures only. Use reduced rates for species mixtures.*Some characteristics listed may vary greatly by location.

Mulching

The moisture loss can also be minimized by spreading suitable mulches on the tree basin. It prevents the loss of soil moisture by reducing evaporation. After a drought period, field capacity is attained sooner under mulch than natural sod, probably due to increased infiltration capacity of mulched soil. Besides increasing yield and fruit quality, organic mulches are helpful in improving the physical conditions of soil. Various organic (straw, hay, organic manure, tree leaves) and inorganic (black polyethylene, white polyethylene) mulches are available which can be used for growing temperate fruits under water scarcity areas. Mulching with straw, sawdust, oak leaves or other organic matter increases the humus content and water holding capacity of soil. The higher moisture content is found throughout the growing season beneath hay or black plastic mulch than unmulched basins.

Mulches can make the difference between a decent looking crop and one that does not make the grade for finish and volume. It can be costly to move large volumes of various organic materials like straw or spoiled hay into the orchard but the results can be comparable to supplying minimal amount of irrigation. The added advantage of mulching is in soil temperature regulation, which could be critical for some rootstocks. Mulches can help level out those wild swings in soil moisture, which can contribute to bitter pit problems in susceptible cultivars (Gardner,2003). Different types and method of mulching are as under:

Grass	:	10 cm beneath the tree
Grass + Herbicide followed by	:	10 cm thick grass beneath trees glyphosate
Composted conifer leaf	:	10 cm beneath the tree
Composted conifer + Herbicide	:	10 cm beneath the tree followed by glyphosate
Black polyethylene	:	1.10 radial distance from the tree

Application of organic manures

Application of organic manures like FYM, green manures and compost, not only improves the soil tilth but improves the infiltration capacity of the soil and also improves the fertility, thus reducing the runoff and soil loss.

Irrigation management

Water, carbon dioxide and sun light interact in plants to form simple sugars in a process known as photosynthesis. Water deficiency may reduce photosynthesis by 40% before leaf actually shows wilting in peach (Frecon, 2002). The irrigation water and fertilizers are important input and their adequate and timely supply play a very crucial role in enhancing production and productivity. Lack of moisture during growth period can cause a great setback to fruit size. Size once lost due to dry weather can not be fully recouped, even if heavy rain follow thereafter. Lack of moisture can increase preharvest drops and fruit drops later in the season after June drop. Drought is likely to accentuate the problem of alternate bearing in apple trees thus resulting in lower production (Verma, 2001). Water requirements increase with increased air movement and decreased relative humidity.

In water deficit areas, growers can adjust by applying water when trees are more sensitive to stress (critical stage) and by taking measures to minimize water losses that occur during irrigation. So water deficit irrigation can be a good strategy and can be a potential tool to manage irrigation. Almond can tolerate drought stress fairly well during the two months prior to harvest which allows successfully use of deficit irrigation strategies. By providing less than full water requirement during this period, minimal impact on kernel weight has been noticed. A peach attain about 66% of its final fruit volume during its last 30 days on the tree. Water is critical during this period partially because the evapotranspiration rate is very high (Frecon, 2002).

For managing irrigation, water use should be minimized, and maximum efforts should be directed towards the increasing the percentage of applied water stored in the root zone for use and towards applying irrigation water as evenly as possible throughout the orchard. It has been found that drip irrigation method is technically feasible, economically viable and socially acceptable. This method is highly suitable for undulating terrains, rolling topography and hilly slopes. Such areas are accessible to drip irrigation without any investment on land leveling and shaping. This system has shown an inbuilt capacity to save water and increase production. Drip irrigation is slow, frequent and precise, and is based on the simple concept of supplying water to plant directly on the basis of the their daily water requirement. In peaches, trickle irrigation on mature trees can save between 30-50% of water needed to irrigate (Frecon, 2002). He suggested that a reasonable design objective for drip irrigation had to wet 25-60% of root zone in a mature tree.

Newer technologies like sub surface drip irrigation help the water conservation. Since the lines are buried, the soil evaporation factor is largely eliminated. Water leaves the orchard ecosystem by way of transpiration only after being picked up by the tree roots. The disadvantage of this system is that you can not readily see what is going on underground. These systems should not be installed where excessive compaction of soils is likely to occur (Gardner, 2003). The application of fertilizers and some pesticides can also be made through the drip system.

Simple water harvesting technology consists in storage of rain water during monsoon season and than utilizing it when there is need. It may be adopted under limited water conditions. It is best suited for valley areas and less sloppy lands. The following engineering methods can also be opted for water management in temperate fruits.

- Interplant harvesting runoff water
- Terracing
- Harvesting of gully water
  - Diversion of ditches
  - Gully head control structures
- Ponds and reservoirs

Nutrition

One of the most important aspects of fertilizer usages is to know when fertilizers should be applied. This depends primarily on the crop and on the mobility of the particular nutrient applied to the soil. Potassium and phosphates, which hardly leach in medium to heavy textured soils are often applied in autumn and incorporated into the soil ready for the crop growth in the spring. On the other hand, nitrogen fertilizers, which are generally susceptible to leaching are applied in spring. Where irrigation facilities exist or rainfall is well distributed, half dose of N should be applied after fruit set.

The NPK fertilizers should be broadcast on the soil surface under the spread of trees and slightly mixed with soil. Fertilizers should be applied 30 cm away from the trunk. For soils of poor nutrient status, the application of fertilizers in a band can often yield better results. The band application must be given at a place where the roots activity is concentrated. In temperate fruit species, this zone lies between 1.0m of tree trunk and up to 60 cm in depth, therefore, the band placement should be 15-30 cm deep in this zone, so that bulk of the feeding roots may help in the uptake of the nutrients in that area. In case of P, the soils having high P fixing capacity particularly soils which are high in Fe, Al, Ca and acidity, should be given band application instead of broadcasting.

Fertilizers should never be applied under dry and excessive wet conditions. Fertilizer dose may be reduced by 1/3-1/2 under continuous conditions and the nutrients requirements should preferably be met through foliar application. Foliar application is particularly useful under conditions where nutrients uptake from the soil is restricted. These nutrients are frequently fix by soil particle and for this reason are scarcely available to plant roots. On such soils, foliar application in the form of inorganic salts or chelates is a valuable tool in combating nutrient deficiencies. As micronutrients are only required in small quantities, a foliar spray applied once or twice and correctly timed, is adequate to meet the demand of the crop. The schedules of foliar nutrition in pome fruits are given in Table 9 and 10.

Table 9: Foliar nutrition in apple

Element	Formulation	Concentration (%)	Time of spray
N	Urea	0.5	After petal fall
		0.5	Postharvest (Oct-Nov.)
Ca	$CaCl_2$	0.5	45 and 30 DBH
Zn	$ZnSO_4$	0.5	After petal fall
Mn	$MnSO_4$	0.5	After petal fall
B	H_3BO_3	0.1	At green tip to pink bud and petal fall
Cu	$CuSO_4$	0.1	After petal fall

Foliar application of urea (25% of soil dose) is equally effective to full soil application.

Table 10:Foliar nutrition in pear

Element	Formulation	Conc. (%)
N	Urea	0.5
Ca	$CaCl_2$	0.25
Fe	$FeSO_4$	0.06
B	H_3BO_3	0.05
Zn	$ZnSO_4$	0.04
Cu	$CuSO_4$	0.02

Time of spray is same to apple.

Diseases and pests management

The diseases and pests management along with nature of damage/ symtoms are given in Table11 and 12, respectively.

Table 11: Major diseases of temperate fruits

S. No.	Diseases and their causal organism	Symptoms	Management
Pome fruits (Apple and Pear)			
1.	**Scab** (*Venturia inaequalis*)	**Leaf:** symptoms first appear in the spring as small velvety, olive green in color and have unclear margins lesions on the lower surface. As they age, the infections become darker and more distinct in outline. Lesions may appear more numerous closer to the mid-vein of the leaf. When infection becomes heavy, the leaf is distorted and drops early in the summer. **Fruits:** small superficial lesions or large patches developed. The margins of the spots are more distinct and lesions becomes darker (black or scabby) with age. Badly scabbed fruit becomes deformed and may fall before reaching good size.	Spray apple trees with urea (5kg/100L) at pre-leaf fall. Spray of carbendazim (50g/100L) or thiophonate methyl (75g/100L) before initiation of leaf fall. Adopt the apple scab spray schedule.

2.	**Premature leaf fall** (*Marssonina coronaria*)	Dark green circular patches on the upper leaf surface, which become dark brown. As the disease progresses, black pinhead-like fruiting bodies (acervuli) develop and when the lesion are numerous, they coalesce to form large dark brown patches (blotches) and the surrounding area turn yellow. The infected leaves drop off and produced the characteristic symptoms of premature leaf fall.	Carbendazim (0.05%) spraying in orchards when disease symptoms appeared followed by a mancozeb (0.3%)/ propineb (0.3%)/ dodine (0.075%)/ fluquinconazole (0.02%).
3.	**White root rot** (*Dematophora necatrix*)	The **above ground symptoms** include premature yellowing of foliage, couples with bronzing and defoliation. The initial infection starts from the root hair and progresses through tertiary roots, secondary roots and then extends slowly to the primary roots. The latest roots turn dark brown and are covered with a greenish grey or white mat having flocculent web of whitish strands or ribbons during monsoon season. Appearance of pear shaped swellings near speta in the hyphae, the cortical cells are ruptured resulting in disruption of the plant system leading to the death of the tree and hence expressing disease on above ground plant parts.	The excess water should be drained. The acidic soils should be amended by applying lime for some years. Soil amendment with neem cake should be used. The fungicides should be applied through deep holes (15-20 cm) made in the basin of tree at 30 cm distance from each other. Bioagents viz., *Pseudomonas fluorescens*/ *Bacillus subtilis* or both or *Trichoderma harzianum* should be applied in basin.
4.	**Alternaria leaf spot** (*Alternaria alternata*)	Dark brown to black elongated lesions appears on leaf petioles and mid ribs in early part of the season, later coalescing to form bigger blotch causing severe yellowing and dropping of the infected leaves	Protective sprays with dodine (0.075%) or zineb (0.3%) or mancozeb (0.3%) at 15 day interval.
5.	**Powdery mildew** (*Podosphaera leucotricha*)	Symptoms first appear in the spring on the lower surface of leaves, usually at the ends of branches. Small, whitish felt-like patches of fungal growth appear and quickly cover the entire leaf. Diseased leaves become narrow, crinkled, stunted and brittle. The fungus spreads rapidly to twigs, which stop growing and become stunted. In some cases the twigs may be killed back. Diseased fruit has a fine network type surface blemish called russetting.	Pruning/ removal of affected twigs, mildewed shoots and leaves in late May. Spray of sulphur fungicides like cosan, solbar, sulforix and thiovit. Recently, the use of ergosterol biosynthesis inhibitor fungicides such as triadimenol (0.18 l/ha), and bitertanol (0.4l/ha). Apple scab spray schedule also reduce intensity of powdery mildew.

6.	**Sooty blotch and fly speck** (*Gloeo des* and *Schizo thyriu m pomi*)	Sooty blotch appears as sooty or cloudy blotches on the surface of the fruit. The blotches are olive green with an indefinite outline. Minute blotches may coalesce to cover much of the fruit. Fly speck Groups of a few to 50 or more slightly raised, black and shiny round dots that resemble fly excreta, appear on the apple fruit. They consist of definite circular black dots.	**1.** Prune trees annually to open up the tree canopy and go for thinning to separate fruit clusters open center for maximum air circulation and exposure to sunlight. **2.** Remove or destroy nearby wild or neglected apple trees. **3.** Post-harvest dips (1 minute) in stable bleaching powder (5kg/100L) or sodium chlorate (3 kg/100L) to remove signs. **4.** Apple scab spray schedule also reduces the spots quite effectively. **5.** Two Sprays of zineb, captan and carbendazim (each at 0.1%) at 40 and 20 days before harvest.
7.	**Mosaic**	Creamy white or yellow patches appear on the leaf lamina and later on, leaf turns yellow. The virus is transmitted through bud-wood by natural root grafting and budding.	1. Maintain seedling at 40^{0}C for 20 days to inactivate the virus. 2. Virus free mother plants be indexed for graft and bud-wood material.
Stone & nut fruits			
8.	**Leaf Curl** (*Taphrina deformans*)	The leaf blade becomes thickened and puckered along with midrib leaf curls and turns yellowish. Finally, the affected leaves dry and drop. Infected twigs become swollen and may exude a gum like substance and ultimately die. Flowers and fruits, when infected, drop prematurely.	Spray of carbendazim (100g/100L) or mancozeb (200g/100L) at leaf fall, bud-swell, petal-fall and fruit-let stage.
9.	**Leaf Blotch** (*Marssonina juglans*)	Small, circular and light brown spots develop on leaves finally, these spots develop into large blotchs having irregular outline and often involve the entire lamina.	1. Collect and burn fallen leaves in autumn. 2. Spray of copper oxychloride (300g/100L) in late spring at 15 days intervals.
10.	**Gummosis complex**	Production of mucilaginous gum like material from any plant part. Affected areas on bark are depressed, cracked and gum comes out in the form of globules or sheets. Cankers develop in the affected plant parts.	1. Fungicidal pests applied on infection site. 2. Foliar sprays of streptocycline (0.01%) plus copper oxychloride (0.2%) after appreance of symptom. 3. Sprays of silver nitrate at 150 and 200 ppm after petal fall.
11.	**Canker Complex**	The bark of almost entire length of the trunk initially exhibit depressions which turns dark brown to black in colour. Bark dries up and splits, cracks also develop in the underneath wood which leads to the drying of foliage and twigs, resulting in defoliation. As the disease progresses fruiting bodies of causal organism develop.	1. After pruning and training practices ,cut ends should be treated with fungicide pastes like Bordeaux paste/paint and Chaubatia paste. 2. Application of copper oxychloride paste on diseased parts scrapping the affected tissues followed by foliage spray of the same fungicide.

Table12: Major pests of temperate fruits

S.No.	Pests	Nature of damage	Management
1.	**San jose scale** (*Quadraspidiotus perniciosus* Comstock) **Damaging Stage:** Nymphs and Females	Sucking of plant juice from all the aerial parts of the plants by the pest showing low vigor and poor growth. In case of light infestation, small disc shaped greyish specks appear on the bark and fruit. Infested parts are covered with ash grey scale. Infested fruits show round red spots with white circular raised specks in the middle at the pedicel and calyx end. In extreme case of attack, plant dies within 3 years.	1. Follow winter spray (December) with diesel oil emulsion (diesel oil 4.5 l + ordinary soap 1.0 Kg + 1.0 Kg $CuSO_4$ + water 72 l) @ 8-12 l / tree. 2. Apply miscible oil @ 2% during Feb-March. 3. Use natural predators like *Aphytis procila, Chilocorus bijugus, Pharoscymnus flenibilus* and *Coccinella septempunctata* 4. Spray chlorpyriphos (0.04%), fenitrothion (0.05%) or dimethoate (0.03%) at peak crawler emergence stage in severe cases.
2.	**Apple wooly aphid** (*Eriosoma lanigerum*) **Damaging Stage:** Nymphs and Adults	Serious pest of pome and stone fruits. Whitish cottony patches appear on stem and branches. It migrates from root to shoot and *vice versa* and suck the sap throughout the year. Characteristic galls or knots develop on roots. The chief mode of dispersal is through infested nursery plants.	1. Treat nursery bed with phorate @ 0.5 g a.i. per plant and treat the nursery plants with chlorpyriphos (0.05%) before planting. 2. Use resistant root stocks like Merton stock 778, 779, 789 and 793, M 21, M 25 etc. 3. Apply imidacloprid 0.4 ml or dimethoate 1 ml/l water during spring but not during summer because of harmful effect on natural parasites. 4. Apply phorate and chorbofuran (each at 3 g a.i. / tree) around the tree in spring and fall season, respectively. 5. Use natural parasites like *Aphelinus mali* (at lower altitude) and *Chrysoperla carnea* (at high altitude) during rainy season.
3.	**Apple stem borer** (*Aeolesthes holosericea* Fabricius) **Damaging Stage:** Grub	Serious pest of apple and stone fruits and can be identified by the piles of brown pallets mix with excreta adhering to the bark or lying on the ground near the stem. It feeds on conductive vessels of stem and on the sap wood of the tree. Tissues start rotting. The hole (5-7 mm) is visible from outside the stem.	1. Insert cotton wick soaked in insecticide emulsion of about 10 ml of rogor or phosphomidan inside the hole and plugging it by moist soil or mud. 2. Place 0.2 g aluminium phosphide tablets inside the hole and then plug with mud. 3. Cover tree trunk with muslin cloth/ dry grass in March till Oct.

4.	**Codling moth** (*Cydia pomonella*) **Damaging Stage:** Larvae	Serious pest of cold arid regions. Larvae bore into fruits generally from the calyx end and adult emergence take place in June. Infested fruits drop prematurely and are unmarketable.	1. J&K Govt. has imposed strict restrictions on the movement of fresh fruits from Ladakh region under SRO 397 dated 8th Sept. 1981 issued under Plant Disease Act, 1973 to avoid its entry in other areas of the country (Dwivedi *et al.*, 2003) 2. Use light traps, sex pheromones, burlapping of tree trunk in autumn, and remove dead bark which provide shelter to larvae during winter months. 3. Release egg parasite (*Trichogramma embiothagum*). 4. Spray cypermethrin (2ml/l) twice at 15 days intervals.
5.	**Peach leaf curl aphid** (*Brachycaudus helichrysi* Kattenback) **Damaging Stage:** Nymphs and Female adults	Serious pest of peach, plum and almond which is confined to the growing shoots and leaf, and suck the cell sap during spring and early summer, causing curling of new leaves with poor fruit set.	1. Spray methyl demeton (0.025%) or dimethoate (0.03%) at pink bud stage 2. Prune the current growth during Dec. - Jan, carrying eggs.
6.	**Peach fruit fly** (*Batrocera* spp.) **Damaging Stage:** Larvae (maggots)	It is characterized by the dark punctures, oozing of liquid and rotting or dropping of the fruits which become malformed, under sized and unmarketable. The flies lay eggs inside the fruit skin and the maggots (dirty white, head less and leg less) can be seen inside the fruits.	1. Collect and destroy fallen fruits. 2. Use Male Annihilation Technique (methyl euginol and Cue Lure Traps for females) and Bait Annihilation Technique (poision baiting of 20 ml malathion + 200 g gur in 2 l water for male and female flies trapping). 3. Spray the surrounding vegetations with malathion @ 2 ml/l of water.
7.	**Anar butterfly** (*Deudorix isocrates* Fabricius) **Damaging Stage:** Caterpillars	The females lays egg singly on the calyx and small fruits and catterpillars bore inside the developing fruits which feed on pulp and seed just below the rind. Offensive odour and excreta of the larvae come out from the entry holes. Infested fruits further attacked by fungi and bacteria causing rotting of fruits. The extent of damage is 40-90%.	4. Destroy infested fruits. 5. Bagging of fruits with muslin cloth bags before fruit maturity is effective. 6. *Brachymeria euploeae* Westw. Parastizes the larvae of this pest. 7. Spray carbaryl @ 2 g/l of water or phosphomidon @ 3 ml/l of water fortnightly from the starts of fruit formation.
8.	**Walnut weevil** (*Alcidodes porrectirostris* Marshall.) **Damaging Stage:** Grubs and adults	This serious pest of walnut has two generations in a year (April-July and June-Sept.) and passes winter as adult. Pupation occurs in the fruit. First generation is more damaging as its grubs feed on developing kernel and can cause 100% premature fruit drop having grub inside. The adult feeds on petioles, green twigs, buds, flowers and young fruits.	1. Collect and destroy the fallen fruits in May and June. 2. Apply two sprays of endosulfan @ 2ml/l of water fortnightly starting from emergence of female flowers.

References

Albrecht, H. J. and Schuldt, C.(1997). Productive cultivars Beitr. Int. Wild frutchttagung. Humbdolt, Berlin: 3-6

Ananda, S. A.(2001). Production and distribution of quality planting materials of temperate fruits. In: Productivity of Temperate Fruits Issues and Strategies (Eds. K K Jindal and D R Gautam), Dr YSPUHF, Solan, HP.Pp. 43-54

Belykh, A. M.(1998). Results and prospects of using raw material sources of seabuckthorn. In: Int. Symp. on Seabuckthorn , Novosibirsk, Ulan Ude, Russia. Pp 178-179

Dwivedi,S. K.; Narboo, Sonam; Dwivedi, D. H. and Kareem, A.(2000). Codling moth-A major threat to the fruit industry of Ladakh. In: Nat. Sem. Sustainable Pest Management, organized at BHU, Varanasi during Jan. 2-4, 2003: 79-80

Frecon, J. L. (2002). Best management practices for irrigating peach. Rutgers Coopråtive Extension, N. J. AES, Rutegers, The State Univ of New Jerse, New Brunswick

Gardner, J. (2003). Water conservation in apple orchards .ag.info.omafra@ontario.ca

Ingles, C.(1994). Selecting the right cover crop gives multiple benefits. Calif. Agric. (Sept-Oct.)

Sounduri, A. S. and Sharma, A. K.(2005). Pomological characterization of some walnut selections. The Hort. J., 18:157-160

Sounduri, A. S.; Sharma, A. K. and Mir, M.A. (2005). Performance of indigenous walnut selections compared to exotic cultivars. Env. Ecol., 23:441-444

Verma, H. S.(2001). Irrigation/ fertilization technology for increasing productivity of temperate fruits. Productivity of Temperate Fruits Issues and Strategies (Eds. K K Jindal and D R Gautam), Dr YSPUHF, Solan, HP. Pp. 245-258

CHAPTER

11 Resource Management for Sustainable Crop Production In Rainfed Agro-ecosystem

Amarjit S. Bali[1], Vikas Sharma[2] and Brij Nandan[3]

[1]*Division of Agronomy, Faculty of Agriculture, Sher-e-Kashmir University of Agricultural Sciences and Technology of Jammu, Chatha, Jammu-180009.*

[2]*Junior Scientist, Regional Agricultural research Station, SKUAST-Jammu, Rajouri.*

[3]*Junior Scientist, Pulses Research sub-station, SKUAST-Jammu, Samba*

In spite of low productivity dry land agriculture in the country it supports 40 per cent human, 60 per cent cattle population and contributes 44 per cent in the total food grain production (*Singh et al., 1999*).

Crop failure is a common feature either due to inadequacy of rainfall or shortage of soil moisture to meet the water requirements during crop phenophases. Conservation, improvement and efficient utilization of natural resources are a pre-requisite for successful farming under rainfed conditions. To achieve this goal, farming should be in harmony with natural processes building up resources, promoting agro-ecosystems, resilience, minimizing the adverse impact on environment, up-gradation and improvement in farm productivity and profitability (*Mac Rae et al., 1990*).

Several workers have indicated the possibility of increasing crop yield by about 150 per cent especially in assured areas receiving a rainfall more than 750 mm (*Gautam, 1982*).

Achieving high yield levels under rainfed conditions demands the conservation of basic resources like soil and rainfall in addition to effective use of inputs.

Characteristics of rainfed agriculture

- Uncertain, ill distributed and limited annual rainfall with higher coefficient of variation.
- Occurrence of extensive climatic hazards
- Undulating soil surface.

- Prevalence of monocropping
- Very low productivity
- Poor market facility
- Poor economy of the farmers.
- Poor health status of the farmers.
- Poor health of the cattle.
- Poor soil status.

Adoption of sustainable agriculture under rainfed conditions

Rainfed agriculture a complex phenomenon, being influenced by extremes of climate, location and terrain conditions need support through adoption of several measures of crop management which can help to get best advantage from the prevailing rainfall and agronomic situations. Some of the measures can be adopted by the following approaches:

A. Crop management approach

1. Selection of crop

Crops with higher water use efficiency or producing more with less amount of water provides better out put from these areas. Crops like millets, barley, lentil, chickpea are considered more stress tolerant than crops like wheat and rice. The crops having profuse root growth and early seedling vigor are considered good to with stand moisture stress. In a crop rotation under rainfed situations, it is not only the total productivity but it is valid that preceding crop should leave enough moisture for the germination of succeeding crop as after harvest of cowpea crop, higher soil moisture was recorded than that of sorghum crop (*Moroke et al. 2005*).

2. Selection of crop varieties

Under rainfed situations a variety having less test weight, owing to more seed surface area germinates better than variety having higher test weight. The variety should have characteristics like short duration in maturity with profuse tillering with higher root to shoot ratio. Differential response to moisture stress has been observed by *Mugabe and Myakatwa(2000)*. SKUAST-J has also tested and recommended wheat varieties like PBW-175, PBW-297 and RSP-81 (Jitto) for early sown conditions and Raj -3077, PBW-226 for late sown rainfed conditions and DGS-1 for Gobhi Sarson and RSPT-I and RSPT-11 for toria of oilseed crops.

3. Weed management

Weeds if not properly controlled can take away as much as 30 per cent moisture and nutrients leading to substantial losses in yield (*Venkateshwarlu, 2001*). Weed control not only improves crop yield but it also imparts stability in crop productivity.

4. Nutrient management

Low supply of plant nutrients is also a major constraint for low productivity of crops in rainfed farming as these soils are not only thirsty but hungry as well. In rainfed system conjunctive use of organic and inorganic fertilizers is beneficial for improving soil fertility and water holding capacity. Proper crop reside management can also help to improve soil health, supplement plant nutrient for crop growth and conserve soil moisture through mulching effects.

A study under dry land conditions of Jammu also indicated an increase of 16.4 per cent in soil N, P and K status with 50 per cent replacement of N through FYM as compared to farmers practice in maize-wheat cropping system. Under Banglore conditions integrated supply of nutrients helped in improving productivity of ragi and maintaining a stable yield measured through sustainable yield index (SYI) developed by Singh *et al* (1990).

$$SYI = \frac{\bar{Y} - \sigma}{\bar{Y}_{max}}$$

Where is the average yield, ó is the standard deviation and is the maximum yield over years.

5. Plant protection management

There is a need to evolve need based plant protection modules which do not have adverse effects on eco-system degradation.

B. Conservation approach

This approach includes:

- Contouring across the slope
- Scooping of land
- Opening of ridges and furrow
- Compartmental bunding
- Bedding system
- Deep summer ploughing

C. Physiological approach

Under this approach the aim is to reduce evaporation and transpiration for which various chemicals like ASA (Alkenyl succinic acid) and DSA (Decenyl succinic acid) have been used for improvement in cell membrane permeability. Antitranspirants like PMA (Phenyl mercuric acetate), HS (Hydroxy sulphonate) and ASA have shown promising results with respect to reduction in water losses.

D. Genetic approach

The approach aims at development of ideal plant type with the characteristics like:

- Early in growth duration and early vigour
- Deep root system with maximum branching at deeper zone
- Dwarf plant type with lesser number of erect leaves
- Moderate tillering
- Good expression of ear heads
- Resistance to diseases
- Bolder grains with moderate dormancy
- Effective photosynthetic behavior with greater sink capacity

E. Risk minimization approach

Rainfed farming often considered a gamble in nature can well managed by minimizing risk through strategies like:

- Careful analysis of the soil and climatic data and generation of database
- Preparation of contingency planning
- Diversification in agriculture through integrated farming systems
- Alternate land use system by adoption of Agroforestry, Silvipastoral and Agri-horti systems

Future thrust area

- Identification of technologies for conservation of natural resource base
- Development of viable and profitable farming systems
- Climatic change studies in context to rainfed farming
- Development of appropriate season-wise contingency plans
- Identification of niche areas
- Development of suitable infrastructure for sound market base

Conclusions

Under rainfed agro-ecosystem adoption of soil and water conservation practices along with appropriate agricultural practices matching with moisture availability should be considered as essential pre-requisites for sustainable crop production.

References

Mac Ra, R.J., Hills , S.B., Mehuys, G.R. and Haming, J. 1990. Farm scale agronomic conversion from conventional to sustainable agriculture. Adv . Agron. 43: 155-198.

Moroke , T.S., Sehwartz, R.C., Brown , K.W and Juo, A.S.R. 2005 . Soil moisture depletion and root distribution of three dry land crops. Soil Sci. Soc. Am. J. 69, 197-205.

Singh , H.P. and Venkateshwarlu B. 1999. Turning grey areas green. The Hindu sources of Indian Agriculture pp 25.

Singh , R.P. , Das, S.K., Bhaskara Rao, U.M and Reddy, M.N. 1990. Towards sustainable dry land agriculture practices. CRIDA pp 106.

Venkateshwarlu, J. 2001. Technical manual on water shed management III, MANAGE, Hyderabad.

CHAPTER 12

Crop Improvement in Food Legumes Under Limited Moisture Conditions

R. K. Salgotra[1] and Anil Kumar[2]

[1] *Division of Plant Breeding and Genetics, Sher-e-Kashmir University of Agricultural Sciences and Technology of Jammu, Chatha, Jammu-180009.*

[2] *Division of Agronomy, Faculty of Agriculture, Sher-e-Kashmir University of Agricultural Sciences and Technology of Jammu, Chatha, Jammu-180009.*

Water scarcity is one of the 21st Century's major challenges. In the dryland farming systems of semi-arid regions, drought stress is common and unpredictable. According to climate-change models, annual fluctuations in rainfall are likely to increase in most of Asian countries in the coming decades. To counter this serious threat to agricultural production, agricultural research systems in India focus much of their effort on breeding for improved drought tolerance and increased water-use efficiency. To be successful in breeding drought tolerance and increased water-use efficiency varieties, there needs to be a ready supply of genetic diversity from which to select needed traits (Cutforth and McConkey. 1997).

Food legumes, being rich in protein, are important ingredients in the Indian vegetarian diet. They are often grown in marginal lands and constitute an important component in the dryland farming system. In spite of their importance, the efforts done into the research and development of food legume crops have been rather marginal whereas problems confronting pulses improvement are enormous. This is the reason that the food legumes production in the country has not increased in the same dimension as in the case of cereals and millets.

Regarding world food legume production, common bean (*Phaseolus vulgaris*) stands first followed by field pea (*Pisum sativum*), chickpea (*Cicer arietinum*), lentil (*Lens culinaris*), field bean (*Vicia faba*), cowpea (*Vigna unguiculata*), pigeon pea (*Cajanus cajan*), lupin (*Lupinus spp.*) and Vetch (*Vicia spp.*). Among the various pulses grown in India, chickpea or gram stands first contributing 29.37 and 37.98 percent of the total area and production under pulses, respectively. This is followed by pigeonpea, black gram (urdbean) and green gram (mungbean) (FAO, 2008).Water deficit is one of the major limiting factors of food legume production in India, because these crops are mostly grown as spring sown and rainfed. Even in irrigated crops such as common bean, water deficit may be responsible for a 30% reduction in yield. Heat stress is frequently associated with water deficit, and it may also induce significant reduction of biomass production and grain yield in food legumes.

The productivity of food legume crops is far below compared to that of food cereal crops. In India even though it is cultivated over 1/5th of total cultivated area, its production is only 1/12th of total food production. Among the many reasons attributed for its lower productivity, lower yield potential of varieties, cultivation in marginal lands, below average management efforts, non availability of quality seeds, prevalence of higher temperature in its growing environment, susceptible to pod borers and wilt diseases are important (Paroda, 1989). The challenge of a quantum jump in pulses production in India is formidable, as it requires addressing important developing and research issues. The demand for food legumes during 2030 AD would be around 26 million tonnes in India with an expected annual growth rate of 3.3 per cent per annum. To achieve this, the present productivity level of 0.6 tonnes/ha has to be increased to 0.99 tonnes/ha. The limitation is its popularity with the farmers as intercrops rather than as sole crops.

Yield-limiting factors and crop improvement objectives

Grain yield of food legume crops on farmers' field is very low. A number of biotic and abiotic constraints contribute to reduce the gap between the potential and actual yields. Correspondingly, the major objectives of crop improvement programmes in different regions include:

- Grain yield
- Early maturity and reduced height
- Tolerance to drought
- Resistance to diseases
- Resistance to insect pests, especially pod borers, pod suckers, and podfly
- Suitability for intercropping
- Enhanced nitrogen fixing potential and survival in infertile soils
- Seed characteristics, especially size and colour depending upon requirements.

Grain yield potential

Until the early 1980s no improved varieties were available to farmers mostly in developing and under developing countries. Although several improved varieties are now available, adoption is limited and most farmers grow low-yielding, late-maturing landraces. Significant yield increase have been reported in food legumes crops like pigeonpea (upto 4.6 tonnes/ha) on farm trials of new varieties, indicating that productivity can be substantially improved with new varieties and better crop management.

Early maturity

Late maturing varieties leave farmers with little time to prepare the field for the next crop and conservation of residual moisture content under dryland farming system (Ney and Turc, 1993). To avoid such delays farmers often plant widely spaced rows and sow other fast-maturing crops in between (wide row spacing also facilitates land preparation and weeding). The long duration may be a disadvantage for subsistence farmers who have to wait nearly a year to harvest. Varieties with different maturity durations should be developed to fit different cropping systems and agro-ecological zones, but early-maturing cultivars should be given priority (Johnston *et al.*, 1999).

Plant height

Most local varieties of food legumes are tall, and are thus difficult to harvest and spray. They cannot be grown in close association with shorter plants due to shading effects unless wide spacing is used or by exploiting their slow early growth. They also tend to lodge-therefore short or medium-statured varieties may be more desirable.

Diseases and insect pests

Several diseases and insect pests have been reported in food legume depending upon different locations and regions. Consequently, yield losses will depend on infestation levels and the natural tolerance of the plant. Efforts to develop disease and insect-pest resistant varieties have met with little success; and there is still limited understanding of insect-host relationships and control methods. Chemical control is often not viable for subsistence farmers; the best option is probably the development of resistant varieties. Moreover, there is an urgent need to develop food legume varieties tolerant of both field and storage pests as a component of integrated pest management strategies.

Drought stress

Food legumes are mainly grown in semi-arid areas with unreliable rainfall, where crop failures are frequent. Although food legume crops are drought tolerant, they grow best with rainfall of 600-1000 mm. Yields are substantially reduced under drought. This can be overcome by developing either early-maturing varieties or varieties with tolerance to drought.

Water use efficiency (WUE)

Lower water use efficiency (WUE) is the main limiting factor to food legume crop production in the Great Plains (Farahani *et al.*, 1998). The mean WUE values for all other legume crops appeared lower than that of spring wheat though the maximum values observed for chickpea and lentil were similar to that for spring wheat. The observation of chickpea and lentil WUE values approaching that of spring wheat indicates that there is much to learn about attaining optimal yield response in the food legume crops because actual yields are often much lower than spring wheat yields. The two warm-season pulse crops, dry bean and soybean, had the lowest mean (and range of) WUE values. This reflects a typically delayed seeding date and longer plant maturity, extending crop growth into the mid- and late-summer season when peak evapotranspiration demand occurs. As such, dry bean and soybean generally are not acceptable to be included into dryland cropping systems in semi-arid regions of the northern Great Plains, especially in western regions of India where rainfall is normally concentrated in the late spring and early summer (Peterson *et al.*, 1996).

Adaptability

Adaptation is the process by which plants/varieties/organisms become more suited to survive and function in a given environment. The need for adaptation to environments is modified by a need to yield well across a range of seasons and changing micro-environments that can lead to large genotype-environment interactions. These interactions may be linked to specific physiological or other traits of the plant which are under genetic control and may be understood. Consequently, different breeding schemes (e.g., farmer participation or research station directed) may be needed in different situations. Similarly under different agro-ecological situations different types of plants may need to be selected (e.g., well watered vs. rainfed). A range of possible factors that affect the ideal adaptation of food legume crops as a means to better understand the process of adaptation that has taken place and continues to take place around the world. It is, therefore, important to test new varieties over as wide an area as possible to determine the areas of optimum productivity. It would also be useful to define and characterize legume-growing environments in a region so that breeders can target cultivars to specific environments (Padbury *et al.*, 2002).

Crop management

Late planting, inadequate weeding, poor land preparation, and low plant populations contribute to the low yields (Johnson, 1987 and Auld *et al.*, 1988). Future efforts must concentrate on disseminating the improved technologies already available.

Varietal improvement through conservation of genetic diversity

Over the millennia, numbers of varieties of food legume crops have been developed through selection by farmers through various adaptive mechanisms for drought stress tolerance. Crop wild relatives represent an even richer source of genes for stress tolerance and adaptation, because they have been around much longer, and have survived periods of very harsh climate (Paroda *et al.*, 1988).

Conservation of genetic diversity

Landraces in farmers' fields and populations of crop wild relatives in low-rainfall and drought-affected sites are natural reservoirs of genes for drought tolerance, but they are now increasingly threatened by genetic erosion. The principal cause is changes in traditional agro-ecosystems, such as replacement of genetically diverse landraces with uniform modern varieties, cultivation of rangeland, and overgrazing leading to the loss or degradation of habitat for crop wild relative populations. To save the rich genetic diversity and make it available to researchers and breeders, plant collectors carry out collection missions, in which landrace and wild relative populations are sampled and conserved in genebanks.

Documentation of existing abiotic stress germplasm

The high genetic diversity for the components of abiotic stress response, existing in large germplasm collections, must be properly evaluated and documented in order to be useful in meeting the challenges posed by abiotic stresses, and drought in particular. In stressed environments, individual stress effects usually cannot be separated from the multiple-stress response. So researchers attempt to characterize germplasm in controlled conditions using morpho-physiological criteria, which can reveal new insights into different components of abiotic stress response. Moreover, advances in molecular biology are providing new means of understanding the physiological and genetic mechanisms involved in plant response to different stresses. This field is also providing tools for molecular screening and genetic engineering applications in crop improvement for increased abiotic stress tolerance. A thorough evaluation and exploration of genetic diversity existing in the crop wild relative and landrace collections might be a strategic first step in germplasm development to counter the negative impact of climate change on rainfed farming systems in the extensive semi-arid and arid drylands of developing countries.

Some of the most significant achievements for development of food legume varieties of widely grown species for various abiotic and biotic stresses using different varietal improvement programmes are discussed below.

About 200 improved varieties of different pulses have been released in the country during the last 40 years. These varieties are high yielding and also possess several desirable characters. Some of the varieties though evolved earlier are still popular among farmers (Table 1). During recent years, emphasis is being given to develop varieties having resistance/tolerance to major diseases. Quite a large number of varieties have, therefore, been developed with satisfactory level of resistance to major diseases. Besides this, a large number of varieties have also been recently released on account of their specific characteristics, such as ideal plant type, early maturity and grain character.

In pigeonpea, the maturity period has been reduced from 300 days to less than 150 days and the varieties having short maturity period are now available to the farmers. These are UPAS 120, Pusa 74, Pusa 84, Sagar, Manak and ICPL 151. With the availability of these varieties, pigeonpea-wheat rotation has already become popular in the northern irrigated belt. Resistance to sterility-mosaic disease has been incorporated and varieties Bahar, Hy 3C and Maruthi have been released for cultivation.

In case of chickpea, resistance to wilt (Avrodhi, JG 315) and *Ascochyta* blight (Gaurav, GNG 146, GNG 588) has been incorporated. Development of short-duration varieties has made mungbean a catch crop during summer season in the irrigated belt of northern India. Varieties PDM 11, PDM 54, Pusa 105, Pant Mung 2 and Pant Mung 3 are not only resistant to yellow-mosaic virus but are also suitable for spring/summer planting. Similarly, for *kharif* season, varieties, such as ML 131, ML 267 and ML 337 have also been developed. Old varieties K 851, T 44 and Pusa Baisakhi are still quite popular with the farmers for the summer season.

Table 1. List of promising varieties of pulse crops released in India

Crop	Still popular old varieties	Recently developed promising varieties
Chickpea	T3, Radhey, H 208, Kabuli L 550, C 235, Annigeri, Chaffa and Dohad Yellow	K 468, K 850, Pant G 114, BG 256, Phule G 5, Gaurav JG 315, ICCC 32 (Kabuli), BG 261, RSG 2, BDN 9-3 Jyoti, Mahamaya 2 and B 108
Pigeonpea	T 7, T 17, Gwalior 3, No. 148, C 11 and T 15-15	UPAS 120, Bahar, Manak, AL 15, BDN 1, HY 3C, ICPI 87, LRG 30, PT 221, Pusa 85, Vishakha 1, Sweta and Chuni
Uridbean	T 9, No. 55, B 76, CO 2, TMV 1, Gwalior 2, Khargone 3 and Mash 48	Pant U 19, Pant U 30, LEG 17, (*Rabi season*); ADT 2, ADT3, CO3, CO, CO5, KM1, KM 2, Krishna and PDU 1, (Spring season)
Mungbean	T 44, Pusa Baisakhi, PS 7, PS 16, Jawahar 45, Kopergaon, Khargone 1, BR 2, B 1, MI 5, S-8 and Shining mung No. 1	Pant Mung 2, Pant Mung 3, ML 267, K 851, ML 131, Pusa 105, Sunaina, CO3, CO4, BO 105, Gujarat 2, Sabarmati, Hy 12-4, PDM 54, (*Kharif* and Summer) and PDM 11 (Summer/Spring)
Fieldpea	T 163, BR 2 and Swarnrekha	Rachna, Pant P 5, HUP 2, JP 885, DMR 11 and HFP 4,
Lentil	L 9-12, T 36, BR 25, B-77, and JLS1,	Pant L 639, Lens 4076, PL 77-2, Malika, Pant L 406, B 256, and Ranjan
Mothbean	T1, Baleswar 12, and Jadia	Jwala
Cowpea	C 152, CO1, T 5269, K 11, RS 9, S 488, NO. 74 and PTB 1	V 240, RC 19, Gomti, JC 10, CO3, and CO4
Horsegram	CO 1, PDM 1 and VZM 1	HPM2, HPK 4, Hebbal Hurali-1 Hebbal Hurali-2 and Madhu
Rajmash	VL63 and Jwala	PDR 14
Lathyrus	B-1	Pusa 24

In urdbean, yellow-mosaic virus-resistant varieties (Pant U 19 and Pant U 30) have also been developed. For spring planting, PDU1, and for winter planting in central and southern parts of the country, LBG 17 have been developed. The latter is resistant to powdery-mildew and therefore, it has become quite popular in rice fallows. In peas, powdery-mildew is a serious problem. An old variety T 163 has become highly susceptible to this disease. Newly evolved varieties Rachna, Pant P 5 and HUP 2 are highly resistant to this disease.

In lentil, varieties Pant L 209 and Pant L 406 for resistance to rust, and Malika and Lens 4076 for bold seededness, have been developed. Among the minor pulses Jwala for mothbean, HPK 4 and Madhu for horsegram, V 240 for cowpea and PDR 14 and HUR 15 for *rajmash* are recently released varieties. PDR 14 (Udai) *rajmash* has been released for northern India, especially the eastern part for October-November planting. Its yield potential ranges between 20 and 30 q/ha.

In addition to varietal improvement work as enumerated above, disease screening techniques for wilt, sterility-mosaic and *Phytophthora* blight in chickpea; yellow mosaic virus, powdery mildew and *Cercospora* leaf spot in urdbean and mungbean; powdery mildew in pea; and rust in lentil have been evolved. These techniques are being used for screening the germplasm at various centres in the country. The donors for resistance to these diseases have been identified and are being used in crossing programmes. Similarly, the agronomical research has led to the successful introduction of many non-traditional pulses in different regions of the country, such as *rabi* pigeonpea in Bihar, *rabi* urdbean and mungbean in rice fallows in the coastal region of Andhra Pradesh, Orissa, Karnataka and Tamil Nadu, and *rabi rajmash* in plains of the northern India. Under *utera* cultivation, lentil has become an alternative to *Lathyrus* in Madhya Pradesh, Bihar and Orissa.

Specific approaches to enhance production under limited water conditions

Significant progress has been made during the last 5 years in the understanding and modelling of plant response to soil water deficit, especially in peas. Despite significant research effort to identify traits and screening techniques for tolerance to drought in food legumes, the outputs of this research in terms of cultivar release are still very weak. Similarly the progress on stress physiology is still poorly mobilized in research programmes on cropping systems designed to improve crop adaptation to drought. On the other hand significant progress has been made by breeders and agronomists on adaptation of legume crops to dry regions by working on plant reaction to stresses other than water deficit and heat. The clearest example in food legumes is probably the development of winter sowing of chickpea for which drought escape and avoidance was improved by selecting on frost tolerance and disease resistance. The same type of progress could be obtained by selecting on adaptation to specific soil conditions of dry regions. Plant phenology and architecture also plays a major role on adaptation to water deficit and heat stress but their impact on yield is depending on the pattern of stress development (intensity, duration, position in the plant cycle).

i) Multi-criteria approach to enhance yield under dryland system

The development of genotypes and cropping systems with stable yield and quality in variable conditions of rainfall and temperature (over years and soil type) remains one of the major goals for the food legume research community. For the driest regions an optimal balance with straw yield must be maintained, because this by-product is sold for animal feeding and is an important component of the farmers' income. Nitrogen fixation must not be neglected, but depending on the local conditions, it will have to be considered as a limiting factor for yield or a by-product to improve the nitrogen budget of the crop especially so in stressed situation where plants cannot afford to spend their energy lavishly on workload other than assimilating for economic products.

ii) Quantification of drought stress

The symptoms of drought can be the result of several abiotic stresses (soil water deficit, air water deficit, heat, mineral deficiency or toxicity, salinity, disease) which are frequently combined under field conditions. There is a need for more surveys in farmers' fields in order to determine the probability of occurrence of all these stresses and their contribution to yield reduction. Soil water deficit itself cannot be considered as a single stress because its effect on yield, quality and nitrogen fixation are dependent on its intensity and position in the plant cycle. The work conducted on pea has shown that rapid progress can be made in analysis and modelling of plant response to drought if soil water deficit and heat stress experienced by the plant are quantified and placed ahead of the development of each organ of the plant.

iii) Optimisation of carbon and water balance

In rainfed conditions agronomists and plant breeders have to face a difficult challenge because most of the characters allowing high yield potential require maximum light interception and stomatal conductance, thereby increasing the amount of water used and the risk of water deficit. In Mediterranean countries the high variability for carbon and water balance over the years would imply that repeating the experiments over a decade in order to complete the necessary risk analysis for each new cropping technique or cultivar. For that carbon and water balance models of food legume crops are also required.

Future needs and opportunities

Food legume crops appear to have enormous potential for increasing the sustainability of dryland cropping systems (Drinkwater *et al.*, 1998). For some legume crops, acceptable drought and heat tolerance, efficient water use, and moisture-conserving growth habits are obvious adaptive advantages for dryland system. However, food legume crops range widely in growth habits, maturity requirements, and yield response to climatic conditions. Comparative adaptation among food legume crops within the highly variable geographic and interannual climatic gradients remains poorly understood. The advent of no-till systems adds further confusion about climatic relationships with productivity by conserving greater amounts of precipitation for crop growth and by changing thermal patterns within the crop canopy.

The development of cultivars and cropping systems on the basis of food legume adaptation to abiotic stresses remains as an important challenge. Significant research progress has been made in recent years but an integrated and oriented programme is now required based on the following elements: (i) Field surveys in the main regions of production, in order to identify the various abiotic and biotic stresses experienced by food legume crops, and their contribution to yield losses. (ii) An ambitious co-ordinated research programme to analyse the plant responses to these stresses and to identify genetic and crop management techniques to improve food legume adaptation. This programme should use the most recent progress in molecular biology, genetics and crop physiology, in order to insure rapid transfer of knowledge between the various food legume species. (iii) Several engineering programs for specific cultivar and cropping systems development, based on field surveys and research output combinations, and conducted in collaboration between extension services, agro-industries and the research institutes. International networks of trials and crop models would be key methodological elements to link the various partners of this program and to ensure a more rapid transfer of knowledge between species and between the regions.

References

Auld, D. L., Bettis, B. L., Crock, J. E. and Kephart, K. D. 1988. Planting date and temperature effects on germination, emergence, and seed yield of chickpea. Agron. J. 80: 909–914.

Cutforth, H.W. and McConkey, B. G. 1997. Stubble height effects on microclimate, yield and water use efficiency of spring wheat grown in a semi-arid climate on the Canadian Prairies. Can. J. Plant Sci. 77: 359–366.

Drinkwater, L.E., Wagoner, P. and Sarrantonio, M. 1998. Legume-based cropping systems have reduced carbon and nitrogen losses. Nature 396: 262–265.

FAO, 2008. Food and Agriculture Organisation, Bulletin of Statistics, 1: 20.

Farahani, H.J., Peterson, G. A. and Westfall, D.G.1998. Dryland cropping intensification: A fundamental solution to efficient use of precipitation. Adv. Agron. 64: 197–223.

Johnson, R.R. 1987. Crop management. In J.R. Wilcox (ed.) Soybeans: Improvement, production and uses. 2nd ed. Agron. Monogr. 16. ASA, CSSA, and SSSA, Madison, WI. pp. 355–390.

Johnston, A., Miller, P. and McConkey. B. 1999. Conservation tillage and pulse crop production—western Canada experiences. In Momentum in northwest direct seed farming. Proc. Northwest Direct Seed Cropping Syst. Conf., Spokane, WA. 5–7 Jan. 1999. Univ. of Idaho, Moscow pp. 176–182.

Ney, B., and Turc, O. 1993. Heat-unit-based description of the reproductive development of pea. Crop Sci. 33: 510–514.

Paroda, R.S. 1989. Research advances in crop sciences, In 40 years of agricultural research and education in India. ICAR Publication, New Delhi, pp. 35-78.

Paroda, R. S., Arora, R. K. and Chandel, K. P. S. (Eds.). 1988. Plant genetic resources: Indian perspective. NBPGR Publication, New Delhi. 545 pp.

Padbury, G., Waltman, S. Caprio, J. Coen, G. McGinn, S. Mortensen, D. Nielsen, G. and Sinclair, R. 2002. Agroecosystems and land resources of the northern Great Plains. Agron. J. 94:76-85.

Peterson, G. A., Schlegel, A. J. Tanaka, D. L. and Jones, O. R. 1996. Precipitation use efficiency as affected by cropping and tillage systems. J. Prod. Agric. 9:180-186.

CHAPTER

13 Problems and Prospects of Vegetable Production in Deficit Water Conditions

Sandeep Chopra, Satesh Kumar and Deep Ji Bhat

Division of Vegetable Science & Floriculture, Faculty of Agriculture, Sher-e-Kashmir University of Agricultural Sciences and Technology of Jammu, Chatha, Jammu-180009.

India has about 108 million hectares of rainfed area which constitutes nearly 75% of the total arable land. In such areas crop, production practically becomes difficult as it mainly depends upon intensity and frequency of rainfall. The crop production, therefore, in such areas is called rainfed farming as there is no facility of giving any irrigation, and even protective or life saving irrigation is not possible. These areas get an annual rainfall between 400 mm to 1000 mm which is unevenly distributed, highly uncertain and erratic. In certain areas the total annual rainfall does not exceed 500mm. The crop production, depending upon this rain, is technically called dry land farming and such areas are called as dry lands. So dry farming or dryland farming may be defined as "A practice of growing profitable crops without irrigation in areas which receives an annual rainfall of 500 mm or even less.

Indian perspective

India has about 47 million hectares of dry lands out of 108 million hectares of total rainfed area. Dry lands contribute 42% of the total food grain production of the country. These areas produce 75% of pulses and more than 90% of sorghum, millet, groundnut and pulses from arid and semi-arid regions. Thus, dry lands and rainfed farming constitutes an important part in Indian Agriculture production. By 2010 A.D., India will have to produce 300 million tonnes of food grains to feed her 1.5 billion population (approx.). This target cannot be realized from irrigated areas alone as we have irrigation potential for 178 million hectares only. Therefore, we will have to evolve an appropriate technology for dry land farming. On the other hand, we can say that second 'green revolution' in Indian agriculture can find its place in rainfed/ dryland agriculture.

Characteristics of Dryland Agriculture

Dry land areas may be characterized by the following features:

- Uncertain, ill-.distributed and limited annual rainfall;
- Occurrence of extensive climatic hazards like drought, flood etc;
- Undulating soil surface;
- Prevalence of mono-cropping system;
- Similarity in types of crops raised by almost all the farmers of a particular region;
- Very low crop yield;
- Poor market facility for the produce;
- •Poor economy of the farmers; and
- Poor health of cattle as well as farmers.

Problems and prospects of dryland farming in reference to vegetable cultivation

The major problem which the dryland farmers have to face very often is to keep the crop plants alive atleast and to get some economic returns from the crop production. But this single problem is influenced by several factors which are briefly described below:

1. Moisture stress and uncertain rainfall

According to definition the dry farming areas receive an annual rainfall of 500 mm or even less. The rains are very erratic, uncertain and unevenly distributed. Therefore, the agriculture in these areas has become a sort of gamble with the nature and very often the crops have to face climatic hazards. The farmers also take up farming half heartedly as they are not sure of being able to harvest the crops. Thus, water scarcity becomes a serious bottleneck in dry land agriculture. Under such conditions efficient irrigation scheduling has got immense importance e.g., in brinjal the critical water requirements varied from 592 mm to 772 mm based on the optimum irrigation adopted. Irrigation at IW/CPE 0.80 is the optimum irrigation regime as it resulted in highest optimum yield (Lourduraj *et. al.*,1997).

2. Effective storage of rain water

According to characteristics of dry farming, either there will be no rain at all or there will be torrential rain with very high intensity. Thus, in the former case, the crops will have to suffer a severe drought and in the latter case, they suffer either flood or water logging and they will be spoilt. In case of very heavy downpour, the excess water gets lost as runoff which goes to the ponds and ditches etc. This water could be stored for providing life saving or protective irrigation to the crops grown in dry land areas. The loss of water takes place in several ways namely runoff, evaporation, uptake through weeds etc. The water could be stored for short period or long period and it can be preserved either in soil, pond or ditches based on situation and utilized for irrigation during dry periods.

3. Heat and Wind

The major effects of heat and wind are to increase the rate of evaporation of water, and thus to increase the effects of aridity. Wind may also cause mechanical damage to crops. The effects of winds can be reduced by windbreaks like tall species of tamarisk, casuarina, and eucalyptus. However, the allelopathic effect of the border row trees should also be kept under consideration.

4. Soils

Soils of the arid regions are highly variable ranging from poor to medium in fertility. Thus vegetables like cassava, pumpkin and gourds should preferably be opted under such conditions.

5. Disposal of dry farming products

In dry farming all the farmers grow similar crops which are drought resistant. These crops mature at the same time and the growers like to dispose off their product soon after the harvest. This results in a glut of products in the market and the situation is badly exploited by the vegetable traders and middlemen. Therefore, marketing of perishable products such as leafy vegetables, tomatoes, okra etc. becomes a serious problem in dry farming areas.

6. Careful and judicious manurial scheduling

In case of irrigated farming, the farmers are at a liberty to apply [manures and fertilizers according to their availability and facility but in case of dry farming they have to be very careful in fertilizer application. Due to lack of available moisture, broadcasting or top dressing becomes wasteful and meaningless. These can be applied only by deep placement and foliar spray for an improved crop production. Studies conducted in chilli by Uma Maheshwari *et. al.*, (2004) revealed that amino acid application @ 0.75% with complete doses of NPK (175:120:130 kg/ha) resulted in highest nitrogen and phosphorus uptake (59 and 31 kg/ha) and micronutrient uptake i.e., Fe, Zn, Mn, and Cu (0.23,0.04,0.22 and 0.69 kg/ha) respectively.

7. Utilization of preserved moisture

Judicious and purposeful utilization of preserved moisture depends upon soil type, plant type and other factors. The amount of available water to the plants depends upon the depth of plant roots, their proliferation and density. In case of limited moisture condition, the yield directly depends upon the rooting depth (Table - I).

Table 1. Feeder root depth of various vegetable crops.

Shallow rooting depth (< 0.67m)	Medium rooting depth (0.68 m to 1.0 m)	Deep rooting depth (1.0 m to 2.0m or more)
Beans	Celery	Tomato
Artichoke	Cucumber	Lettuce
Brussels'sprouts	Pumpkin	Pepper
Beet	Egg plant	Watermelon
Asparagus	Winter squash	Summer squash
European Carrot	Garlic	Potato
Chard	Mustard	Radish
Muskmelon	Sweet potato	Spinach
	Leek	Sweet corn

Fulton and Tan,1986

Certain varieties pertaining to above mentioned crops shows drought resistance due to extensive root system e.g., Red Rock tomato variety. (Stoner,1978). Therefore mixed planting of crops having variable root depth can not only increase productivity of the land per unit area and time but also prevents the tough competition among roots for moisture if same kind of rooting among the planted crops prevails. Therefore, utilization of preserved moisture is an art in dry farming. The water collected in ponds or ditches may be used to give protective or life saving irrigation. The widely spaced crops like okra and cucurbits can be intercropped with leguminous vegetables. Therefore, the water collected during the rainy season need special technique and skill for its efficient utilization.

8. Quality of the produce

The quality of the produce from dry farming areas is often found to be inferior as the economic parts are not fully developed. So providing the life saving irrigation at the critical stages of the development of the economic part must be kept in mind as the lack of moisture will reduce the market value of produce and the farmers do not get the profit of their labour and investment (Table 2).

Table 2. Critical growth stages of vegetables

Crop	Critical Stages
Asparagus	Fern growth
Beans	Flowering and pod formation
Carrot	Root enlargement
Cucumber	Flowering and fruit enlargement
Brinjal	Flowering and fruit enlargement
Lettuce	Head development
Muskmelon	Flowering and fruit development

Chilli	Fruit setting and fruit development
Pumpkin	Flowering and fruit development
Radish	Root enlargement
Summer squash	Flowering and fruit development
Tomato	Flowering and fruit setting
Watermelon	Blossoming to harvesting

Fulton and Tan,1986

9. Disease and Pest Problems

Arid regions are infested heavily with termites, nematodes, cutworms and diseases like powdery mildew and wilting etc. Hence the selection of resistant varieties/ sources is the only solution to tackle the biotic stresses under dryland vegetable farming (Table 3).

Table 3. List of vegetable varieties resistant to biotic stresses

Crop	Variety	Resistance to
Tomato	SL-120,Hisar Lalit	Nematode
	HS-24 (Hisar Anmol)	TLCV
Brinjal	PPC, Kt-4, Hisar Shayamal, Pant Samrat, Pant Brinjal Hybrid-1	Bacterial wilt
	Pant Samrat, Punjab Neelum, Punjab Barsati	Little leaf, Fruit and Shoot Borer
Chilli	Pusa Jwala	Thrips, mites and aphids
	Pusa Sadabahar, Punjab Lal	CMV, TMV and Leaf curl
Okra	Co-1, Punjab Padmini, P-7, Varsha Uphar, Hisar Unnat, A.Anamika	YVMV, Jassids and cotton boll worm
Onion	PBR-2, PBR-6, A.Niketan, P.Ratnar	Thrips
	A.Kalyan, P.Ratnar, Punjab Selection-3	Purple Blotch
Round Gourd	A.Tinda	Fruit fly
Pumpkin	A.Suryamukhi	Fruit fly
Bitter gourd	Hisar-2	Fruit fly
Watermelon	A.Manik and A.Jyoti	Powdery mildew

Ram,1998

Recommendations for effective vegetable cultivation under dry land conditions

Some of the recommendations from the outcome of studies undertaken in the dryland farming areas in relation to vegetable cultivation are as follows:

1. Tillage practices

Bunding across the slope and leveling the land should be done before onset of monsoon. Deep summer ploughing should be followed by surface tillage during monsoon months and also during rest of the year. The cultural management like zero-tillage is one of the most important factors for higher crop production in rainfed agriculture. On different vegetable crops in hill slopes (10-20%) it was observed that zero-tillage gave significantly higher yield of ginger but cabbage and tomato performed well in pit methods of planting.

Table-4: Effect of tillage on the yield of different vegetable crops

Crop	Yield (q/ha)			
	Zero-tillage (dibbling)	**Furrow (minimum)**	**Pit**	**Conventional (Spading)**
Cabbage	76.33	-	97.62	80.3
Ginger	8.47	7.38	-	-
Cowpea	1.39	1.56	-	-
Tomato	17.73	25.57	28.11	-

Miah Md. and Saheed,2000

Different vegetables such as sweet corn, squash, cucumber, beans and cabbage etc. perform well under no preparatory tillage. Table 5. shows that the superiority of no till performance of cabbage was due to the increased soil moisture in no till plantings during the drought period. Moreover, no till has been successful on the soils that tend to crust and where weed growth is suppressed.

Table.5. Influence of tillage on total yield of cabbage

Tillage system	**Yield (t/ha)**	
	Spring planting	**Autumn planting**
Conventional tillage	21.9	18.0
No-till	23.6	33.5

Source: Vegetable Growers News 1984: 38(6)

2. Fertilization

Application of organic manures like FYM compost. etc. @ 15-20 tonnes/ha or green manuring should be done. These manures should be applied about 20-25 days before sowing and should be well mixed in the soil. Studies also show that in Chilli, the compost @ 80kg/ha combined with 133 kg N/ha increased the soil carbon and nitrogen content as well as N mineralization under drip irrigation conditions. Fertilizers should be basal placed at a depth of 7.5 to 10 cm in the soil and the seeds should be sown in the same furrows about 3 cm. above the fertilizers. This is important especially during winter season. The nitrogen (20-50% of total) should be top dressed by side or band placement method at about 10- 15 cm apart. The crop rows should be done soon after the rains but if there is not sufficient moisture in the soil, the nitrogen should sprayed over the foliage with urea solution containing 3-5% nitrogen. Zinc and sulphur should be applied as basal dose if needed.

3. Selection of crops/varieties

Selection of suitable crops and their varieties should be done according to their suitability to a particular region/ micro climate. e.g., There are plants with a short life cycle that can germinate, grow and produce the economic part during a very short period of available moisture like Silvery fir tree tomato, small carrot leafed plant that takes 75 days from the seed or 55 days after transplanting, plants with deep or extensive root systems which have the ability to gather water over a wide area e.g., okra, asparagus, sweet potato, rhubarb, pumpkin and corn etc. There are plants which store up water in their tissues and release it very slowly like cassava, colocasia, yams etc. that are protected from water loss by wax or other impediments like onion and garlic.There are plants with very small or narrow leaves, thus reducing water loss like chilli, garlic, onion, etc. There are plants in which the tissues themselves can withstand much desiccation without dying like Newzealand spinach, Sugar baby watermelon, Swiss chard, Purslane, Pearson tomato and Amaranthus. Plant vegetables that produce a copious amount of edibles like tomatoes, squash, peppers, eggplant etc. Vegetables like broccoli and cauliflower which consume more water and space should be avoided for drylands.

4. Seed treatment

Seeds must be treated with a suitable fungicides and that of leguminous vegetables like fenugreek and beans with rhizobium culture before sowing. Soaking seeds in plain water for rabi sowing helps in getting higher germination, better seedling vigour and an early maturity.

5. Crop rotation

Proper crop rotation should be followed which should preferably have at least one legume every year.

6. Moisture conservation practices

It is a common perception that vegetables cannot be grown under rainfed situation in arid and semi arid regions which are otherwise dominated by cereals, pulses and other field crops. However several vegetables such as Clusterbean, Okra , Onion and Chilli have great potential in diversification in drylands. Moisture being the most limiting factor for growth and development in these lands need to be conserved. In this regard research information is rarely available in vegetable crops. The beneficial effects of moisture conservation on different crops have been reported. Radder *et al.*(1991) has observed higher soil moisture due to compartment bunding while Selvaraju *et al* (1999) reported similar views due to adoption of tied ridges and furrows and compartment bunding as compared to flat bed method. So conservation of *in-situ* still holds more relevance in vegetables. Alloli *et al (2008)* conducted on farm adaptive research trials on chilli with different moisture conservation practices and reported that ridge and furrows along with mulch enhanced the vigour of the crop as manifested in higher plant height, leaf area and dry matter production. Further moisture conservation practices (ridges and furrow + mulch) helped to promote productivity of chilli as evident in significantly higher yield per unit area. The pooled data indicated the higher yield (38.40q/ha) due to ridges and furrows + mulch followed by ridges and furrows as compared to flat bed method (30.80 q/ha).

7. Intercropping

Intercropping should be a routine practice under the dry land conditions for the purpose of making best use of soil and inter row moisture harvesting. Singh (1993) studied the production potential and economics of vegetables intercropped with rainfed okra. Intercropping of cowpea reduced the yield of okra but the okra-equivalent yield increased to a maximum of 34.98t/ha, raising the productivity as compared with other intercropping combination as well as sole crop of okra (6.31t/ha)

8. Method of sowing

Line sowing at a depth of 7.5 to 10 cm or even more, depending upon the situation, should be practiced because it helps in better seed germination. This also helps in stabilizing the required plant population and thereby in getting better yield.

9. Interculture

Proper weed management practices should be followed by adopting integrated weed control measures. Mulching should be done by providing frequent interculture and pulverizing the soil. If intercultural operations are not possible then use of artificial mulches like covering the surface with tree leaves, uprooted weeds, sugarcane leaves, saw dust or polythene sheets are used to check the evaporation of water from the soil. In tomato, Chakarvarti and Sadhu (1994) observed that polyethene mulches irrespective of colour were superior to rice straw or water hyacinth. Higher soil temperature (2-3°C above the control), greater soil moisture conservation (31.5 to 67.7%) and lower salinity levels (36.7-59.8%) were observed with polyethene mulches but the mulch colour had no appreciable effect. Considerable yield advantages in various vegetables due to surface mulch under rainfed farming system have been obtained by many investigations (Table 6).

Table 6: Response of different vegetables to mulching

Crop	Yield (t/ha)	
	Mulch	No mulch
Cabbage	68.82	56.01
Tomato	34.29	12.17
Ginger	13.58	1.31
Cowpea	1.56	13.70

Hopen and Oebker, 1975

10. Water management

Water harvesting between the rows should be done by growing some pulse crops and runoff water should be collected in some nearby located ponds and used as life saving irrigation.

11. Plant protection

An efficient plant protection measure should be adopted to protect the crop from insect pests and disease damage. Under dry conditions, integrated approach of pest and disease management should be the mandate and practiced.

12. Harvesting stage

The crop should be harvested at proper physiological maturity so that the following or succeeding crop may be sown slightly earlier than the scheduled time and best use of rain water or residual moisture may be made for crop production.

13. Use of chemicals

Crops like Chillies, Okra and Cucurbits etc. should be sprayed with CCC or cycocel for modified growth, higher drought resistance and better yield. Use of chemicals which reduces rate of transpiration on its application to the plant surface (anti-transpirant) e.g., Phenyl mercuric Acetate (PMA), Hydroxy Sulphonates (HS), Alkenyl Succinic Acid (ASA), Adol-52 (a formulation of alcohol), and S-600 (a plastic transplanting spray). Likewise, use of calcium chloride solution (0.25%) for 20 hours as seed treatment impart drought resistance and results in better germination. Boron solution is also used for soaking seeds. Agrosan is a fungicide but also induces drought resistance in vegetables when seeds are treated with this chemical.

14. Start cultivation before the onset of summer

Instead of doing a heavy summer planting, practise your planting in spring with short season vegetables like lettuce, peas, radishes, spinach, beets, onions, garlic and broccoli all which thrive in the cooler spring weather.

15. Water at Night

Most vegetables grow at night and therefore irrigation during evening hours is most beneficial otherwise most of it will evaporate during morning or mid day application and will not benefit the plant at all.

16. Close planting

There are many advantages for planting vegetables close together like formation of a canopy layer over the soil, which shades it and prevents evaporation.

17. Use drip irrigation facilities

Using drip irrigation method, vegetables with very less quantity of water can be grown besides liquid fertilizers can effectively be added into the irrigation water. In tomato, Dhake *et.al.,*2009 reported that the yield contributing parameters were significantly higher under liquid fertilizer through drip irrigation than solid fertilizer applied through drip or surface irrigation. In vegetables, 35 -50 percent water saving along with yield increase by 15-40% have been reported (Table 8.)

Table 7 : Effect of drip irrigation on vegetables

Crop	Water saving (%)	Yield increment (%)
Tomato	45.26	15.81
Brinjal	44.37	18.46
Bottlegourd	45.27	20.00
Onion	35.26	39.32

Source: ISVS Souvenir: Silver Jubilee National Symposium, Dec 12-14,1998. pp98-99

Conclusion

Even after utilizing all the available water resources, about 50% of our cultivable area will still depend on rains. Therefore, our agricultural scientists, policy makers and farmers should appropriately realize the magnitude of role that rainfed agriculture or dryland farming can play. They should thoroughly examine the problems of dry land agriculture from different view points and evolve appropriate technologies, crop varieties, etc. for these areas to better the economic condition of the farmers. Farmers should utilize well in time whatever improved technology and varieties suitable for dry farming are available. They should be extra careful about the utilization of available rain water, selection of crops and protection of crops from different harmful physiological or biological agencies. Only then can they make vegetable cultivation profitable under dry farming conditions.

References

Allolli, T.B., Hulihalli, U.K. and Athani, S.I. 2008. Influence of *in-situ* moisture conservation practices on the performance of dry land chilli. *Karnataka J. Agric. Sci.,* **21**(2):253-55

Chakravarti, R.C. and Sadhu, M.K. 1994. Effect of mulch type and colour on growth and yield of tomato. *Indian J. Agric. Sci.,* **64**(9): 608-612.

Dhake, A.V., Adsul, G.G. and Subramaniam,V.R. 2009. Studies on liquid fertilizer application through drip irrigation in tomato cv.Dhanshree. *Green Farming.* **2**(5): 280-82

Fulton. J.M. and Tan, C.S. 1986. Irrigation of vegetable crops. Ontario Ministry of Agriculture and Food. *Fact Sheet Agdex* 250/560

Hopen, H.J. and Oebker, N.F. 1975. Mulch effects on ambient carbon dioxide levels and growth of several vegetables. *HortScience* **10**:159-161

Lourduraj, C.A., Padmini, K., Pandiarajan, T. and Sreenarayanan, V.V. 1997. Effect of mulching and irrigation regimes on brinjal. *South Indian Hort.,* **45** (5&6):228-234

Radder, G.D., Itnal, C.J., Surkod, V.S. and Biradar, B.M. 1991.Compartment bunding: an effective in-situ moisture conservation practice on medium deep black soil. *Indian J. Soil Conserv.* **19**:1-5

Ram, H.H. 1998. Vegetable breeding: principles and practices. Kalyani Publishers. 421p.

Selvaraju, R., Subbian, P., Balasubramanian, A. and Lal, R. 1999. Land configuration and soil nutrient management options for sustainable crop production on allifisols and vertisols of southern peninsular India. *Soil tillage Res.* **52**:203-206

Singh, R.V. 1993. Production potential and economics of vegetables intercropped with rainfed okra. *Indian J.Hort.,***50**(1)73-76

Stoner, A.K. 1978 Breeding Vegetables tolerant to environmental stress. *Hort Science* **13**:684

Umamaheshwari, Hari, P. and Kamala. 2004. Impact of foliar application of organic nutrients on the nutrient uptake pattern of chilli cv .K-2. *South Indian Hort.,* **51** (1&6):168-172

CHAPTER

14 Organic Vegetable Production Techniques in Dry Land Farming Systems

Sanjeev Kumar and Puja Rattan

Division of Vegetable Science & Floriculture, Faculty of Agriculture, Sher-e-Kashmir University of Agricultural Sciences and Technology of Jammu, Chatha, Jammu-180009.

India has about 108 million hectares of rainfed area which constitutes nearly 75% of the total 143 million hectares of arable land. In such areas crop production becomes relatively difficult as it mainly depends upon intensity and frequency of rainfall. The crop production, therefore, in such areas is called rainfed farming as there is no facility to give any irrigation, and even protective or life saving irrigation is not possible. These areas get an annual rainfall between 400 mm to 1000 mm which is unevenly distributed, highly uncertain and erratic. In certain areas the total annual rainfall does not exceed 500 mm. The crop production, depending upon this rain, is technically called dry land farming and areas are known as dry land.

India has about 47 million hectares of dry lands out of 108 million hectares of total rainfed area. Dry farming or dry land farming may be defined as, "a practice of growing profitable crops without irrigation in areas which receive an annual rainfall of 500 mm or even less".

Characteristics of Dry land Agriculture:

Dry land areas may be characterized by the following features:

- Uncertain, ill-distributed and limited annual rainfall.
- Occurrence of extensive climatic hazards like drought, flood etc.
- Undulating soil surface.
- Occurrence of extensive and large holdings.
- Practice of extensive agriculture i.e. prevalence of mono cropping etc.
- Relatively large size of fields.
- Similarity in types of crops raised by almost all the farmers of a particular region.
- Very low crop yield

- Poor market facility for the produce
- Poor economy of the farmers, and
- Poor health of cattle as well as farmers.

Further the crop production is highly risky in arid and semi-arid climates. In such conditions generally two types of agriculture is practiced. One is crop production or arable farming and the other is mixed farming i.e. animal husbandry together with crop production and pasture management.

2. Prospectus of Organic Farming in rainfed areas:

In traditional India, the entire agriculture was practiced using organic techniques, where the fertilizers, pesticides, etc., were obtained from plant and animal products. Before the Green Revolution, it was feared that millions of poor Indians would die of hunger in the mid 1970s. However, the Green Revolution, within a few years, showed its impact. As time went by, extensive dependence on chemical farming has started showing its darker side. The land is losing its fertility and is demanding larger quantities of fertilizers to be used. Pests are becoming immune requiring the farmers to use stronger and costlier pesticides. Due to increased cost of farming, farmers are falling into the trap of money lenders, who are exploiting them to no end, and forcing many to commit suicide. Both consumer and farmers are now gradually shifting back to organic farming in India. It is believed by many that organic farming is healthier. Though the health benefits of organic food are yet to be proved, but even then the consumers are willing to pay higher premium for the same.

As per Food and Agriculture Organization (FAO) study of mid-2003, India had 1,426 certified organic farms producing approximately 14,000 tons of organic food / produce annually. In 2005, as per Govt. of India figures, approximately 190,000 acres (77,000 hectares) were under organic cultivation (table 1). The total production of organic food in India as per the same reference was 120,000 tons annually, though this largely included certified forest collections. However, the data collected from certifying agencies like SKAL, Eocert, SGS, IMO, etc, the certified organic area has increased up to the tune of 173682.50 ha in 2005-2006 (Bhattacharyya, 2007).

Table 1. Statewise status on certified organic area (2005-2006)

States	Area (ha)	States	Area (ha)	States	Area (ha)
Andhra Pradesh	1661.42	Karnataka	4117.07	Rajasthan	22104.91
Arunachal Pradesh	557.76	Kerala	15474.47	Sikkim	177.64
Assam	1817.50	Manipur	347.65	Tripura	20.87
Chhattisgarh	293.16	Maharashtra	18786.69	Tamilnadu	5423.63
Delhi	1658.71	Madhya Pradesh	16581.37	Uttar Pradesh	3033.97
Goa	5555.07	Mizoram	300.40	Uttaranchal	5915.85

Gujarat	1627.06	Meghalaya	378.89	West Bengal	6732.43
Haryana	3437.52	Nagaland	718.76	Others	824.73
Jammu and Kashmir	22315.92	Orissa	26387.86		
Jharkhand	5.00	Punjab	3779.31	Total	173682.50

Majority of India's cropped area is not irrigated and it can be safely assumed that high-input demanding crops are not grown on these lands. Fertilizer use on dry lands is always less anyway as chemical fertilizers require sufficient water to respond. Pesticide use in these lands would also be less as the economics of these hardy or "not-so profitable" crops will not permit expensive inputs. These areas are at least "relatively organic" or perhaps even "organic by default". While neither of these terms necessarily denotes a healthy farm or a recommended agriculture system, it would at least imply a non-chemical farm that can be converted very easily to an organic one providing excellent yields and without the necessity and effort of a lengthy conversion period.

Since the organic food is getting popular and results into high remunerative, so organic vegetable production can be one of the options in the dry land to increase the returns in such areas with slight modifications in their cultural practices. Furthermore, the mixed farming practices i.e animal husbandry along with crop production can be best utilized in an economic way in the organic farming as the animal waste can be successfully used as organic manure for raising the crops.

Organic vegetables production in the rainfed areas is a strategic work and requires keen observation on the water management along with the avoidance of chemical use.

3. Techniques to be Pursued for the Water Management Under Organic Farming

- Growing quick maturing, drought resistant crops.
- Following farm practices which conserve and utilize the available rainfall to the fullest extent. To obtain maximum storage of moisture under any rainfall condition, the soil must absorb as much water as possible when it rains and losses by evaporation or transpiration must be kept to a minimum.
- Wind and heat must not cause excessive evaporation at critical stages of plant growth.
- Soil should be deep (preferably 10 feet - 3 meters) with no clay, sand, or gravel seems to interfere with capillary movement of water. The minimum feasible soil depth is 18 inches (450 mm) but water storage capability and drought resistance increases with increasing soil depth.

Growing Vegetables with Less Water Requirement

Since water requirement is the main criteria of growing vegetables in the dry land or rainfed areas, so vegetables, which are adapted to drought conditions at a very basic or cellular level, are preferred for growing in such area. Vegetables which reportedly require less water for their growth and development are enlisted below:

Rabi Season	Spring Season	Rainy Season
Pea, leaf mustard, chick pea, radish, spinach, kale, leek, beans.	Water melon, musk melon, amaranthus, tomato.	Okra, bottle gourd, pumpkin, bitter gourd, tomato, cowpea, chilli, baby corn, cucumber, sponge gourd, and ash gourd.

Tomatoes, squashes and melons establish deep root system quickly and are able to draw moisture from the deeper soil layers after the surface has become dry in mid summer. As long as these crops have water in their early growing season, they're tolerant to terminal drought. In fact, many tomatoes actually do better if you cut off irrigation in mid-to late summer. Okra withstands drought well. By and large, cool-season crops are not drought-resistant. Many of them originated in the Mediterranean region, where they germinate when the rains fall, grow slowly through the mild but cool winter, and flower and fruit in the spring. These are the cool-season legumes like peas, lentils and faba beans, and the crucifer crops, including Brussels sprouts, cabbage, cauliflower, collards, kale, kohlrabi, mustard, broccoli, turnips and watercress.

Farming Techniques

This includes the techniques which conserve the moisture, by avoiding the formation of crust on the soil surface and reduce run off of water. The soil crust formation can be prevented by following ways:

- Crop land should be as level as possible.
- All tillage and plantings must run across (or perpendicular to) the slope of the land. Such ridges will impede the downward movement of water.
- For every two feet of vertical drop or 250 feet of horizontal run, the field should either have bunds or contour strips.

Reducing Soil Evaporation

As water near the surface of the soil evaporates, water is drawn up from below to replace it, thinning the water film. When it becomes too thin for plant roots to absorb, wilting occurs. This can be reduced by following ways:

- Shelter belts of trees or shrubs reduce wind speeds and cast shadows which can reduce evaporation 10 to 30 per cent by itself and also reduce wind erosion.
- Mulching reduces the surface speeds of wind and reduces soil temperatures.
- Shallow tilling can create a dirt mulch 2 to 3 inches deep which dries out easily but is discontinuous from the subsurface water, preventing further loss. Tillage must be repeated after each rain to restore the discontinuity.

This is most workable where rainfall occurs in a few major rainfalls with relatively long intervals in between.

Reducing Transpiration

All growing plants extract water from the soil and evaporate it from their leaves and stems in a process known as transpiration. Important points taken into consideration for reducing transpiration rate are:

- Weeds compete not only for soil nutrients, but water as well and so their control is essential.
- Selection of crop is significant as well.
- In dry farming, the number and spacing of plants is reduced so that fewer plants compete for soil moisture.
- Post harvest tillage will create stubble and dirt mulches and destroy weeds before the onset of the dry spell.

4. Systems-based Organic Production Practices

These conserve nutrients, protect water quality, by a combination of the following:

- Increasing soil organic matter by returning organic materials to the soil and choosing practices that support a biologically active humus complex.
- Composting animal manure and other organic residues to form a more uniform and chemically stable fertilizer material.
- Timing the release of nutrients from organic-matter mineralization to coincide with the times when plants are actively growing and taking up nutrients.
- Using inter-cropping practices to diversify crops in the field, enhance soil fertility, increase the efficiency of nutrient use, and decrease pest pressures.
- Planting catch crops or cover crops to recover nutrients that may otherwise leach into the subsoil.
- Using conservation practices that reduce the potential for water runoff and wind and water erosion.
- Managing and monitoring irrigation practices to enhance nutrient uptake, decrease leaching of nutrients, and minimize root and stem diseases.

5. Different Organic Ways to Conserve Soil Moisture

Cover crop:

Cover crops are crops which are not usually grown for harvest, but which serve multiple functions in crop rotation systems. Although cover crops are commonly grown to prevent soil erosion and for improvement of soil tilth, other important roles include the enhancement of soil structure, improvement of soil fertility, enhancement or preservation of environmental quality, and contributions to the management of weeds, insect pests, and plant pathogens. Not all of these benefits are associated with a specific cover crop use, however many of these benefits can occur simultaneously.

Purposes

- To control erosion during periods when the harvested crop does not furnish adequate soil cover.
- To recover phosphorus and other nutrients in the root zone by plant uptake.
- To improve infiltration, thus reducing surface runoff from the soil. Cover crops establish during the non-crop period, usually after the crop is harvested, but can be inter-seeded into a crop before harvest by aerial application or cultivation.

Major benefits of cover crop are

1. Increase evapo-transpiration, increase water infiltration and decrease soil bulk density.
2. Reduce sediment production, decrease impacts of raindrops, decrease runoff velocity.
3. Increase soil quality by improving the biological, chemical, and physical soil properties.
4. Increase organic carbon, carbon exchange capacity, aggregate stability and water infiltration.
5. Grass and *Brassica* species are great nitrogen scavengers and increase carbon inputs.
6. Cover crops improve soil and water quality and may reduce nutrient and pesticide run-off by 50% or more, decrease soil erosion by 90%, reduce sediment loading by 75%, reduce pathogen loading by 60%.

Green Manuring

Green manuring involves the soil incorporation of any field or forage crop while green or soon after flowering, for the purpose of soil improvement. A cover crop is any crop grown to provide soil cover, regardless of whether it is later incorporated. Cover crops and green manures can be annual, biennial, or perennial herbaceous plants grown in a pure or mixed stand during all or part of the year. In addition to providing ground cover and, in the case of a legume, fixing nitrogen, they also help suppress weeds and reduce insect pests and diseases. When cover crops are planted to reduce nutrient leaching following a main crop, they are often termed "catch crops." As these suppress weeds, these reduce the competition for available water. Furthermore, these provide cover on the soil surface which reduces the evapo-traspiration and thus reducing the formation of soil crust. Green manuring can be used in various types of soils. It is appropriate for infertile soils and areas with low rainfall, where artificial fertilizer cannot be used effectively because of the lack of moisture. It is useful on sandy soils, as deep-rooted legumes help recycle nutrients. It also helps improve the structure of heavy soils. It can help control erosion in hilly areas or places with strong winds.

Advantages of green manuring

- Improves the soil fertility by the increasing the humus content of the soil,
- Adds nutrients and organic matters,
- Improves the soil structure,
- Improves soil aeration,
- Increases the water holding capacity of the soil,
- Increases the nutrient holding capacity of the soil,
- Helps control the soil born insects/ mite, nematodes and diseases,
- Used as a soil binder in the sloppy areas,
- Conserves and recycles plant nutrients,
- Protects the soil from erosion,
- Promotes habitat for natural enemies,
- Serves as a good food for the earthworms,
- Increases soil's biodiversity of beneficial microbes by stimulating their growth.

Mulches

Mulch prevents loss of moisture from the soil, suppresses weed growth, reduces fertilizer leaching, cools the soil, and keeps vegetables off the ground. Fruit rot sometimes occur when vegetables touch the ground and mulching serves as a barrier between the plant and the soil, helping prevent the disease.

Mulching has several advantages. It reduces labor required in cultivation, since emerging and small weeds perish under their dark barrier. Therefore, mulching reduces the need for tillage and the use of weed-control chemicals. Water is conserved because mulches reduce the evaporation of soil moisture by lowering the soil temperature. Water absorption by a mulched soil is greater than that of an unmulched soil. Mulch also prevents the formation of soil crusts. In addition, soil loss from heavy rain and wind is decreased. In fact, mulches are excellent conservation agents.

Mulch is an excellent insulator and prevents drastic fluctuations in soil temperature. Mulch keeps the soil cooler in summer and warmer in winter, improving both root growth and nutrient availability. At the end of the growing season, organic mulches can be tilled into the soil to further increase the organic matter content and the water-holding capacity of the soil. Finally, mulches impart a neat, trim look to gardens and reduce the incidences of mud-splashed vegetables after heavy rains, which could lead to disease problems.

Ecological Pest Management Techniques

Direct Control Methods

- Quarantine: This involves isolation of infected crop or animal from the farm
- Biological control: This involves use of a natural enemy of a pest to control its population, for instance a lady bird on a farm will control the population of aphids in a tomato garden, others include; wasp Vs caterpillar larva, spiders Vs flying insect pests.

Indirect Control Methods

This refers to change in the farming technologies so as to interfere with a pest's life cycle and include the following:

Mixed cropping

This includes growing of more than one crop on the same piece of land, for instance interplanting of simsim with cassava to control Morrow rats. The simsim has very bitter roots that ward off the rats. Agroforestry technologies also lie in this category for this case the various trees on a farm interfere with odours from the host plant thence making it hard for the pest to locate the plant.

Hygiene methods

These involve ensuring that the fields are kept clean and free from infected and rotting plants. For instance the infested plants should be removed from the fields as soon as they are observed. It should be ensured that the field is free of weeds and thus crop inspections at regular intervals are necessary to remove the weeds and infected plants from the field.

Physical management

This includes the following specific measures:

a) Hand picking
b) Scare crow
c) Trapping
d) Crop rotation
e) Double fencing
f) Ditches (control wild pigs)
g) Encourage pests' natural enemies
h) Increase soil fertility
i) Plant tolerant crop varieties
j) Mixed cropping
k) Early cropping

6. Long-term Benefits of Organic Management

Organic farming practices promote the formation of soil humus and the accumulation of nutrient reserves in the bodies of soil organisms and in the readily decomposable form of soil organic matter (Ryan, 1999; Wander et al., 1994). As communities of soil organisms become larger and more diverse, the decomposition of added organic matter will be enhanced, as will the ability of this biological community to temporarily store mineralized nutrients (Drinkwater et al., 1998). As the ability of soils to store nutrients increases, crop nutrient demands will be met from a combination of applied and stored nutrient sources. Careful management of the types of organic residues added to the soil can also control nutrient mineralization and immobilization processes. In the fall, you can either apply carbon-rich organic materials to the soil or leave woody crop residues on the soil. Soil organisms decomposing these materials will use excess soil nutrients to meet their nutrient demands. Nutrients immobilized in the bodies of soil organisms can be made available to crops in the spring by adding a nitrogen-rich form of organic matter to the soil shortly before the onset of the growing season. This will stimulate the decomposition of the high-carbon material and the mineralization or release of nutrients held in the bodies of soil organisms. Remember that climate conditions affect the time needed for either immobilization or mineralization processes to occur. In cold or arid climates, these processes will be much slower than in warm humid climates.

Readily available forms of nutrients can be applied to crops to meet high nutrient demands or to stimulate mineralization of nitrogen-poor organic materials. These nutrient sources and application methods are expensive; careful monitoring of nutrient additions in relation to plant uptake needs can save money, enhance plant production, and reduce the risks of nutrient leaching and run-off. Use of these readily available nutrients without proper management can increase the potential for nutrient leaching and runoff.

References

Bhattacharyya, P. 2007. Status of organic farming in India www.oranicexchange. org/ meetings/presents/ind7_drp.pdf.

Drinkwater, L.E., Wagoner P., and Sarrantonio M.. 1998. Legume-based cropping systems have reduced carbon and nitrogen losses. Nature. Vol. 396. p. 262–265.

Ryan, M. 1999. Is an enhanced soil biological community, relative to conventional neighbours, a consistent feature of alternative (organic and biodynamic) agricultural systems? Biological Agriculture and Horticulture. Vol. 17. p. 131–144.

Wander, M.M., Traina S.J., Stinner B.R., and Peters S.E.. 1994. Organic and conventional management effects on biologically active soil organic matter pools. Soil Science Society of America Journal. Vol. 58. p. 1130–1139.

CHAPTER 15

Rainfed Floriculture - A Shot in the Arm for Dryland Farmer

Sheetal Dogra, R. K. Pandey, Satesh Kumar and Jag Paul Sharma

Division of Vegetable Science & Floriculture, Faculty of Agriculture, Sher-e-Kashmir University of Agricultural Sciences and Technology of Jammu, Chatha, Jammu-180009.

Water has been prioritized to be the most crucial resource for sustainable development of agriculture in India as on average 70% of its arable area is rain-fed. The life of mankind and almost all the flora and fauna on the earth depends on the availability of fresh water resources. It is an extremely important component of biological systems. The most of the plant cells and tissues contain 80-90 per cent water by weight. Plant water status influences all the physiological processes of plants directly or indirectly. Highly variable global distribution of annual average rainfall of about 1000 mm on the earth surface is responsible for the disparities in agriculture production and socio-economic conditions. Fifty five per cent or more than one-half of the total land surfaces of the earth, receives an annual precipitation of less than 500 mm and must be reclaimed, if at all by dry farming practices. Area with 500-750 mm rainfall, which accounts for 10 per cent of the total land area, also need dry farming measures for successful crop production.

In India also, distribution of the annual average rainfall of about 1200 mm, is highly variable, irregular and undependable with widespread variations among various meteorological sub-divisions in terms of distribution and amount. The spatial distribution of rainfall varies from 100mm/annum in Rajasthan to about 11000 mm/annum in Cherrapunji in Meghalaya. Agriculture uses almost 85 per cent of the total water available in the country. Of the total geographical and net cultivated area about 92 and 33 million hectares receive less than 750 mm rainfall annually.

Table 1. Distribution of dryland area in different rainfall zones in India

Mean annual rainfall (mm)	Climate zone	Dry land area (million ha)
<500	Arid	10.9
500-750	Dry semi-arid	22.2
750-1000	Wet semi-arid	21.2
1000-1250	Sub-humid	19.6
>1250	Humid	27.7

Virmani *et. al* (1991)

Agriculture continues to be mainstay of the Indian economy. It contributes 25 per cent of the national gross product. In this contribution, rainfed agriculture in India is important today and will continue to be so in future. Currently, about 63 per cent of agriculture in India is rainfed. There can be an enormous shift in crops and cropping patterns in rainfed areas as the irrigation water is becoming scarce and the quality of available irrigation water is deteriorating.

Floriculture in India, is being viewed as a high growth Industry. Commercial floriculture has gained importance over the years. The total area under flower crops in 2007-08 was estimated around 160.7 thousand hectares, which included traditional flowers. Indian floriculture industry has been shifting from traditional flowers to cut flowers for export purposes. In India, Maharashtra, Karnataka, Andhra Pradesh and Haryana have emerged as major floriculture centres in recent times. Production flowers are estimated to be 870.4 Mt of loose flowers and 43417.5 million (numbers) of cut flowers in 2007-08. According to the report of APEDA, India exported 36,240.78 MT flowers and planting material of worth Rs. 34,014.42 lakhs in 2007-2008.

Floriculture has a great potential to improve the resource utilization efficiency of current cropping patterns. The total water requirement of many drought tolerant ornamental and native plants is lower than that of existing crop alternatives. In addition most ornamental plants are propagated by vegetative cuttings, so the larger the plants become with time, the more cuttings become available. This is contrary to vegetables and fruit trees, whose productivity declines with time. The most appealing factor for introducing drought tolerant native and ornamental plants into local cropping patterns is the export potential. Since ornamental and native plants are not consumable goods, most countries have less rules and regulations on importing these items. Exporting ornamental plants has the potential to become one of India's primary income generating activities.

Ornamental plants having less water requirement and can be successfully established in dry land conditions by using certain technologies like selection of suitable plants, moisture conservation, improved production technologies, rain water harvesting, etc.

2. Choice of ornamental plants

A plant is drought tolerant if it can survive a dry spell of more than 2 or 3 months without supplemental watering. This means that most native plants from regions that occasionally experience 2-3 months without rain, are drought tolerant. Drought tolerant plants should have following characteristics:-

- Deep root system
- Resistance/tolerance to biotic stresses
- Resistance/tolerance to abiotic stresses
- Low rate of transpiration
- Less sensitive to photo-period
- Greater phenotypic plasticity

Most established ornamental trees and shrubs have such wide-ranging roots that they are drought proof and need no irrigation. Plants such as the Rock Daisy (*Pachystegia insignis*), often have leaves with wax coated upper surfaces and felting on the undersides. These coating provide protection from salt spray and moisture loss. Other drought tolerant plants, such as *Artemisia arborescens, Convolvulus cneorum, Stachys byzantina* and *Euryops pectinatus*, have silver leaves to reflect harsh sunlight. Many grasses have low moisture requirements. In drylands, cacti and succulents can also be extensively used. They require little or no irrigation once established and rain may, at times, even be a disadvantage. *Aloe, Crassula, Agave, Kleinia, Sedum* and the myriad of cacti species have very bold forms and are used for rainfed conditions.

Table 2. Drought Tolerant Plants

ANNUALS	
Ageratum	*Ageratum houstonianum*
Blanket flower	*Gaillardia pulchella*
Calendula	*Calendula officinalis*
California poppy	*Eschscholzia californica*
Cockscomb	*Celosia cristata*
Cosmos	*Cosmos bipinnatus,*
Dusty miller	*Senecio cineraria*
Foxglove	*Digitalis purpurea*
Geranium	*Pelargonium* x *hortorum*
Globe amaranth	*Gomphrena globosa*
Marigold	*Tagetes erecta, T. patula*
Moss rose	*Portulaca grandiflora*
Pansy	*Viola* x *wittrockiana*
Petunia	*Petunia* x *hybrid*
Salvia	*Salvia splendens, S. farinacea*

Snapdragon	*Antirrhinum majus*
Statice	*Limonium*
Sweet alyssum	*Lobularia maritime*
Verbena	*Verbena spp. and hybrids*
PERENNIAL PlANTS	
Parrotlily	*Alstroemeria psittacina*
Iron plant	*Aspidistra elatior*
Madagascar periwinkle	*Catharanthus roseus*
Whorled coreopsis	*Coreopsis verticillata*
Grand crinum lily	*Crinum asiaticum*
Tall cigar plant	*Cuphea micropetala*
Angel's trumpet	*Datura inoxia var. quinquecuspida*
Cassava	*Manihot esculenta*
Four o' clock	*Mirabilis jalapa*
Daffodil	*Narcissus spp.*
Pink evening primrose	*Oenothera speciosa*
Aasparagus fern	*Protasparagus densiflorus*
Bracken	*Pteridium aquilinum*
Sage	*Salvia lyrata*
Goldenrod	*Solidago odora*
SHRUBS	
Glossy abelia	*Abelia x grandiflora*
Aucuba	*Aucuba japonica*
Red barberry	*Berberis thunbergii var. atropurpurea*
Butterfly-bush	*Buddleia davidii*
Harland boxwood	*Buxus harlandii*
Rockspray contoneaster	*Cotoneaster horizontalis*
Japanese euonymus	*Euonymus japonicas*
Gardenia	*Gardenia augusta (G. jasminoides)*
Himalayan indigo	*Indigofera heterantha*
Virginia sweetspire	*Itea virginica*
Showy jasmine	*Jasminum floridum*
Italian yellow jasmine	*Jasminum humile*
Winter jasmine	*Jasminum nudiflorum*
Crapemyrtle: New Orleans, Centennial, Chica, Tonto, Acoma, etc.	*Lagerstroemia indica and hybrids*
Lantana,	*Lantana camera*
Common lavender	*Lavandula angustifolia*
Japanese privet	*Ligustrum japonicum*
Little Gem, Pendula, etc.	*Picea abies (dwarf cultivars)*
Shrubby podocarpus	*Podocarpus macrophyllus var. maki*

Flowering almond	*Prunus glandulosa*
Pomegranate	*Punica granatum*
Formosa pyracantha	*Pyracantha koidzumii*
Needle palm	*Rhapidiophyllum hystrix*
China rose	*Rosa chinensis*
Rugosa rose	*Rosa rugosa*
Rosemary	*Rosmarinus officinalis*
Butcher's broom	*Ruscus aculeatus*
SMALL TREES	
Mimosa	*Albizia julibrissin*
Smoke tree	*Cotinus coggygria*
Italian cypress	*Cupressus sempervirens*
Chinese juniper: Wintergreen, Spartan, Hooks, etc.	*Juniperus chinensis*
Golden rain tree	*Koelreuteria bipinnata*
Golden rain tree	*Koelreuteria paniculata*
Crapemyrtle	*Lagerstromia indica*
Choctaw, Muskogee, Natchez, Tuscarora, etc.	*Lagerstromia indica x fauriei*
TALL TREES	
Red maple	*Acer rubrum*
Deodar cedar	*Cedrus deodara*
Cryptomeria	*Cryptomeria japonica*
China fir	*Cunninghamia lanceolata*
Cypress	*Cupressus arizonica*
Eucalyptus	*Eucalyptus spp.*
Ginkgo	*Ginkgo biloba (male)*
Southern magnolia	*Magnolia grandiflora*
Pine	*Pinus spp.*
London plane tree	*Platanus acerifolia*
Oak	*Quercus spp.*
Sassafras	*Sassafras albidum*
VINES	
Mexican creeper	*Antigonon leptopus*
Bougainvillea	*Bougainvillea spp*
Railroad vine	*Ipomoea pes-caprae*
Japanese honeysuckle	*Lonicera japonica*
Passionflower	*Passiflora caerulea*
PALMS	
Coconut palm	*Cocos nucifera*
Queen sago	*Cycas revolute*
Virgin palm	*Cycas circinalis*

Dragon tree	*Dioon edule*
Chinese fan palm	*Dracaena draco*
Canary Island date palm	*Livistona chinensis*
Date palm	*Phoenix canariensis*
	Phoenix dactylifera
Date	*Phoenix roebeilinii*
	Phoenix rupicola
California fan palm	*Rhaphiolepis x delacourii*
	Washingtonia filifera
GRASSES	
Red Top	*Agrotis alba*
Creeping Bent fine Grass	*Agrotis palustris*
Colonial Bent Fine Grass	*Agrotis tenuis*
Carpet Grass	*Axonopus affinis*
Bermuda Grass	*Cynodon dactylon*
Centipede Grass	*Eremochloa ophiuroides*
CACTI AND SUCCULENTS	
Agave	*Agave Americana*
	Agave fillifera
Aloe	*Aloe vera*
Yucca	*Yucca spp.*
Mother in law's tongue	*Sansevieria marginata*
Sedum	*Sedum spp.*
	Kalanchoe spp.
	Euphorbia spp.
	Opuntia spp.
Golden Barrel	*Echinocactus grusonii*
Crab cactus	*Zygocactus spp.*

3. Moisture conservation

In dry land areas, soil moisture is the most limited resource for plant growth. Soil moisture is lost as evaporation from the soil surface and as transpiration from plant surface. For prolonging the use of limited amount of soil moisture by the crop plants, evaporation losses of soil moisture has to be arrested, as these are not related to productivity of the crops, whereas transpiration can be reduced to some extent without affecting productivity of crop plants. This reduction in soil moisture losses for increasing the productivity and water use efficiency through various measures is termed as moisture conservation.

Moisture conservation practices

- Mulches
- Anti-transpirants
- Weed control
- Optimum plant population

- Wind break and shelterbelts
- Shade nets

Mulching

Mulch is simply a protective layer of a material that is spread on top of the soil. Mulches can either be organic such as grass clippings, straw, bark chips, and similar materials or inorganic such as stones, brick chips, polyethylene and plastic.

Mulches prevent rapid evaporation from the soil surface and reduce rapid drying. Mulching also suppress weed infestation effectively. Furthermore, it stimulates microbial activity in soil through increasing soil temperature, which improves agro-physical properties of soil. Mulching is used as a means of successful crop production mainly in place where irrigation facilities are scanty. The effect of these is to cut off the upward flow of underground moisture at a point below the actual surface and to prevent its rapid escape into the air during dry weather (Macmillan, 1989 ; Suh and Kim, 1991). It has a unique character of reducing the maximum soil temperature and increasing the minimum temperature (Singh *et al.*, 1987). Photodegradable and biodegradable mulches made of plastic, paper, or other materials have been tested for their utility for annual crop production with varying degrees of success (Greer and Dole, 2003). As early as the 1920s, researchers developed acceptable production practices with paper mulches in annuals. Solaiman *et al* (2008) obtained maximum plant height and highest number of flowers in china aster (*Callistephus chinensis*) with black plastic mulch (Fig. 1 & 2). Lippert *et al* (1964) and Waggoner *et al.* (1960) reported that moisture distribution in the upper soil layers in mulched is more uniform compared to unmulched soil, root development is better in the upper soil layer, which usually is rich in nutrients and useful microorganisms. Lal (1989), also reported that biological activity of soil fauna increased with mulching thereby increases soil fertility. Mulching has been reported to increase yield by creating favorable temperature and moisture regimes.

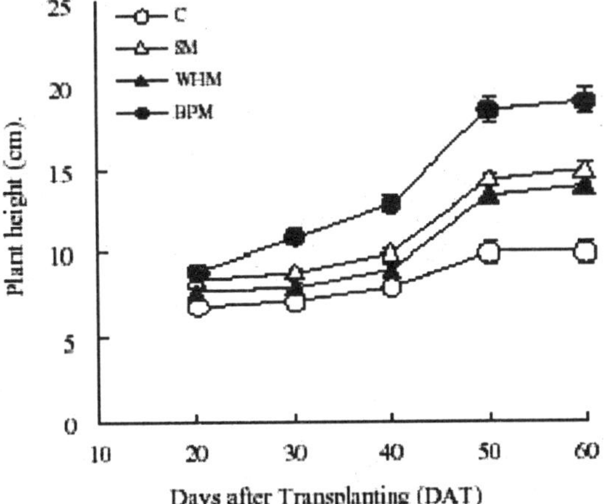

Fig 1.Variation of plant height of Aster due to different mulches. C, control; SM, Straw mulch; WHM, Water hyacinth mulch; and BPM, Black plastic

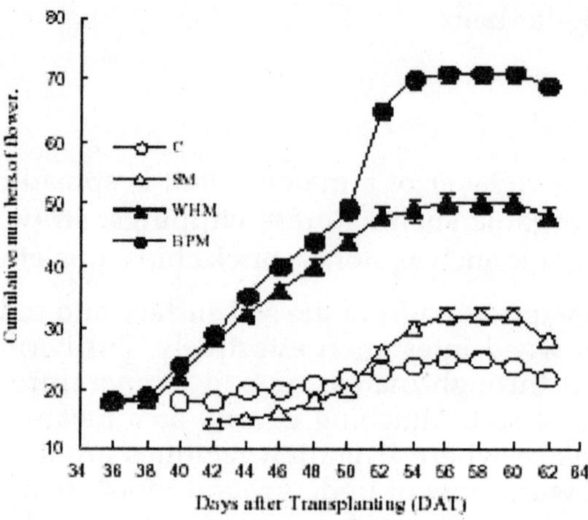

Fig 2: Influence of different traditional mulches on flower initiation and total number of flowers in aster. C, control; SM, Straw mulch; WHM Water hyacinth mulch; and BPM, Black plastic mulch.

Laying of Mulches

For new beds, planting through sheet mulches is very effective. Single plants, such as trees and specimen shrubs, are best mulched to the radius of the canopy whereas beds and borders are mulched in their entirety. Mulching in conifers maintains the relatively cool soil temperatures that most conifer prefers. The mulch should be 2-3 inches deep and should never come in contact with the trunk of the plants (Roger *et al.*,1997). When babul is grown in dry areas, mulching is generally recommended in the first year. Mulching in babul needs to be carried out during November-December for optimum results. With new plants, apply mulch after thorough watering.

Anti-transpirants

Plants transpire water vapours continuously from all the above ground parts particularly through leaves. This process of evaporation of water from the aerial parts of plants is termed "transpiration". Any material that is applied to plant surfaces with the aim of reducing or inhibiting water loss from plant surface is called "anti-transpirant". Antitranspirants are used on cut flowers, newly transplanted trees and shrubs, to preserve and protect plants from drying out too quickly. They have also been reportedly used to protect leaves from salt burn and fungal diseases.

Types of anti-transpirants:- Based on the mode of action, anti-transpirants are of four types.

i) Stomatal closing type

Spray materials used for various purposes such as certain fungicide (phenylmercuric acetate (PMA), herbicide (atrazine) and metabolic inhibitors have been found to cause the closure of guard cells of the stomata and thereby reduction in transpiration, when sprayed in low concentration on the leaf surface of crop plant. Kamp (1985) controlled powdery mildew of zinnia with antitranspirant more effectively than a commercially used fungicide, acti-dione [cycloheximide]. It also showed that plants treated with the antitranspirant had closed stomata, and presumably this had an effect on water uptake as well as plant host interactions. Antitranpirants reduced the water requirements of oleanders and turf grass by spraying plants with PMA.

ii) Film-forming type

These form a colorless film on the leaf surface which allows diffusion of gases but not of water vapour. Examples include silicone oil and waxes. The effect of the 10 and 125St Dimethyl silicone materials lasted more than 16 days, decreasing both photosynthesis and transpiration of trees (Parkinson, 1970).

iii) Increasing leaf reflectance type

White reflecting materials such as whitewash or kaolinite spray form a coating on the leaves and increase the leaf reflectance. By reflecting the large amount of radiation, they reduce leaf temperature and vapour pressure gradient from leaf to atmosphere and thus reduce transpiration. Application of 5% kaolin spray has been found to reduce transpiration losses markedly.

iv) Growth retardants

Foliar application of chemicals such as cycocel reduces shoot growth, increases root growth and induces stomatal closer.

Scope of using anti-transpirants

- Under dry land area to reduce water losses through transpiration.
- For extending the irrigation interval.
- In areas having poor quality of soil water or irrigation water to reduce the uptake of salts.
- For reducing transplanting shock of nursery plants.

Weed control

Weed is a plant which grows out of place. Weeds are unwanted, useless, prolific, competitive and often harmful to the environment. They interfere with agricultural operations, compete with crop plants for water, nutrients, space and light and thus reduce their production potential. Weeds can cause yield reduction up to 90 per cent in many crops. Critical weed free periods of 30 days for short duration crops and 45-60 days for long duration crops are observed for higher yield. Weeds frequently transpire greater amounts of water per unit of dry matter produced than do the crop plants with which they grow in association. Weeds can remove more quantity of soil moisture and they can survive better than crops under drought conditions. Weed management practices under dry land include weed prevention and control measures.

Weed prevention: It includes prevention of weed seeds spread, growing crops and varieties which grow faster than weeds, checking weed seed production, adjusting plant population to suppress weed, using weed seed free crop seed, using well decomposed farm yard manure and compost etc.

Weed control measures

- Cultural control envisages inter/mixed cropping, mulching, selection of suitable variety/crop, proper crop rotation, adjusting time of sowing etc.
- Physical methods include hand weeding, tillage, hoeing and other inter-culture operations.
- Chemical weed control measures include application of selective and non-selective herbicide for the control of weeds on cropped and non-cropped land. Crop specific herbicides are available for effective and economical weed control under dry farming *e.g.* Trifluralins and Oryzalin as pre-emergence herbicide are mostly used for annual grasses and few broad leafed weeds in marigold and calendula and number of other crop specific herbicides are available, which kill the weeds on the cropped land.
- Biological control includes the use of insects, animals and pathogens to control or limit weed infestation.

Plant population

In dry land ecology, moisture is the most limiting factor. Therefore, plant population must be adjusted to available soil moisture levels, either within rows or between rows. Under rainfed system, the optimum plant density is usually less than recommended under irrigated agriculture. The excessive foliage results in quicker loss of soil moisture through transpiration and higher plant density becomes counter productive if conserved moisture exhausted before the seed formation and development or rainless period prolongs during critical stages of crop growth and development. Moreover, added soil fertility in rainfed conditions is not high and relatively a higher plant density would have individual plant under nutrient stress, leading to lower yield (Rana, 2007).

Shelter from Drying Winds

Good wind protection goes a long way towards minimising moisture loss from both the soil and the plants' foliage and also makes for a more pleasant living environment. The shade from shelter belts and carefully sited trees creates microclimates that are suitable for less drought tolerant shrubs and perennials that would suffer if planted in full sun. Use small-leafed trees, such as the vanilla tree (*Azara microphylla*) or the mayten (*Maytenus boaria*), that won't suck up too much of precious soil moisture to keep their foliage turgid.

Shade nets

The crop productivity is greatly influenced by the growing environment and management practices. Under open field cultivation, it is not possible to control the growing environment of plants, thereby affecting the quality and productivity of the crop. The main purpose of protected cultivation is to create a favourable environment for the sustained growth of crop so as to realize its maximum potential even in adverse climatic conditions. Protected cultivation in the forms of greenhouses, net houses, low tunnels, cloches, mulches etc. offers several advantages to grow crops of high quality and yields, thus using the water and other resources more efficiently.

Shade nets are used in cultivation of flowering plants, foliage plants, and for raising nurseries. In pot plants and greens namely, Palms, Cactii and succulents ranges, *Acalypha, Eucalyptus, Ficus* range, Hedera range, *Maranta* familygroups, Monstera group, *Peperomia, Pilea* range, *Sensviera* range, *Scindapsus* range, *Schefflera* range, *Strobilanthus, Syngonium* range, *Tradescantia* groups, Yucca, *Cissus* varieties can be grown under shade houses. In the flowering segment, *Bougainvillea* group, *Calliandra, Clerodendrum, Crossandra, Episcia, Euphorbia, Hibiscus* group, *Ixora* group, *Jatropha, Jasminum* group, *Lantana* group, *Musa* families, *Pentas* group, *Geranium* groups, and *Strelitzia* varieties can be grown.

4. Rain Water harvesting and recycling

Rain water harvesting is the collection, storage and recycling of run-off water for irrigation and other uses. The surface runoff water harvesting can be achieved through dugouts ponds, tankas, *khadins, havelis,* diversion bunds and roof-top rain water cisterns. Stream flow runoff harvesting is practiced through *nala*-bunding, check-dams, stop dams, percolation tanks/ponds and *nadi.* Rajasthan people have used the scarce and little rainfall through rainwater harvesting structures such as tankas (dugout and lined circular holes, 3-4 m diameter), *nadi* (small excavated or embanked village ponds), *khadins* (runoff from hilly and wasteland is imponded by constructing earthen bunds and crops are grown) *etc.* For efficient use of stored water, it is necessary to consider conveyance of water, time of irrigation, methods of irrigation, quantity of irrigation water applied and selection of crops and cropping systems. Drip and sprinkler methods of irrigation are recommended to minimize wastage of stored water and to bring more area under command. For collection of higher amount of rainfall, under arid region, runoff is induced either by compacting the soil surface or using ground cover of plastic films, rubber and metal sheeting material or using concrete and asphalt, or chemical treatments with soil flocculants (sodium salts) and water repellent (silicone). In the hilly areas, use of low density polyethylene films (250 micron) has been found suitable as a low cost measure for lining of small tanks to check seepage losses.

5. Production Technology of Rainfed Ornamentals

Land preparation

Land preparation (preparatory as well as post-harvest tillage operations) influences the infiltration, runoff flow and rate of soil erosion. Deep ploughing or chiseling after crop harvest has been found to increase the infiltration rate especially in soil, where hard pan formation is very common. Since soil moisture is the limiting factor of crop production in dry land, tillage practices should be aimed to fulfill the need of conservation of rain water and moisture. Major consideration in the land preparation is timeliness and precision of operations. As the rainfall period is very short, the field preparation and sowing have to be completed within short span of time. Any delay in these operations can reduce the availability of soil moisture, which in turn seriously affects the germination stand and yield of the crop.

Soil fertility management

Indian dry lands are not only thirsty but also hungry. The fertilizer consumption is approximately 15-18 kg/ha under dry farming conditions as against the average consumption of more than 100kg NPK/ha under irrigated conditions. Dry land soils are mostly deficient in nitrogen, often deficient in phosphorus and medium in potassium. A great scope for increasing the productivity exists under dry land conditions through efficient nutrient management practices. Crops in dry land areas suffer not only from moisture stress but also from nutrient stress. The method and time of application of fertilizer dose is an important recommendation to optimize the use of nutrients and available water. As the quantum and distribution of rainfall are highly unpredictable, fertilizer dose and timing have to match with rainfall events. Two to three splits have been recommended depending on the crop duration and rainfall pattern. Basal dose of nutrient should be placed below the seed at appropriate distance for their efficient use.

Important recommendations to improve the fertilizer use efficiency in rained dry farming are:-

i) Scheduling the fertilizer application based on soil test;
ii) Selecting the type of fertilizer based on soil reaction;
iii) Placing the fertilizer at 3 to 4 cm by the side or below the seed zone;
iv) Adjust N dose depending upon moisture availability;
v) Spraying of fertilizer nutrients on need basis;
vi) Adding organic manure/green manure at least once in three years;
vii) Including leguminous crops either in rotation or as an intercrop;
viii) Application of bio-fertilizers like *Rhizobium*, *Azospirillum*, *Azotobacter*, PSB and VAM as a component of integrated nutrient management.

Drip Irrigation

Drip irrigation is the slow, deep and regular application of water directly to the root zone of plants. This method of irrigation requires 30-70 percent less water than conventional sprinklers by providing water directly to the plants, and reducing waste from runoff and evaporation. The water flows at a low pressure through polyethylene ("poly") pipes or soaker hoses laid in rows or serving groupings of plants. The water slowly enters the soil from emitters (also known as drippers), pre-punched holes or porous pipe. Drip irrigation is used to water everything from annuals to full-scale trees.

There are many advantages of using drip irrigation including water conservation, reduced soil erosion, reduced plant disease and weed growth, increased plant vigor and protection of our groundwater. It maintains a constant level of soil moisture, while not saturating the root zone. This reduces the stress on plants caused by the traditional flood-to-drought cycles of overhead sprinklers. By reducing the amount of water splashed on plant foliage, drip irrigation discourages fungus and bacterial plant diseases. For example, powdery mildew can be significantly reduced on roses with the installation of soaker hoses or drip irrigation. This reduces the costly and time-consuming application of fungicides. Farmers are cultivating marigold and rose in Andhra Pradesh by drip irrigation and earning more money than other conventional methods of irrigation.

Applying water directly to the root zone of plants reduces the amount of water lost to surrounding soil. In a field nursery, only about 75% of the water applied by overhead sprinkler irrigation becomes available to the plant, remaining lost due to wind, evaporation, and spraying water on areas not intercepted by roots. Micro- irrigation delivers water directly to the root zone and can be 90% to 95% efficient (Ayala, 2009).

Every farmer has experienced the frustration of a freshly weeded field that sprouts a new crop of weeds after the first irrigation. Drip irrigation controls the flow of water to areas where individual plants, shrubs and trees are growing, while leaving open spaces dry. This combined with generous amounts of mulch will deter weeds and conserve water.

Spacing of soaker hoses or drip emitters

Spacing of soaker hoses or drip emitters depends on the soil, plant types, and spacing. Water spreads just 6 inches from a soaker hose or drip emitter in sandy soil, but a foot or more in clay (Fig 3). It is good to wet at least 50% of the root area of each perennial or shrub. Table 4. shows optimal spacing for drip and soaker hoses in various situations.

Watering schedules

Watering schedules depend on the weather, plants, soil, and emitter or soaker hose flow and spacing. Table 5. below suggests watering times for various situations, but the best advice is to observe plants and soil moisture before and after watering.

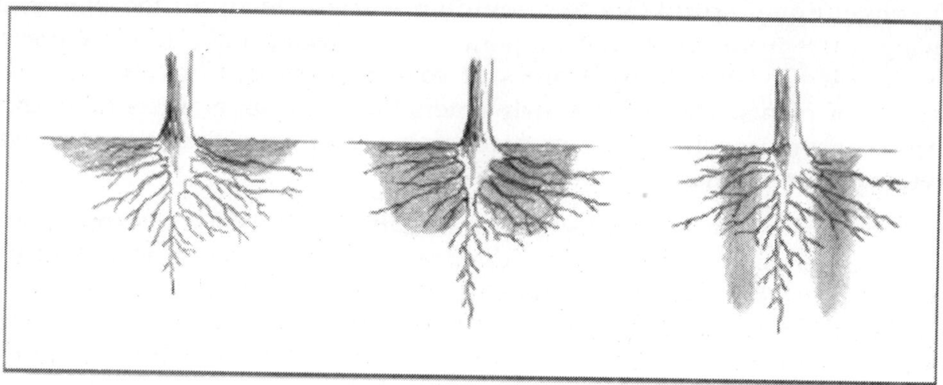

Fig. 3 Water spread from drip or soaker hose in clay(left), loam (middle), and sandy (right) soils. Terrell, (2005)

Table 4. Spacing of soaker hoses or drip emitters

Plant type	**Space Between Drip Lines or soaker Hoses**		**Number of Emitters Per Plant**	
	Clay Soil	*Sandy Soil*	*Clay Soil*	*Sandy Soil*
	Emitters 15-18″ apart in line	Emitters 6-12″ apart in line		
Shrubs and Trees	3-4′ between lines or hoses	2′between lines or hoses	One emitter/2′ plant circumference	One emitter/1′ plant Circumference
Annuals	One soaker or emitter line per row of plants		One per plant	One per plant
Perennials	One soaker or emitter line per row of plants		One per plant	One or two per plant

Terrell, 2005.

Table 5. Watering schedules for annuals and perennials

Plant Type	**Layout on Loam Soil**	**Water Need**	**Weekly Run Time**
Annuals and Perennials	One 1/2 gallon/hour emitter per foot of row	3/4 to 1″ per week	One and a half hours (split over 2-3 days)
	One soaker hose per row	3/4 to 1″ per week	45-60 minutes (split over 2-3 days)

References

Ayala, L. 2009. Grow Healthier Plants with Half the Water and Less Work than Sprinklers. Drip Irrigation for home landscapes www.olympiawa.gov

Greer, L. and Dole J.M., 2003. Aluminum foil, aluminum painted, plastic and degradable mulches increase yields and decrease insect-vectored diseases of vegetables. Hort. Technol., 13: 276-284.

http://www.apeda.com/apedawebsite/six_head_product/floriculture.htm

Kamp, M.1985. Control of Erysiphe cichoracearum on Zinnia elegans, with a polymer-based antitranspirant HortScience 20:879-881

Lal, R., 1989. Conservation tillage for sustainable agriculture: Trop. versus temperate Environ. Advs. Agron., 42: 147-151

Lippert, L.F., Takatori E.H. and. Whiting F.L, 1964. Soil moisture under bands of petroleum and polyethylene mulches. Proc. Amer. Soc. Hort. Sci.,. 85: 541-546.

Macmillan, H.F., 1989. A handbook for tropical planting and gardening. Scientific Publisher. Maan York. Bhavan, India, pp: 261. 15.

Terrell, M. 2005. Drip Irrigation. Masters Garden. Washington State University, Havana.http://spokane-county.wsu.edu/spokane/eastside/ mastergardener@spokanecounty.org

Parkinson, K. J.. 1970.The Effects of Silicone Coatings on Leaves. Oxford University Press. Rothamsted Experimental Station Harpenden, Herts

Rana D.S. 2007. Management of rainfed agriculture. In : Integrated Water Management. Applied Zoology.

Report on micro-irrigation- A Technology for Prosperity by International commission on Irrigation and Drainage www.icar.org

Roger, W. T., Susan, F. M. and Kim E T. 1997. Growing conifers. In: Gardening.111pp.

Sadaphal M. N. and Rajat De. 1994. Resource Management for Sustainable Crop Production. Indian Society of Agronomy, New Delhi.

Singh, P.N., Joshi B.P. and Singh G, 1987. Effect of mulch on moisture conservation, irrigation requirement and yield of potato. Ind. J. Agron., Agric. Exp. 32: 451-452

Solaiman A.H.M., Kabir M.H., Jamal Uddin A.F.M. and Hasanuzzaman Mirza. 2008. Black plastic mulch on flower production and petal coloration of aster (Callistephus chinensis American-Eurasian Journal of Botany, 1 (1): 05-08.

Suh, J.K and Kim Y.S., 1991. Studies of improvement of mulching, cultivar and method in onion. Res. Rep. Rural Dev. Adm. Hortic., 33: 31-36.

Virmani, S.M., Pathak, P. and Singh, R. 1991. Soil related constraints in dryland crop production in vertisols, alfosols and entisols of India. In: Soil Related Constraints in Crop Production, Bulletin 15. Indian Society of Soil Science, New Delhi, pp. 80-90.

Waggoner, P.E., Miller P.M. and. Deroot, H.C 1960. Plastic mulching: principles and benefits. Connecticut Agric. Exp. Station Bull., pp: 534-544.

CHAPTER

16 Prospects of Agroforestry in Drylands

Sandeep Sehgal and N S Raina

Division of Agro-forestry, Faculty of Agriculture, Sher-e-Kashmir University of Agricultural Sciences and Technology of Jammu, Chatha, Jammu-180009.

Drylands cover about 40% of earth's total land surface (UNDP, 1997). About half of the world's countries have some portions of lands in dryland environments. Over 2.3 billion people live in drylands, accounting for nearly 40% of the world's population. According to the Millennium Ecosystem Assessment (MA), about one billion of these people live below the poverty line, accounting for almost half of the world's poor. Africa is relatively driest of the continents. Even in South America, known for its rainforests, about a third of the land is in dry zones. Asia also contains substantial drylands which has the largest population concentrated in the drylands. Drylands vary in terms of their climate, soils, flora, fauna, land use and people. Dryland environments are fragile with very sparse vegetation cover. The soils are shallow and stony with poor water holding capacity. Droughts and crop failures are a common phenomenon in drylands, making rainfed agriculture problematic. The problems of drylands are numerous and are further being complicated by the rapid increase in human as well as cattle population. Rising population has resulted in an increase in the demand for food, fodder and fuel. The limited arable land resources are therefore being put under enormous pressure by the intensification of agriculture, coupled with poor farming practices, excessive tillage of soils and overgrazing. These activities leave exposed soils, which quickly erode and loose fertility, resulting into excessive runoff. Erosion is a major problem in Drylands. Rainfall is inadequate and erratic and the growing period is short. Rainfall intensity and variability are very important. Since, the soils of drylands often cannot absorb all of the rain that falls in intensive storms; water is frequently lost in runoff and seepage processes. At the opposite extreme of the spectrum, the water from a rainfall of low intensity can be lost through evaporation and evapotranspiration when the rain falls on a dry soil surface. Rainfall intensity also relates to the risk of soil erosion. Individual raindrops carry enough energy capable of removing topsoil upon impact, causing splash erosion, which can degrade or destroy the soil structure over time (Brooks et al. 2003). Moreover, rates of deposition, accumulation and decomposition of organic matter are low in dryland environment. All these factors lead to poor moisture holding capacity and ultimately moisture deficit which is a primary factor controlling plant growth processes and productivity (Kramer and Kozlowiski, 1979). Drylands are constrained by factors related to natural causes, threatened and accelerated by factors related to anthropogenic causes (IUFRO, 2004).

Food insecurity and extreme poverty are most challenging in drylands. Poverty in the drylands is debilitating as compared to the inhabitants of irrigated areas (Sehgal *et al*, 2008). Moreover, poverty does not allow certain costlier interventions to improve the conditions of the drylands, as the income and saving potential of dryland farmers is low and uncertain. Tackling dryland degradation and promoting sustainable resource use are key challenges for both environment and development goals because:

- Some of the world's poorest and most vulnerable people live in drylands.
- In many drylands, environmental resources such as land, water and forests are being degraded or are under pressure.
- Many poor people are critically dependent on such resources for their livelihoods.

Despite the importance of woody species in traditional farming systems, only limited information is available on their potential contribution, when used in agroforestry systems. To improve traditional agricultural systems, alternative farming methods with improved soil management technologies are being investigated by various research institutions worldwide. Emphasis is being given to technologies that promote sustainability and serve as intermediate steps to more permanent cultivation through the use of soil conservation, crop rotation, and intercropping techniques. Agroforestry is one such option, which has the potential to significantly improve the livelihoods, economic viability, and agricultural production of small farmers. Trees are highly regarded by the farmers of the drylands. To evade or minimize the adverse affects of frequent droughts, the native peoples in arid zones have often developed production systems in which woody perennials have a very important role, both from a productivity as well as a resource conservation point of view (Tewari *et al.*, 2000). Deliberate integration of trees on farms leads to a variety of functions that directly or indirectly contribute to the livelihood security. Trees play an important role in sustaining the crop production by restoring the soil nutrients, improve the physical properties, increase crop yields, increase the availability of nitrogen in the soil, increase the water-holding capacity of soil and reduce soil erosion (Young, 1989). They also improve the household economy by providing products which can be used directly by the rural households such as food, fuel and construction material and small timber for making agricultural implements (Michael *et al* 1999). Agroforestry, based on thousands of years of farmer's innovation, has been developed into a continuously evolving interdisciplinary science (Lassoie and Buck, 2000). Scientifically speaking, agroforestry is the combination of woody perennials with crops/animals on the same land management unit. The practice of agroforestry involves the combination of agriculture, livestock production, forestry and other types of production systems on the same piece of land rotationally or simultaneously. Regardless of the nature of these combinations, the goal is attaining ecological stability and sustainability.

Traditional agroforestry systems

All over the drylands, farmers traditionally practice agroforestry. The trees in the drylands are intricately associated with the life of human beings. They are grown scattered in agricultural fields for numerous uses like fodder, fuelwood, shade, small timber, vegetables and medicinal uses. Some of these practices are very extensive and highly developed (Table-1). Trees in drylands may be observed in different farming systems and spatial arrangements in the agrarian landscapes in the form of

- agricultural or silvipastoral systems
- scattered on farmlands
- along the farm boundaries
- as windbreaks and shelterbelts

Table-1 Main woody components and associated crop/grass species in combined productive-protective systems (extensive agroforestry systems) of the hot Indian arid zone.

Tree/Shrub Species association	Annual rainfall (mm)	Habitats	Associations of crops/grasses
Calligonum-Haloxylon-Leptadenia	100–150	Sand dunes, interdunes	Pearl millet, cluster bean *Lasiurus scindicus*
Ziziphus-Capparis	150–200	Rocky, gravelly pediments	Pearl millet, green gram, moth bean, cluster bean *Cymbopogon jwarancusa, Aristida sp., Cenchrus ciliaris*
Calotropis-Calligonum-Clerodendrum	200–250	Sandy undulating plains	Pearl millet, green gram, moth bean, sesame, *Cenchrus ciliaris* with *C. setigerus*
Prosopis-Ziziphus-Capparis	250–300	Alluvial plains, soils, often with kankar pans at 80–150 cm soil depth	Pearl millet, cluster bean, green gram, moth bean, sesame, *Cenchrus ciliaris* with *C. setigerus*
Salvadora-Prosopis-Capparis	250–300	Alluvial plains but soils are moderately saline	Cluster bean, pearl millet and sesame and wheat (irrigated areas) with *Cenchrus setigerus, Sporobolus* sp.

Prosopis-Tecomella	275–325	Sandy undulating plains	Pearl millet, cluster bean, green gram, moth bean -*Cenchrus ciliaris* with *Cenchrus. setigerus*
Prosopis	300–350	Alluvial plains	Pearl millet, cluster bean, green gram, moth bean-*Cenchrus ciliaris*
Prosopis-Acacia	300–350	Alluvial plains (irrigated)	Sorghum, cumin, pearl millet, mustard, wheat

Source: Saxena (1997).

The practice of growing agriculture crops under scattered trees on farmlands is quite old and seems to have scarcely changed for centuries. Some of the tree species common in these types of arrangements include *Prosopis cineraria, Faidherbia albida, Acacia tortilis, Mangifera indica, Azadirachta indica, Emblica officinalis, Zizyphus mauritiana, Aegle marmelos, Tamarindus indica, Acacia nilotica and Leucaena leucocephala* etc. The diversity of products and services considerably vary from one site to another, making species and their uses, site specific and agroforestry systems complex by nature (Depommier, 2003). Though the worldwide list of such trees is very long, some of them have received more attention than others, for example, in India, *Prosopis cineraria* in drylands is commonly found scattered and sometimes on boundaries in association with cereals and pulses at a density of 5-80 trees per hectare (Tejwani, 1994). The leaves are rich in proteins and other nutrients and constitute quality livestock feed (Bhandari *et al.*, 1979). The tree also provides fuel wood of high calorific value. Fruits are used as vegetable, flowers and bark are used in medicine and gum is used in sweets. The wood is used for making cot frames, agricultural implements, racks and bullock cart frames etc. In summer during scorching heat and dry periods it provides shade to human beings, livestock and birds. The tree has deep root system and grows well under water scarce conditions. *Prosopis cineraria* also enriches the soil (Table-2), by increasing the nutrient content of soils below it.

Table-2 Nutrient contents of soil under two *Prosopis* species in arid regions of India.

Species	Organic matter	Nitrogen	Macro-nutrients (kg/ha)			Micro-nutrients (ppm)			
		(%)	N	P	K	Zn	Mn	Cu	Fe
Prosopis cineraria	0.57	0.042	250	22.4	633	0.6	10.0	0.5	0.3
Prosopis juliflora	0.39	0.033	212	10.3	409	0.5	7.5	0.5	2.6
Open field	0.37	0.020	203	7.7	370	0.2	6.9	0.3	3.0

Source: Singh and Lal (1969).

Experiments conducted by Shankarnarayan *et al.* (1987) show that *P. cineraria* has no adverse effect on grain production of mung bean (*Vigna mungo*) and guar (*Cyamopsis tetragonoloba*). Under different spacing treatments during 1985, when the general level of crop yield declined due to moisture stress, the yield of guar improved due to association with *P. cineraria*. *Prosopis cineraria* and *Zizyphus numularia* naturally come up in the rangelands and make a silvipastural system where cows, goats and sheep are the part of the farming system. In Africa, *Faidherbia albida* is a widely distributed leguminous tree. It is present in Sudanese and Sahelian zones that are affected by a long dry season. This species is distributed over territories with an annual rainfall ranging from 50 to 1500 mm (Fagg & Barnes, 1990). In eastern, central and southern Africa, it occurs naturally along riverbanks on alluvial soils. The best known peculiarity of *Faidherbia albida* is its reverse phenology (Wickens 1969; Roupsard *et al.*1999). Trees are in leaf, growing and fruiting during the dry season, whereas leaves are shed after the first rains and growth resumes only at the end of the wet season. This phenology is advantageous for agroforestry, because competition with associated crops growing during the wet season is minimized. *Faidherbia albida* trees produce abundant fodder for livestock during the dry season. Farmers value the tree for fodder and green manure and it is harvested for tannis and gums, charcoal, and wood for carpentry (Hocking 1987). *F. albida* significantly contributes to maintaining crop yield through biological nitrogen fixation and provision of a favourable microclimate while minimizing tree–crop competition. A study on an *F. albida*–millet parkland system in Niger demonstrated that shade-induced reduction of soil temperatures, particularly at the time of crop establishment, is critical for good millet growth (Vandenbeldt and Williams 1992). The traditional *Acacia senegal* agroforestry system practiced for centuries in the Sudan-Sahelian zone is another such example. Neem (*Azadirachta indica*), a tree famous for its medicinal and insecticidal properties, is a part of many folklores in the dryland areas of India (Rawat, 1995). Similarly, the inherent ability of dryland acacias to restore the land productivity is a well established fact. These systems have low productivity but are an example of an excellent synergism between the nature and the natives and that's why they have been sustained for centuries.

New agroforestry systems

The practice of agroforestry has been prevalent for many centuries in different parts of the world. However, it was during the late 1970s that the efforts were initiated to bring these traditional practices into the science of agroforestry. Realizing its potential benefits, agroforestry became quite popular, especially in the developing countries (Garity, 2004). Some agroforestry systems were developed for the drylands to improve the overall productivity and sustainability. New systems were developed and management of existing systems was studied in order to reduce the negative effects of trees on the associated crops. Many new agroforestry systems have been developed including both fruit and timber trees. Horti - pastoral system is an important land use system for the drylands. Its establishment depends upon the lands available for cultivation, fertility status of the soils and the economic condition of the farmer. Lands with gentle to moderate slopes and soils with poor fertility are preferred. The extent of land required for such land use is normally more than that for agricultural use. A list of suitable fruit trees for these environments is presented in table-3.

Table-3 List of fruit trees cultivated under poor (500 mm) and moderate (500 mm to 750 mm) annual rainfall distribution patterns

Scientific name	Common name
Annona squamosa	Custard apple
Psidium guajava	Guava
Tamarindus indica	Tamarind
Zizyphus mauritiana	Ber
Moringa oleifera	Drumstick
Grewia subinequalis	Phalsa
Emblica officinalis	Amla (Aonla)
Carissa carondas	Karonda
Sizygium cuminii	Jamun
Mangifera indica	Mango

Source: Sulladmath, 1983; Ismail, 1999

Similarly, silvipasture systems include trees and grasses in some combination on the same unit of land and, wherever possible, multipurpose trees should be an integral component of the system. The grass species should be tolerant to drought and extremes of temperature, and the fodder should be highly palatable and nutritive. In table 4, some of the feasible tree grass combinations found in different regions of India are presented.

Table-4 : Suitable species for silvipasture systems in different regions of India

Trees/Shrubs	Grasses/Legumes
Hot Arid Desert (Western Rajasthan, parts of Gujrat and Harayana)	
Acacia tortilis, Acacia nilotica, Albizia amara, Eucalyptus terticornis, E. camaldulensis, Prosopis cineraria, Prosopis juliflora	Grasses- *Cenchrus ciliaris, C.setigerus, Lasiurus sindicus* Legumes- *Atylosia scarabaeoides, Stylosanthes Hamata, Clitorea ternatea*
Semi-arid, Rocky & Gravely (Parts of UP, MP, AP and Karnataka)	
Acacia catechu, Ailanthus excelsa, Albizia lebbek, Cassia siamea, Dalbergia sissoo, Dendrocalamus strictus, Prosopis juliflora, Hardwicikia binata	Grasses- *Cenchrus ciliaris, C.setigerus, Panicum antidotale, Chrysopogan fulvus, Dicanthium annulatum* Legumes- *Macrroptelium atropurpureum, Desmodium trifolium, Stylosanthes spp.*
Cold desert (Laddakh-J&K)	
Hippophae spp., Juniperus communis, Juniperus wallichiana, Populas spp., Salix spp.	Hay species- *Pranges pabularia*

Ravine lands (UP, MP, Rajasthan & Gujarat)	
Acacia nilotica, Acacia catechu, Acacia tortilis, Albizia lebbek, Dalbergia sissoo, Dendrocalamus strictus, Prosopis juliflora, Eucalyptus terticornis, E. camaldulensis	Grasses- *Cenchrus ciliaris, C.setigerus, Chrysopogan fulvus* Legumes- *Atylosia scarabaeoides, Stylosanthes gracilis S.humilis*
Cultivable wastelands	
Acacia spp., Albizia spp., Hardwickia binata, Leucaena leucocephala, Sesbania spp.	*Chrysopogan fulvus, Dicanthium annulatum, Pennisetum polystachyuon, Sehima nervosum* Legumes- *Clitoria ternatea, Glycine javanica, Macropetelium atropupureum Stylosanthes spp.*

Source: Faruqui and Kumar, 2009

Another new agroforestry system developed for over more than a decade called hedgerow- intercropping, also called as alley cropping, has been tested and promoted in various regions all over the world. It was developed keeping in view the important role that soil organic matter and other biotic factors play in maintaining the land productivity. Scientists at the International Institute of Tropical Agriculture (IITA) in the 1970s incorporated woody species in crop production systems. This ultimately led to the development of the alley cropping system (Kang *et al.* 1981). In alley cropping, food crops and woody species are intercropped, food crops are grown in the alleys formed by hedgerows of planted trees and shrubs, preferably legumes. The hedgerows are cut back at planting and periodically pruned during cropping to prevent shading and to reduce competition with the food crops. The prunings are used as green manure or mulch. The hedgerows are allowed to grow freely to cover the land when there are no crops. Success of alley farming system depends on: the right choice of woody species; successful hedgerow establishment, and proper hedgerow and crop husbandry. Desirable characteristics of woody species for these types of agroforestry systems are:

- ease of establishment
- coppicing ability
- withstand pruning
- deep rooting

The general practice under alley cropping type of agroforestry world over has been to integrate conventional agricultural crops with different tree species. Although various tree species have been tried in alley cropping, *Leucaena leucocephala* being the most widely studied so far, in all kinds of environments. The choices of tree species as well as management of the hedgerows and alley width are very important, for any successful hedgerow intercropping (Singh, 2004).

Other problems of drylands include high temperatures and hot desiccating winds. These affect the establishment, growth and yield of crops in arid areas. A mixture of trees and shrubs planted across the wind direction help in reducing the wind speed (Table-5). Establishment of windbreaks and shelterbelts help in combating these problems. Shelterbelts of trees like *Acacia tortilis, Cassia siamea, Prosopis juliflora* and *Albizzia lebbeck* can help in reducing wind speed and wind erosion besides improving the microclimate.

Table-5 Design and species for shelterbelt plantations

Purpose	Design	Suitable species
Road side	3-5 staggered rows	*Acacia tortilis, Albizzia lebbeck, D. sissoo, Prosopis juliflora* and *Tamarix articulata*
Railway lines	6 rows	*Prosopis juliflora, P. acculeata, T. articulata*
Canal side	6 rows	*Acacia nilotica, Eucalyptus spp, T. undulata, A. tortilis, P. juliflora, D.sissoo, A nubica, P. cineraria*
Farm boundary (rainfed)	1/2/3 rows	*A. tortilis, A. lebbeck, A.indica. D. sissoo, P aculeate, P. juliflora*
Farm boundary (irrigated)	2 rows	*A. tortilis, A. lebbeck, Dicrostachys cineraria, P. juliflora*

Source:Prasad andMertia (2009).

Conclusions

Agroforestry is a promising low investment option for subsistence farmers in the dryland areas of the world. The resiliency of agroforestry makes it a powerful tool for the sustainable management of the drylands. Dryland agriculture faces specific problems that must be overcome both for improved agricultural production and environmental conservation. In general, agroforestry produces a multitude of benefits, including increased income from selling surplus goods, improved environments and enhanced biodiversity, increased access to education through income generation, shelter from harvesting trees for building materials, improved health when medicinal plants are incorporated, and a greater diversity of food sources. These benefits are desirable in creating a sustainable agro-ecosystem in the drylands of the world. In the drylands, there are a number of indigenous agroforestry systems. The existence of these systems has a great potential for further development and the introduction of new agroforestry systems. However, except for a general description, no in depth study has been carried out. It needs focus on the evaluation, improvement and encouragement of sustainable traditional agroforestry systems. A lot of trees suitable for dryland agroforestry have been screened and tested but there are many others that have hardly been investigated. The International Agroforestry Centre ranked the following as priorities for domestication in the Sahel: *Adansonia digitata, Vitellaria paradoxa, Parkia biglobosa, Tamarindus indica* and *Zizyphus mauritiana* (Leakey 1999). Screening of indigenous trees for their inclusion in drylands as well as the design and management of new agroforestry systems with meticulous understanding of the nature of interactions among components of the system are the key factors influencing the system productivity and sustainability. Efforts should also be made to popularize some of the potential systems among the farming community. For designing and developing new agroforestry systems for these areas, socioeconomic study of the intended users is a major necessity to develop viable agroforestry systems for the drylands. It has been established that food security and income are among the primary motivations that influence the farmers to adopt certain agroforestry system (Brown, 2003).

The comparison of traditional and new agroforestry does not intend to judge which option is the better, but aims to understand the strengths and weaknesses of each system, to work out the interaction and complementarity between the two. Like anywhere else, the acceptability of agroforestry systems and practices in the dryland/rainfed environments will improve if interactions that exist between trees, crops, and/or livestock remain largely beneficial; the negative interactions stay minimized, so that productivity per unit area of land is increased whilst reducing environmental risks associated with monocultural systems.

Reference:

Bhandari, D.S., Govil, H.N. and Hussain, A. 1979. Chemical composition and nutritive value of Khejri (Prosopis cineraria) tree leaves. *Annals arid Zone.* 18: 170-173.

Brooks, K.N., Ffolliott, P.F., Gregersen, H.M. & DeBano, L.F. 2003. *Hydrology and the management of watersheds.* 3rd ed. Ames: Iowa: Iowa State University Press. pp. 373-384.

Brown, D M. 2003. Considering the role of landscape farming system and the farmer in adoption of trees in Claveria, Misamis Oriental province, Philippines, MS Thesis, University of London.

Depommier D., 2003. The tree behind the forest: ecological and economic importance of traditional agroforestry systems and multiple uses of trees in India, *Tropical Ecology*,44(1): 63-71.

Fagg, C.W. & Barnes, R.D. 1990. *African acacias: study and acquisition of the genetic resources.* Final Report on ODA Research Scheme R. 4348. Oxford Forestry Institute, Oxford.

Faruqui, S A and Sunil Kumar. 2009. Meeting challenges in forage production through agroforestry in India. In: *Agroforestry: Natural Resource Sustainability, Livelihood & Climate moderation.* (Eds OP Chaturvedi, A Venkatesh, R S Yadav, Badre Alam, R P Dwivedi, Ramesh Singh & S K Dhyani). Satish Serial Publishing House, Delhi. pp. 757-775

Garrity, D.P. 2004. Agroforestry and the achievement of the millennium development goals: *Agroforestry Systems* 61 p. 5-17

Hocking, D. 1987. Acacia albida - the farmers choice for semi-arid and arid zones. Nitrogen Fixing Tree Association NFT Highlights 98-02, Waimanalo, Hawaii, USA

Ismail S. 1999. Agrihorticultural systems for drylands. In: *Sustainable Alternate Land Use Systems for drylands* (Eds R P singh and M Osman). Oriental Enterprises, Dehradun.

IUFRO. 2004. Trees, Agroforestry and Global change in Dryland Africa (TACCDA). Proceedings from VITRI/ETFRN/IUFRO workshop, 31 July- 4 August 2003, Hyytiala, Finland.

Kramer P J, Kozlowski T T. 1979. *Physiology of woody plants.* New York: Academic Press

Kang, B.T., Wilson, G.F., and Sipkens, L. 1981. Alley cropping maize *(Zea mays L.)* and leucaena *(Leucaena leucocephala* Lam.) in southern Nigeria. *Plant and Soil* 63: 165-179.

Lassoie J P and Buck L E.2000. Developing of agroforestry as an integrated land use management strategy. In:, *North American Agroforestry: An Integrated Science and Practice* (Eds. Garrett H E, Rietveld W J and Fisher R F.), American Society of Agronomy, Madison W I, USA, pp 1-30

Leakey, R.R.B., Wilson, J. and Deans, J.D. 1999. Domestication of trees for agroforestry in drylands. *Annals of Arid Zone* 38, 195–220.

MA (Millennium Ecosystem Assessment), 2005, Ecosystems and Human Well-being: Desertification Synthesis, World Resources Institute, Washington, DC.

Michael Arnold J.E. and Dewees P. A .1999. Trees in managed Landscape; factors in farmer decision making. In: Agroforestry in sustainable Agricultural System (Eds. Louise E., Buck, James, P., Lassoie and Erick C.M.. Fernandes. CRC Press. USA, Pp. 277-294.

Prasad R. and Mertia R. S., 2009. Impact of tree shelterbelts in sustainable management of arid farming system. *In: Agroforestry: Natural Resource Sustainability, Livelihood & Climate moderation.* (Eds OP Chaturvedi, A Venkatesh, R S Yadav, Badre Alam, R P Dwivedi, Ramesh Singh & S K Dhyani). Satish Serial Publishing House, Delhi. pp. 447-460

Rawat, G.S. 1995. Neem (*Azadirachta indica*) -Nature's drugstore. *The Indian Forester* 121: 977-980.

Roupsard O., Ferhi A., Granier A., Pallo F., Depommier D., Mallet B., Joly H. I. and Dreyer E.. 1999. Reverse phenology and dry-season water uptake by Faidherbia albida (Del.) A. Chev. in an agroforestry parkland of Sudanese west Africa. *Functional Ecology* 13: 460–472

Saxena, S.K. 1997. Traditional agroforestry systems in western Rajasthan: *In: agroforestry for sustained productivity in arid regions* (Eds. Gupta and Sharma), Scientific Publisher, Jodhpur. 21-30.

Sehgal S., Mohd. Saleem, Sood K K and Raina N S. 2008. Agroforestry for poverty alleviation, soil conservation and improving productivity in the drylands of J&K. Abstract published in National Symposium on Agroforestry Knowledge for Sustainability, Climate moderation and Challenges Ahead (15-17 December, 2008) NRC-Agroforestry, Jhansi, U.P.

Shankarnarayan, K.A., Harsh, L.N. and Kathju, S. 1987. Agroforestry in the arid zones of India. *Agroforestry Systems,* 5: 69-88.

Singh, R P. 2004. Sustainable agroforestry management options for rainfed lands in arid and semi-arid regions of India. *Indian Journal of Agroforestry* 2:1-9

Singh, K. S. and Lal, P. 1969. Effect of *Prosopis spicigera* (or *cineraria)* and *A cacia albida* trees on soil fertility and profile characteristics. *Annals of Arid Zone* 8: 33-36.

Sulladmath, U V. 1983. Fruit crops of dryland agriculture. Presented at the Seminar cum workshop on Dryland Development, 29-30 Oct.1983. Karnataka State Department of Agriculture and University of Agricultural Sciences, Karnataka, India.

Tejwani. 1994. *Agroforestry in India*. Oxford & IBH Pub. Co. New Delhi

Tewari, J.C., Bohra, M.D. and Harsh, L.N. 2000. Structure and production function of traditional extensive agroforestry systems and scope of intensive agroforestry in the Thar desert. *Indian Journal of Agroforestry*, 1(1), 81-94.

UNDP. 1997. Aridity zones and dryland populations : An assessment of population levels in the world's drylands. UNSO/UNDP, New York.

Vandenbeldt, R.J. and J.H. Williams 1992. The effect of soil temperature on the growth of millet in relation to the effect of Faidherbia albida trees. *Agricultural and Forest Meteorology* 60: 93–100.

Wickens, G.E. 1969. A study of Acacia albida Del. (Mimosoideae). Kew Bulletin 23, 181–202.

Young A. 1989. *Agroforestry for soil conservation. Wallingford, UK: CAB International.*

CHAPTER

17 Beekeeping: A Promising Enterprise in Dryland Farming

D.P. Abrol

Professor and Head, Division of Ag. Entomology, Faculty of Agriculture, Sher-e-Kashmir University of Agricultural Sciences and Technology of Jammu, Chatha, Jammu-180009.

The state of Jammu and Kashmir (32^0-17′ to 37-05′ N latitude and 72^0- 40′ to 80^0-30' E longitude) represents one of the most important beekeeping areas in India. At least four agroclimatic zones ranging from low altitude subtropical, intermediate, temperate and cold alpine occur. Temperatures range from -45^0 to 45^0 and above. Such diversity of geographical features plays a dominant role in determining the topography, climate and plant species present in the region. It offers great potential for both migratory and non-migratory beekeeping. Modern intensive agriculture with its diversified cropping patterns and orchards is becoming increasingly popular in the Jammu and Kashmir state. The current agricultural transformation, once linked to apicultural operations, offers much scope for income generation through beekeeping. The state has about 75% cultivable land area totally rainfed. Per unit productivity is low. It is further affected by small land holding, climatic variations and poor types of soil. The marginal and landless farmers, constituting half of total rural population are the worst effected people having to resort for menial labour job during non-agriculture period. The advantage of Jammu division is the availability of a number of subsidiary occupations to the main agriculture not involving any land resources. The major among these are beekeeping. Beekeeping is being done by 600 families producing 700 tonnes of honey having a value of about Rs 3.5 crores. The state has a potential of 600000 colonies, which would increase the production by up to 5 folds and can sustain 12000 families. Another important advantage of beekeeping is that these bees act as excellent pollinators leading to increased seed/fruit formation in cultivated crops, horticulture etc.

In dry hill farming system, the honeybee colonies can be sustained in forests. As the honey bees are found associated with forests globally. Flowers of forest trees provide subsistence for honey bees and the trees physically provide shelter for a swarm or bee hive. Forest management and beekeeping each have a long history both in the India and globally, but have seldom been integrated or studied in a systematic fashion. Purposeful plantings of trees, as in agroforestry systems, could be designed to favour bee forage or hive protection. Tree growing and beekeeping can easily be combined for several reasons. Both are sustainable on land that is hilly or otherwise less desirable for other agricultural purposes. Both can be sustained while the grower/beekeeper is busy with other farming occupations. Bee hives require very little space, while the bees themselves can forage in a radius of 4 to 5 km. Hives may be located within or near a tree plantation, and utilize both the trees and surrounding other flowering plants for forage. Combining forestry and beekeeping provides annual honey bee products (e.g. honey, beeswax) to supplement income from a landowner's long term forest managements. Combining bees and trees is one way of accomplishing this goal. This paper addresses several important known bee-tree interactions which need more systematic study.

Honeybees play an important role as natural pollinators for a wide variety of crops as well as plants growing in the wild. While visiting flowers to collect nectar, the bees transfer pollen from one flower to another. Three quarters of the world's cultivated crops are pollinated by different species of bees, and honeybees are the most effective and reliable pollinators (Abrol, 1993, 2009). They also play an often unrecognized role in maintaining the vegetation cover: more pollination means more seed, more young plants and eventually more biomass, providing food and habitats for birds, insects and other animals.

Beekeeping helps to create sustainable livelihoods

Beekeeping is crucially important for agricultural well-being; it represents and symbolizes the natural biological interdependence that comes from insects, pollination and production of seed. Useful small-scale efforts to encourage beekeeping interventions can be found throughout the world, helping people to strengthen livelihoods and ensuring maintenance of habitat and biodiversity. Beekeeping fits in well as a subsidiary activity alongside many other livelihood endeavors because bees are capable of harvesting nectar and pollen in plants in the same natural resources which otherwise go waste (Abrol, 1997). There is no competition with other insects or animals for these resources that otherwise would be inaccessible to people. Beekeeping ensures the continuation of natural assets through pollination of wild and cultivated plants. Flowering plants and bees are interdependent: one cannot exist without the other. As bees visit flowers, they collect food and their pollination activities ensure future generations of food plants, available for future generations of bees and for people too. It is a perfect self-sustaining activity. Pollination is difficult to quantify, but if it could be measured it would be the most economically significant value of beekeeping.

Beekeeping helps to sustain the natural resource base. Throughout the world, beekeeping has traditionally been part of village agriculture. Now, as farming practices change, it is essential to ensure that beekeeping is retained and encouraged in order to provide continued populations of pollinating insects.

Beekeeping can add to the livelihoods of many different sectors within a society including village and urban traders, carpenters who make hives and stands, tailors who make veils, clothing and gloves and those who make and sell tools and containers.

Beekeeping outcomes

Beekeeping produces a number of quite different outcomes.

- Pollination of flowering plants, both wild and cultivated, is vital for continued life on earth. However, this essential process is difficult to quantify.
- People everywhere like honey, the best-known beekeeping product. Honey is a traditional medicine or
- Food in most societies. Whether sold fresh at village level or in sophisticated packaging, honey generates income and can create livelihoods for several sectors within a society.
- Beeswax is a valuable product of beekeeping, and much of the world's supply comes from developing countries.
- Beekeeping products such as pollen, propolis and royal jelly can be harvested and marketed, although special techniques and equipment are needed for some of these products.
- Beekeepers and other community members can create assets by using honey, beeswax and other products to make secondary products such as candles, skin ointments and beer. Secondary product brings a far better return for the producer than selling the raw commodity. This work strengthens people's livelihoods.
- Products of beekeeping are used for apitherapy in many societies.
- Honey, beeswax and products made from them, such as candles, wine and food items have cultural value in many societies and may be used in rituals for births, marriages, funerals and religious celebrations.
- Beekeepers are generally respected for their craft. Bees and beekeeping have a wholesome reputation. Images of bees are used as symbols of hard work and industry, often by banks and financial institutions.

These outcomes are real and they strengthen people's livelihoods, even though some of them cannot be fully quantified. Beekeeping helps people to become less vulnerable, strengthens their ability to plan for the future and reduces the danger that they will slip into poverty in a time of crisis, for example, especially at a time if a family member becomes ill or crops fail.

Honeybees produce honey and other hive products (e.g. pollen, propolis, beeswax, royal jelly) for which there are unsatisfied local and export markets. Honeybees pollinate flowers, both wild and cultivated, and are crucially important to the environment and commercial crop production. The role the honeybee plays in ensuring the diversity flora and fauna cannot be over emphasized. Honeybees play a more important role than mere honey producers. They assist in the production of a wide variety of commodities, such as apples, melons, cucumbers, almonds, flax, sunflowers, and clover. These and other fruits, vegetables, tree nuts, and field, seed, and forage crops require, or benefit directly from, bee pollination. For agriculture, the value of honeybees as pollinators far exceeds the value of the honey and beeswax they produce. In addition, honeybees are important to the production of plants that provide food and shelter for wildlife, control soil erosion, and beautify the environment. To ensure an ample supply of honeybees for pollination. Because they are kept in observable units (colonies in hives), honeybees may be used to monitor pollution of and/or, changes in the environment. In the future they may be used to distribute fungicides to crops. Beekeeping supports ancillary industries, such as the manufacturers and suppliers of hives and equipment and publishers of beekeeping books and magazines as well as the retailers of products. They keep bees on a part-time basis and come from a wide variety of backgrounds crossing all areas of the social, religious, and political and gender divide. On average they keep five hives, but this varies according to circumstances, from one to over fifty per beekeeper.

Honeybee species

There are four main species of honeybee found in Jammu and Kashmir, i.e., Apis cerana, *A. mellifora, A. dorsata*, and *A. florea. Apis mellifera*, a European species, is becoming. *A. mellifera* and *A. cerana* are domesticated species and serve the purpose of commercial beekeeping.

A. Dwarf Honeybee *Apis florea* Fabricius, 1787

As its name implies, *Apis florea* is the smallest of the true honey bees and is called appropriately the dwarf or the little honey bee. It is the smallest species of honeybee, both in the body size of its workers and in the size of its nest. A nest of *A. florea* consists of a single comb, whose upper part expands to form a crest that surrounds the branch or other object from which the comb is suspended. Dwarf honeybees nest in the open, but not without camouflage: most nests are hung from slender branches of trees or shrubs covered with relatively dense foliage, usually from 1 to 8 metres above the ground. In India it is known commonly as *katua, bhunga, chhoti madhumakhi, chhoti mahal* (Uttar Pradesh), *phulori masa* (Maharashtra), *visanakarra pattu* (Andhra Pradesh). The bee is generally found in plains or low lands in tropics and sub-tropics. It is rarely found in altitudes above 1500 m. The nests are built in bushes, densely leaved small trees in gardens and orchards, eaves of buildings or sheltered boxes or wall niches in urban areas and on closely placed stalks of crop plants like Sorghum (*jowar*). The dwarf bee is able to survive in very hot and dry climates with ambient temperatures reaching 50 °C or more.

The comb architecture is similar to that in other *Apis* species, except for the honey storage portion, that is distinct in *A. florea* comb. The comb shows a distinct honey portion that is situated at the top, and where the support is free from above, the honey cells are constructed around the support. Besides, honey production, *A. florea* can be utilized also for pollination of several agricultural and horticultural crops. The dwarf bee is an important pollinator of crops in hot and dry agricultural plains of India.

Giant or Rockbee, *Apis dorsata* F.

The giant honeybees are found predominantly in or near forests, although at times nests may be observed in towns near forest areas. The bee shares the open air, single-comb nesting habits of *Apis florea*, suspending its nest from the under surface of its support, such as a tree limb or cliff. In general, *A. dorsata* tends to nest high in the air, usually from 3 to 25 meters above the gound. In tropical forests in Thailand, many nests are suspended in Dipterocarpus trees from 12 to 25 meters high: this tree is probably preferred as a relatively safe nesting site because its smooth bark and its trunk rising for 4 to 5 meters before branching out make it very difficult of access to terrestrial predators.

Nonetheless, about three-quarters of the worker population of a colony of giant honeybees is engaged in colony defence, forming a protective curtain three to four bees thick in the same way as *Apis florea.* While birds are common predators of *A. dorsata*, the workers' large body size protects them reasonably well against ant invasion, so that the sticky bands of propolis characterizing the nests of the dwarf honeybee are not found surrounding the nests of *A. dorsata*, nor are the nests hidden by dense foliage. Nests of *A. dorsata* may occur singly or in groups; it is not uncommon to find 10-20 nests in a single tall tree, known locally as a "bee tree". In India and Thailand, tree harbouring more than 100 nests are occasionally seen in or near the tropical forest.

The single-comb nest, which does not have the crest of honey-storage cells typical of *A. florea* nests, may at times be as much as one meter in width. The organization of the comb is similar to that in the other honeybee species: honey storage at the top, followed by pollen storage, worker brood and drone brood. At the lower part of the nest is the colony's active area, known as the "mouth", where workers take off and land, and where communication dances by scouts, announcing the discovery of food sources, take place. This dance takes place on the vertical surface of the comb, and during its progress, the bees must have a clear view of the sky to observe the exact location of the sun. Workers of *A. dorsata* are however able to fly at night, when the light of the moon is adequate.

A. dorsata is well known for its viciousness when its nest is disturbed: the mass of defending workers can pursue attackers over long distances, sometimes more than 100 meters. Notwithstanding its ferocity, however, this bee's honey is highly prized locally, in some places commanding the best prices in local markets, Rockbees can forage even during moonlit nights (Diwan and Salvi 1965). Its flight range is more than 5 km (Koeniger and Vorwohl 1979, Koeniger and Koeniger, 1980). In the normal forage conditions they have been observed to visit sources 2 to 3 km away from the nest. The bees have an average tongue length of 6.683 mm (Ruttner 1988). These factors provide a large foraging range, both in area and variety of plant species, to the rockbees. An annual production of about 2500 tons of beeswax from the wild rock bees was reported in 1969. Rockbee is an important pollinator of several crops.

Eastern honeybee, *Apis cerana* Fabricius, 1793

Apis cerana, or the **Asiatic honeybee** (or the **Eastern honeybee**), are small honeybees of southern and southeastern Asia, In the wild, the oriental honeybees construct their multiple-comb nests in dark enclosures such as caves, rock cavities and hollow tree trunks. The normal nesting site is, in general, close to the ground, not more than 4-5 meters high. The bees' habit of nesting in the dark enables man to keep them in specially constructed vessels, and for thousands of years *Apis cerana* has been kept in various kinds of hives, i.e. clay pots, logs, boxes, wall openings, etc. Despite the relatively recent introduction of movable-frame hives, colonies of *Apis cerana* kept in traditional hives are still a common sight in the villages of most Asian countries. As a result, the feral nests of the oriental honeybee in tropical Asia sustain fewer casualities in being hunted by man than those of the dwarf and giant honeybees. The several combs of an *A. cerana* colony are built parallel to each other, and a uniform distance known as the "bee space" is maintained.

Natural nests of the Indian hive bee occur in tree trunks, rock crevices, ant hills, underground deserted nests of white ants, or any dark enclosure, sometimes even in the open, but quite dark spaces in forests or unused rooms in buildings. The bee occurs in a wide range of geographical areas from tropical coastal areas to the temperate Himalayan ranges at about 3000 m altitude. Colonies are found in forests or agricultural areas in the plains, and even in urban areas with good vegetation. Bees are larger than the dwarf bee, but are much small than the rockbee. This bee is similar in size to the European hive bee in similar latitudes. Two varieties of the bees had been usually recognized - the hill variety, larger and darker, usually dark brown in colour, with a cell size of about 4.8 mm, occurring in the hills and mountains; and the plains variety, smaller and lighter, generally reddish yellow in colour, with a cell size of about 4.15 mm, occurring in the plains. The hill variety is believed to be more aggressive, is more populous, builds larger nests and produces more honey than the plains variety.

The nest consists of several parallel combs with an uniform distance between them. The nests have usually 6 to 8 combs. A wide variation occurs in the number depending upon the period of stay of the nest in the location, space available in the nest site and its shape. Sometimes only 3 to 4 combs which are narrow but about a metre long are found. In natural nests that lived for over 2 years upto 15 normal sized combs were found, that yielded over 10 kg of honey.

The Indian hive bee does not use propolis as the European bees do. This may be an adaptation to tropical climate, where hive ventilation assumes importance. The cracks in the floor board or gaps in the hive walls or frame joints are not sealed. This may attract pests like wax moth. One of the characteristic features of the hive bee is fanning used for ventilation of the hive. During nectar flows large quantities of water have to be removed from the dilute honey in combs and ripen it. The moisture-ridden air has to be removed from the hive. For this purpose bees undertake fanning vigorously, and it is most visible at the hive entrance. The Indian hive bee fans with its head facing away from the entrance. Contrastingly, the European bee fans with its head towards the entrance. In their experiments on introducing queens of the European bee into the colonies of the Indian bee, Dhaliwal and Atwal (1970) observed the workers of both the species fanning side by side, but heads oriented in opposite directions.

Potential of the Indian Hive Bee

- *Apis cerana* is gentle to handle, industrious and well adapted to the ecological conditions of south and Southeast Asia.
- It is less susceptible than *A. mellilfera* to nosema disease, not seriously affected by *Varroa* and is less prone to the attack of predatory wasps.
- To control diseases, parasites and predators, beekeeping with *A. mellifera* requires chemical treatment of colonies. Chemicals are not required in beekeeping with *A. cerana*
- The variety of geographical races/populations of *A. cerana* that exists in south and Southeast Asia provides excellent opportunities for the genetic improvement of this native species through selective breeding.
- Through genetic engineering techniques it may be possible to introduce desirable genes from *A. cerana* into *A. mellifera.*
- *A. cerana* is sympatric in distribution and can co-exist with the two other species of Asiatic honey bees, *A. dorsata* and *A. florea,* without any adverse ecological consequences.
- For pollination purposes, *A. cerana* is superior to *A. mellifera* in certain aspects, e.g., it is more suitable for cross-pollinating entomophilous crops grown in the small holdings of this region because of its shorter flight range and longer foraging hours than the European honey bee. Use of bee hives for pollination of agricultural and horticultural crops is another field that is gaining importance in recent years. There is an increasing demand for bee colo-

nies by orchardists growing apples, litchi, lemon, other citrus fruits, producers of seed of cucumber and other cucurbits, cole crops, spices, onion and vegetable crops and flower seeds, as also by farmers growing sunflower. Payment by the farmer to the beekeeper for pollination service is also becoming a common practice. Bee colonies migrated to farm and orchard areas, to tide over adverse periods, can be utilized for crop pollination. Such migrations can be doubly beneficial to the beekeeper. The overall income from apiculture can thus be quite substantial. Besides honey, the hive bees can produce beeswax, pollen, royal jelly and bee venom. Technologies exist for beeswax, pollen and bee venom production. For royal jelly too, suitable technology can easily be developed. Beekeeping for package bee production, queen rearing and supply, breeding and supply of improved strains of bees, and similar non-traditional products can be introduced. Efficient management methods for production of these items can augment the productivity of the bees.

D. The European, honeybee *Apis mellifera* Linnaeus, 1758

The **Western honeybee** or **European honeybee** (*Apis mellifera*) is a species of honeybee comprised of several subspecies or races. European bees were successfully introduced into India in 1965. The European bee is similar to the Indian hive bee in its biology, nesting, foraging, colony defence and other behaviour features, with minor differences. According to Chahal et al (1995), the European bee possesses definite superiority over the indigenous bee in Punjab. It can colonize areas where the indigenous bee is not present or cannot do well. It can yield 4 - 5 times as much honey as the Indian bee. Chahal listed some features of the two species that have economic implications in beekeeping (table 3.8). However, the performance of *A. mellifera* may not be the same in other agro-climatic conditions. For example, *A. mellifera* doing well in Punjab not may not survive the new climatic and vegetational conditions in other parts of the country.

Colony organization of honeybees

The bee colony

Honey bees live in a home of wax comb. These six-sided wax cells are very strong and house the brood (immature bees) during development and provide storage space for honey and pollen. In nature bees usually live in a sheltered cavity, such as a hollow tree or rock crevice. The colony is composed of a queen, drones, and workers.

The Queen

There is only one queen bee in the colony (family). As mother of the colony, her purpose in life is to lay eggs. She may lay several hundred eggs (approximately 2000) in one day. These eggs may hatch into drones (males), workers, or new queens. The queen can determine type of egg she is going to lay. She lays only the type that she feels the colony needs. It takes sixteen days for queen to develop from an egg into an adult. About the seventh day after hatching, the queen flies from the hive and mates with one or more drones. This is the only time in her life that the queen mates, though she may live four to five years. The queen is larger than the worker and longer than the drone. Her wings are shorter in proportion to her body length than those of the drone or worker. She has a long tapering abdomen. When undisturbed, a mated, laying queen will usually be found on or near the comb containing the eggs in the hive.

The Drone

The number of drone bees in a colony varies seasonally. There may be none when the bees have little food, but up to 1000 during the honey collecting season. When the honey season is over and food and water become scarce, the drones are expelled from the hive.

It takes 24 days for drone to develop from an egg into an adult. The drone does not work in the hive. The duty of the drone that is only male in the hive is to mate the queen and it dies after mating with her. Drones are larger and fatter than the queen or the workers. Their bodies are not long as the queen's. The drone has a short tongue he uses to take food from workers and from stored honey in the hive. He does not have legs fit to carry pollen and he is unable to produce wax. He has no stinger to defend himself.

The worker

There are 5,000 to 75,000 worker bees in a colony. They do the entire house and field work. Some workers go out of the hive to bring in water, pollen, nectar, and propolis (bee glue). Other workers remain in the hive to guard against the enemies. Still others clean the hive, build wax comb, nurse the young, and control the temperature of the hive. Workers eat honey to produce heat in cold weather and fan their wings to keep the hive cool in the hot weather.

It takes 21 days for a worker to grow from an egg into an adult. During the honey-collecting period, workers have special legs equipped with pollen baskets. They also have glands that produce wax and the scent necessary for carrying out their many duties. Workers are smaller than either the drones or the queen. They have the stinger which when it stings the stinger remains behind and the bee dies.

Where to set up an apiary

It is well known that bees can be kept anywhere. However, if the interest is to increase honey yields and profit margin, the place where hives are placed is paramount to the beekeeper. They must be kept away from the public or a place where they cannot sting anyone. Bees also require **food sources** that are *nectar* and *pollen* therefore they must be able to find their food sources within their vicinity - at least 2km. While bees can fly many kilometers to look for the food, this is uneconomical when it comes to honey production. The shorter the distance for the bees the more they collect the food.

The apiary must have **good air drainage** (good air circulation). This means it must have a good airflow. The hives must be **protected** from **strong wind,** which may cause drifting of bees. It must also be protected from **hot sun** by providing a **partial shade**. Bees need **water** and must be able to find the water within 500m. Sometimes water can be provided in containers and to prevent bees from drowning, sticks or stones can be provided for landing and taking off of bees.

To protect the hives from ants and termites, no weeds must be allowed to grow around the hive as they form the bridge for the ants to reach the hive. The scent from weeding usually upset the bees; hence the apiary must be well prepared in advance. During dry spell, a firebreak can be made around the apiary to prevent the hives from being burnt. The apiary must be kept clean and tidy all the time.

Inspecting the colony

The best time for inspecting the colony is a bright, sunny day when the bees are working normally. Bees should not be disturbed on cold, rainy, or windy days or at night. When inspecting the colony, light the smoker and approach the hive from the side to avoid blocking the bees' entrance. Smoke a bit on the entrance holes especially the busiest. Lift the lid and smoke on the surface and place back the lid. Then it should be removed after a short time and placed upside down. The top bars should be loosened part with the knife, taken out, and examined one by one. The top bars should be handled carefully and always holds the combs vertically to avoid them breaking as shown below. As you inspect the colony, always be mindful of the queen. The top bar where she may be located must be put back as soon as possible as losing the queen would be tantamount to murdering the colony. Therefore, the top bars must be handled with care and they should be no crushing of bees. Always smoke reasonably after inspecting each top bar to calm the bees.

After you have finished the inspection of the top bars, return all the hive parts to their original positions in the same order as they were taken to maintain the structure of the brood nest. Movable top bars soften the job of checking of combs in a hive as each can be lifted and turned around. While inspecting the hive, all the pest and insects must be removed.

Management of honeybee colonies

Once the hive is occupied and the bees are busy, then the following simple basics must adhered to:

- Do not stand in the flight path of the bees.
- Work quietly without excessive talking or drumming noises.
- Work quickly but smoothly. Remove lid carefully and puff smoke gently around the entrance of the hive.

- Remove a few empty bars to create a gap at one end of the hive. This should not disturb the bees. Thereafter, remove one bar at a time. Smoke the gap gently and hold the bar vertically so as not to break off the comb.
- Keep the bars in the same order and try not to squash any bees when replacing them in the hive. Squashed bees release a smell (alarm pheromone) that sets other bees on the attack.
- Do not visit the hive in the warm part of the day - about six o'clock in the evening is a good time.
- Do not try and work with too many hives at any one time. Certainly for not more than 45 minutes in an apiary as bees from the first hive worked on will be agitated and attack which will lead to further commotion amongst all the bees.
- Always wear light coloured clothes. Ideally protective clothing should be worn, especially a veil to protect the eyes and face.
- Make sure the top bars are pushed together as they are replaced, so that no gaps exist. Finally, gently replace the lid on the hive.
- Always keep the grass cut and the area around the hive tidy.

Success in beekeeping mainly depends on the sound knowledge of biology, behavious and accordingly the management practices. Nectar and pollen are not available to bees throughout the year. However, during some parts of the year surplus food is available, minor and subsistence food is available during other periods, whereas bees may face dearth period for certain parts of the year. We also have different seasons in a year with greatly varying weather conditions and the weather at times may be hard for bees. A beekeeper must handle his bees in such a manner that the colonies are well prepared for the coming honey-flow. This has to be done by helping the bees in successfully abridging the dearth period and by reducing the effect of severe weather conditions.

Spring Management

In hills the winters are severe and there is lack of flora, the bees remain confined to their hive for most of the time. The bees have also been facing the problem of maintaining nest temperature. At the onset of spring the colonies emerge in considerably weak condition. It should be the earliest attempt to examine the colonies on a bright, warm and calm day to assess the condition of the colony, working of the queen, amount of brood present, honey and pollen stores and to clean the bottom board debris accumulated during winter. It is a useful management to give a stimulant feeding to colonies when very few spring flowers have blossomed. The stimulant feeding with this sugar syrup (30-40%) will help the colonies to rear more brood and raise greater foraging force to avail spring and summer flow. This feeding also raises the morale of the bees. It is an established fact that colonies which receive stimulant feeding produce more honey. The examination should be done carefully and quickly because robbing is easily induced. All manipulations should be stopped once the robbing becomes apparent and reduce all entrances so that colonies are able to guard against the invasion of robbers.

During spring, bee colonies go 'all out' to rear brood and invest all resources in increasing their strength. Queen lays more vigorously after winter egg laying rest. More drawn combs are added for expanding brood next. If the queen is working unsatisfactorily, that is, she is laying sparingly and or laying drone eggs, efforts must be made to replace the queen at the earliest opportunity. It may be pointed out that very weak colonies desert their hives if disturbed unnecessarily and are easily robbed out if they are fed sugar syrup without great care. It is a wise practice to unite them with others and follow the golden rule, 'always help first those colonies that need the least help, leaving the weakest to be helped the last'. During early spring the weather is unsettled and beekeepers are warned against over expanding the brood nest and dividing it into two or more parts by insertion of empty combs or comb foundations because there can be chances of the outlaying brood being left unattended by the worker bees. Consequently the neglected brood gets chilled because of bad weather. Spring is also a swarming season. Swarm prevention and control measures should be taken.

Honey flow period

The swarming season is followed by good honey flow season. Proper management of the bees is essential during these days. It is true that any amount of diligence during the honey flow cannot make amend for the poor management during the preceding months but any oversight during these crucial days is certain to affect the year's work of the bees and the beekeeper. The principle function of the beekeeper during honey flow period is to keep the colony morale high. In other words, he should ensure that the honey - gathering instinct is dominant and that the instinct is not checked. Congestion in the hive must be avoided and surplus house bees are drawn to supers (second story).

Many times the queen goes to the super chamber and lays eggs and honey extraction becomes difficult. At least three weeks before honey extraction a queen excluder should be placed in between brood and super chamber and queen is confined to brood chamber. Apis mellifera (European bee) attains sufficient strength before honey flow and full depth supers on langstroth hives are used in India, while in valley Langstroth hive is used even for Indian honey bee (*Apis cerana*) . In most parts of valley Newton hive is still in use, where half super is the frames which are three-fourth filled with honey or pollen and one-fourth with sealed brood should also be taken out of brood chamber and in its place empty combs or frames with foundation is added. On warm days, bees are noticed to gather in clusters at the entrance. This is a sign of congestion and poor ventilation. This affects the honey gathering instinct. The situation should be remedied promptly by improving ventilation by removing the entrance rod or shaving the supers backward.

Honey Extraction

When the honey flow begins to slow down, the frames containing honey should be removed for extraction. The honey should finally be extracted when bees are still bringing the nectar. To remove honey combs, a colony is smoked, the desired combs taken out and bees brushed off with a soft brush or a bunch of green grass. Depending upon the strength, 5-15 kg of honey should be left with the colony of *Apis mellifera* for summer and monsoon dearth periods. After the job has been done, the place should be swabbed with water and the appliances cleaned. Hive bodies are washed to remove honey drops. The empty wet combs should be returned to the bees for cleaning and the hive entrances be reduced to avoid robbing.

Summer Management

The honey flow is followed by a summer dearth period. With the ceasing of honey flow season develop a strong tendency to protect their stories and become nervous and excited. Bees start throwing out drones and are not allowed to return because they are now useless in the colony. Unmanaged colonies stop brood rearing in order not to starve in future. The best thing to do during this period is to avoid broodlessness in colonies and stimulate them to rear brood. The strong colonies with sufficient store would continue to rear some brood during summer. The colonies cannot be kept in the open in the sun. These should always be shifted to a place with thick shade. Beekeeper can further help bee colonies by placing them under open straw huts. At places the temperature rises as high and gunny bags, moistened twice at noon and in the afternoon, can be spread over top covers. Frequent examination of colonies should be avoided. The principle is 'All efforts should be made to add to the comforts of bees.'

Winter Management

Honey bees live in a thermal environment of their own and maintain colony temperature between 32 and 35 ^{0}C. The bees form a cluster when the atmospheric temperature drops below 100°C. This roughly spherical cluster becomes tighter as the temperature drops further down. The cluster is composed of inner and outer shell, the latter serves as insulation and prevents the heat loss. Bees change their places between the inner and the outer shells. Bees raise the temperature by muscular movements which is possible by the consumption of honey. The colonies should be kept under damp places and under thick grooves of dry grass. Periodic inspection to verify the food stories is essential in winter. Insulation of hives helps to reduce consumption of honey and saves energy of bees. In the cold months namely December, January, February with days and night chilly, need light insulation. Finely chopped dry grass, wood shavings, saw dust, dry leaves, chopped rice straw or wheat straw are the handy packing materials though thermocol and woolen rugs can also be used. Under valley conditions packing can be given in brood chamber, if the colony material is kept in place with strings or water proof tar paper wrapped over the packing material.

Table 1. Comparative morphometric, behavioural and economic characteristics of *Apis mellifera* and *Apis cerana*

Characteristics	*A. mellifera*	*A. cerana*
Body weight (mg)	90 -120	50 -70
Tongue length (mm)	5.7 -7.2	4.39 -5.53
Nectar load (mg)	40 -80	30 40
Pollen load (mg)	12 29	7 -14
Flight range (km)	2 -5	0.8 -2
Egg laying capacity of queen per day	800 -1800	300 -800
Colony build up at honey flow	40,000 -60,000	25000 -30,000
Swarming	Little	High tendency
Absconding	Very little	Very high tendency
Aggressiveness	Usually calm	Mostly furious
Yield under Indian conditions (kg/colony)	25 -30	4 -5

Source: Chahal *et al* (1995)

An important feature in which *A. mellifera* differs with *A. cerana* is its use of propolis. Propolis is used to seal cracks and crevices in the nest and make the hive weather proof. The frames or the inner cover is often joined to the hive body with propolis. This makes inspection or management of colonies difficult. Some races are heavy propolizers. Special techniques have to be adopted to handle the bees without unduly disturbing the nest. Propolis has antibacterial properties. It has several medicinal applications. With the introduction of beekeeping with *A. mellifera* in India it is now possible to develop suitable technology for production of propolis.

Beekeeping is recognized as a low input and high output activity, suitable for rural, tribal and other weaker sections of population. The peculiarity of this agrobased industry is that it does not require any raw material from the artisans like other industries. The raw material is in the form of nectar and pollen from flowers which is freely available in nature. Beekeeping is a decentralized, forest and rural agrobased industry. Beekeeping can be started by any one who takes interest may be skilled, unskilled, man, woman, old and young, working or retired persons. A technology that is simple, easy accessible and at the same time demanding the least capital investment is suitable to this type of industry. Beekeeping may even be started with a single colony which can be increased to thousands of them. It can provide unemployed and underemployed persons with full employment.

Beekeeping requires very little inputs. A bee box is the only essential requirement to start the industry. The output is comparatively quite high. For the illiterate poor in remote areas of the country, this could be an ideal occupation that could be taken up by them, besides their traditional agriculture and forest related main occupations, if any. Traditionally, only few people are involved in beekeeping, perhaps due to ignorance. Beekeeping requires light labour, no permanent infrastructure and exploits orchard flowers beneficially for extra income generation.

Farm women, men can be trained successfully within a week in beekeeping operations such as hiving bees, bee swarms, occasional feeding, division of colonies by mass rearing of queens, uniting the colonies, queen introduction, prevention of absconding, swarm control, and honey extraction. A household can start with as many as keep as many as 10 bee colonies would cost around ten thousand rupees. Peak average production in the second year would annually be 20kg of honey per colony with an additional five kg of beeswax. The initial investment would easily be recovered during the year following the establishment of the bee colonies.

Honeybees perform several ecological and economic functions without competing for scarce land resources. The complementarity of beekeeping with horticulture is scientifically proven, but so far it has not been widely adopted in mountain areas. Beekeeping is also an environmentally friendly activity and it helps promote off-farm employment and high income opportunities.

Stationery beekeeping, due to the limited period of the flowering season in given locations is a constraint to unhindered production of beehive products. This problem of localised/stationery beekeeping is overcome by introducing migratory/mobile beekeeping. There are proven cases, where honeybees have been used as a promising source of off-farm employment and to overcome the seasonal variation of mountain areas through mobile beekeeping.

Why is beekeeping an appropriate activity for particularly disadvantaged poor rural people in remote mountainous areas?

- It requires no land.
- It requires minimal inputs when appropriate technology and indigenous bee strains are used.
- It is environmentally positive for pollination of wild flowers and trees.
- It is not very time consuming.
- It is generally not heavy work, so it can be done by most people.
- It is a sustainable use of forest and pasture resources, actively encouraging forest conservation.
- It is appropriate in areas where agriculture is prevented by mountainous conditions (i.e., sloping rocky land and high altitude conditions).
- It is a source of income generation (through the sale of honey and wax).
- Its main products, honey and beeswax, are of low-volume, high-value and relatively unperishable and hence relatively easy to export from remote mountainous areas.

- It has potential for value-added product manufacture in the remote area itself as a means of extra income over and above the value of the raw material (e.g., candle and hand cream making with wax, honey, medicines, etc).
- It has the potential to benefit communities through livelihood improvement (i.e., keeping honey for home medicinal use, making ones own hand cream for cuts and grazes, improving family nutrition, etc).
- It improves crop yields through pollination services.
- Honey is not only delicious it is also highly medicinal (containing a natural antibiotic) and provides trace minerals and vitamins.
- Bee pollen, which can be harvested, is one of the most nutritious substances known to man, containing all essential amino acids and most vitamins and minerals.
- Bee venom (bee sting therapy) can be used to treat arthritis and other ailments.

Honey is also a sweet base for a number of medicines and bee venom is used in many pharmaceutical applications, especially to cure rheumatic diseases. It is a natural dehydrant and excellent for those on slimming diets. As a proven anti-ageing agent and natural rejuvenator honey has no equal. Honey bee colonies, however can be placed wherever and whenever they are needed. Furthermore, honey bees have additional advantages over other pollinators, as they could be available in large numbers and have instinctive pollen hoarding behavior. In the absence of pollinating service of honey bees, the cost of many fruits, vegetables, legumes, and seeds would be many times more what it is today.

Ancillary industries based on beekeeping for rural development

An ancillary unit is defined as a unit which produces parts, components, sub-assemblies, and tooling for supply against known or anticipated demand of one or more large units manufacturing/assembling complete product and which is not a subsidiary to or controlled by any large unit in regard to the negotiation of contracts for supply of its goods to any large unit. This shall not, however, preclude an ancillary unit from entering into an agreement with a large unit giving it the first option to take formers output

Industries necessary for apiculture

Supply industries

There are a number of specialist industries which operate solely (or predominantly) to supply requisites to the beekeeping industry. While not quantified as a part of the beekeeping industry, their existence is certainly a part of the indirect economic impact generated by the industry. This "supply" sector includes:

i) **Beehive manufacturers** - while only small in number, companies manufacturing hives for the industry tend to specialize in this activity.

ii) **Extractor/uncapping machinery** - any company skilled in the production of stainless steel equipment for the food industry would be capable of manufacturing the machinery required to "uncap" combs, and extract honey.

iii) **Packers' equipment** - bottles and bottling equipment are required by packers. In addition, drum manufacturers provide the special galvanized drum, made with side bung that is not generally available from other sources for producers and packers. A gradual change to plastic drums is taking place, with intermediate bulk containers, of pallet size being used which hold more honey than do the traditional drums.

iv) **Heat source** - every honey producer requires a steam or hot water boiler to generate steam, and hot water, for processing honey and wax. While not requiring a specialized manufacturing activity, the beekeeping industry generates a demand for such equipment.

v) **Transport/handling equipment** - all commercial beekeepers must purchase trucks and utilities for transporting and servicing hives. Many also own front end loaders for loading hives, on pallets, and loading drums or use other forms of mechanisation.

vi) **Other equipment** - beekeepers also have a need for other equipment such as electric generators and mobile extracting units. While the first two items (as for heat sources) do not require a specialized manufacturing activity, expenditure by the beekeeping industry can provide an important source of demand.

vii) **Quality Assurance** - beekeepers and packers are increasingly introducing quality assured premises and equipment for handling honey, as a food product for human consumption. Again, this is not necessarily a specialised activity unique to the industry, but represents an additional demand for existing services.

It is clear that a considerable amount of other economic activity is generated as a result of the activities of the honey bee industry.

Industries dependent on apiculture

Major products of beekeeping are: honey, beeswax. In addition, specialized segments of the industry concentrate beekeeping activities towards the production of: queen bees, package bees - the provision of specialist paid honey bee pollination services for horticultural and agricultural industries. Queen bees and package bees are sold to other beekeepers both within the country and overseas. Honey and beeswax are still produced by queen and package bee producers, but these products are not the prime goal of their beekeeping activities. The economic dimensions of the industry, and the direct and indirect impacts that beekeeping are as given below:

Industry by sector

The overall apiary industry should be considered in terms of a number of sectors, as follows:

i) Honey.

Honey is the prime output of commercial beekeepers, and is produced by bees from plant nectar. The major producers are Russia, China, USA, Mexico, Argentina, Canada, Brazil and Australia. The major exporters are China, Mexico and Argentina, but the highest colony yields are recorded in Australia and Canada which have a favourable environment as well as highly developed colony management. The major consumers and importers are the industrialized countries led by Germany, Japan, USA and UK. The increased consumption over the last few years can be attributed to the general increase in living standards and a higher interest in natural and health products. Western Europe as a whole imported approximately 140,000 tonnes which is about 55% of consumption. The average EU per capita consumption of 600 g per year varies widely amongst individual nations, from Greece with 300 g per capita to Germany with 1,800 g per capita.

In general, light-coloured honeys bring the highest price and dark ones are most frequently used for industrial production. Mild flavoured honeys are preferred, but characteristically flavoured honeys bring top prices in some countries. Large honey packers usually prefer honeys with a low tendency to crystallize. Some unifloral honeys such as Hungarian Black Locust honey bring twice the price of regular, multifloral honey. Small shipments into Switzerland of unifloral honeys such as lavender honey, in most cases already bottled, bring much higher prices. Local prices in most developing countries are higher than the international market prices and prices in neighbouring countries with less honey production or favourable exchange rates may sometimes be quite attractive.

Expansion of markets with honey-containing products should be considered on a national level or for across-the-border trade. Consumer education and of course, spending power will probably be the most important factors influencing the possibility of expanding local markets or for increased product diversity. The examples given in this chapter might serve as ideas for possible modification and adaptation to individual circumstances.

ii) Beeswax.

Beeswax is a substance secreted by the worker bees. It is recovered by beekeepers primarily from honey comb cappings, and also from cull combs and wax pieces. Beeswax is used in certain pharmaceutical and cosmetic preparations, as a base for polishes and some ointments, for candles - and for comb foundation for beekeeping. It has the highest melting point of natural waxes, and can be sold in either the raw or refined form. Commercial beeswax is generally refined for sale by a manufacturer of apiary products.

The cosmetics and pharmaceutical industries have no complete substitute for beeswax. At least small quantities will always be needed to maintain quality and specific characteristics. Like honey prices, prices for beeswax may vary considerably from place to place.

Markets and prices for products made from beeswax vary widely from country to country. In these industries, beeswax forms only a minuscule part both of the manufacturing process and of the final product. It is used in candle making, skin creams, **grafting wax for horticulture,** polishes and varnishes, paste furniture polish, liquid furniture polish, spray polish, floor polish, shoe polish, cream type. crayons, leather preserves, waterproofing textiles and paper, paint, wood preservative, swarm lure, veterinary wound cream, adhesive lotions.

iii) Live bees.

The production of queen bees, and of entire colonies of bees, is the main diversification available to beekeepers. The queen bee industry is dependent on the existence of a profitable honey industry and on an export market to buy queens at a period when little or no sales. For example, In Australia the demand for queen bees is estimated at around 155,670 per annum - at an average price of $9/queen, this represents a farm-gate value of around $1.5 million. This is a conservative figure because export sales - estimated by industry sources to be $0.75 million - are not recorded separately and have not been added. Live bee exports is a potential growth area for the beekeeping industry, as further markets develop.

Package bees and nucleus colonies are other forms of live bee production, and are sold both within the country and overseas. Again, data on total value of production for this sector of the industry is not available, and has been estimated on the basis of known production. The total value of this sector has been assumed to be $2.25 million, which is almost certainly an under-estimate, but which has been used as a conservative minimum. In India where great potential of beekeeping exists marketing for live bees can be much more.

iv) Other products.

In addition to honey and wax, active beehives are also a source of other products. These include:

Royal Jelly - a milky white smooth jelly secreted by nurse bees, used to feed developing queen larvae and young worker bee larvae. The production of royal jelly is a very specialized procedure, and flora conditions must be ideal before production can be considered. Royal jelly is used as tablets, or mixed into creams and shampoos. Royal jelly can be sold in its fresh state, unprocessed except for being frozen or cooled, mixed with other products, or freeze-dried for further use in other preparations. The fresh production and sale can be handled by enterprises of all sizes since no special technology is required. In its unprocessed form it can also be included directly in many food and dietary supplements as well as medicine-like products or cosmetics. For larger industrial scale use, royal jelly is preferred in its freeze-dried form, because of easier handling and storing. Freeze-dried royal jelly can be included in the same products as the fresh form. The production of freeze-dried royal jelly requires an investment of at least US$ 10,000 for a freeze-dryer, sufficient production volume and an accessible market for the raw material or its value added products. Products containing royal jelly should be specially marked or packaged in order to distinguish them from similar products without it.

As dietary supplement

Royal jelly belongs to a group of products generally described as "dietary supplements" These are products which are consumed not for their caloric content nor for pleasure, but to supplement the normal diet with substances in which it might be lacking.

As ingredient in food products

A mixture of royal jelly in honey (1-3% royal jelly) is probably the most common way in which royal jelly is used as a food ingredient. Among the advantages of this product are that no special technology is required and the honey masks any visible changes in the royal jelly. The final product is pleasant-tasting and it provides the beneficial effects of both products.

As ingredient in medicine-like products

In medicine-like formulations royal jelly is generally included for its stimulatory effects. However, it is also used to solve specific health problems. A variety of formulations are available, often containing ingredients otherwise used to alleviate particular afflictions or as medicine.

As ingredient in cosmetics

Except in Asia, probably the largest use of royal jelly is in cosmetics. Royal jelly is included in many dermatological preparations, but mostly in those used for skin refreshing, and skin regeneration or rejuvenation. It is also used in creams or ointments for healing burns and other wounds.

Others

The only other known uses for royal jelly are in animal nutrition. In particular, royal jelly has occasionally been used (fresh or freeze-dried) to stimulate race horses. For experimental purposes it is also used as a food for rearing mites and insects.

Royal jelly collection

Royal jelly is produced by stimulating colonies to produce queen bees outside the conditions in which they would naturally do so (swarming and queen replacement).

It requires very little investment but is only possible with movable comb hives.

Propolis - a by-product of the bee hive. It originates as a gum secretion gathered by bees from a variety of plants, and can vary in colour depending on the plant species of origin. Propolis has remarkable therapeutic qualities, and is much sought after in some countries for the treatment of a range of human ailments, and for cosmetic purposes. It is used by honey bees as an antiseptic to varnish the interior of honey comb cells used by the hive to rear young brood, to seal cracks in the hive from the winter chill, and for general hive cleanliness purposes. The market for raw material and secondary products containing propolis will probably continue to grow as they find more acceptance in medicinal uses and as more cosmetic manufacturers realize their benefits and marketing value.

Bee venom: Collected by stimulating bees with a mild electric current. The venom is processed, and used in the preparation of pharmaceutical materials. It can be used to detect hypersensitivity or allergic reaction to bee stings. Bee venom is a highly specialized product with only very few buyers. The market volume is relatively small too, although there are no comprehensive surveys. The main venom producer is the USA which has produced only about 3 kg of dry venom during the last 30 years (Mraz, 1982) but there is a large producer in Brazil and more or less significant amounts are produced in many other countries. Prices in 1990 varied greatly between US$100 and US$200 per gram of dry venom (Schmidt and Buchmann, 1992). Prepared for injections or sold in smaller quantities, prices can be much higher. However, the beekeeper often does not get this price. The prevailing prices in European and Asian markets are generally slightly lower.

Pollen: Pollen can also be harvested by beekeepers, at a rate of around 7-10 kgs per hive per year. Pollen is used by bee colonies as a source of protein, but harvesting pollen by the beekeeper requires detailed knowledge of resources, hive management, species flowering variations and timing, and hive response to different honeys and pollens. Pollen is collected via specialized traps fitted to the hives, and must be processed rapidly after collection (usually via freezing or drying) to avoid excessive moisture absorption and fermentation. Many beekeepers harvest pollen to feed back to their hives during periods of natural pollen deficiency. The bulk pollen consumer market seems to be growing in industrialized countries, but pollen tablets are still a common feature of health food stores and command an excessively high price. Encapsulation and extraction of pollen lend themselves easily to small scale manufacturing and result in safer consumer products.

Most of the buyers and large scale sellers of pollen are also honey traders. Crane (1990) however, reports that a lot of commercial pollen is not bee collected, but machine-collected from certain wind pollinated plants which release very large quantities of dry pollen. At least in industrialized countries and those with increasing numbers of health conscious consumers, pollen consumption is likely to increase further. On the other hand, there seems to be a wide market for reasonably priced, encapsulated pollen and tablets.

v) Paid pollination services.

Some beekeepers receive payment for placing hives in close proximity to flowering crops, according to contractual arrangements with farmers. For example, rates for pollination services in inland Australia varied between $25 and $35 per hive in 1996, with variations between crops. It has been estimated that at least $2.9 million is received by the industry in this way, based on total payments received for pollination services in Tasmania and multiplied up to an Australian figure by numbers of hives.

Similar concept is picking up throughout the world including India. In Himachal Pradesh, India this practice has already started and is likely to be followed in other states as the awareness about pollination benefits is realized by the farming community.

Evidently, to ensure the country's self-sufficiency in foodstuffs, to receive foreign currency from excess production, the stabilization of rural populations by complementary activities of both a financially rewarding and environmental nature, and there is no doubt that beekeeping fits perfectly within this framework and hence, efforts are required to popularize and increase beekeeping still an enormous potential waiting to be tapped.

Honeybee pollination is essential for some crops, while for others it raises yield and quality. In addition to the crops, a wide range of pastures, including Lucerne and clover, are pollinated by honeybees hence this estimate understates the potential value of the pollination services.

Beekeeping also increases production of fruit and vegetables, particularly cross pollinated crops such as apples, pears, plums, and litchis and seed production for cabbages, cauliflowers, carrots, turnips, radishes, and other vegetables.

Table 1 Crops dependent on bees for pollination

Category of crops/fruits	Name of the crop
Vegetables	Pumpkin, cucumber, bottle gourd, ridgegourd, carrot, radish, cabbage, knolkhol, cauliflower, onion, soyabean,
Oilseeds	Sarson, toria, sunflower, niger, seasame, safflower, linseed
Pulses	Tur, urad, mung, beans, guar, pea, cowpea
Forage legumes	Lucerne, berseem, clovers
Fruit crops	Oranges, pears, apples, peach, plum, almond, cherry, persimon, strawberry, guava, pomegranate, Jamun, fig, craneberry, grapes, lemon, raspberry, blackberry
Other crops	Buckwheat, cotton, coffee, tobacco, sweet clover
Plants of forest importance	Toon, shisham, soapnut, wild raspberry, stain, Wild cherry, shain, Euretia sp., Robina sp., Trifolium sp., Eupatorium sp. Azadirachta sp, maple chestnut, eucalyptus, willow, linden, catalpa and magnolia etc.

Table 2. Percentage Increase in yield of some crops due to bee pollination

Crop	Increase (%)	Crop	Increase (%)
Fruit crops		**Fodders and legumes**	
Apple	18.00-69.50	Alfalfa	23.00-19,733
Almond	50.00-75.00	Berseem	193.00-6,800
Apricot	5.00-10.0	Clovers	40.00-33,150
Cherries	56.00-1000	Vetches	39.00-20,000
Citrus	7.00-223.00	Birds foot	3.00-1000
Grapes	23.00-54.00	**Miscellaneous crops**	
Guava	12.00-30.00	Buck wheat	63.00-100.00
Litchi	453.00-10,246	Coffee	17.00-39.00
Plums	536-1,655	Cotton	2.00-50.00
Vegetable crops		Field beans	7.00-90.00
Cole crops	100.00-300.00	**Oil seed crops**	
Radish	22.00-100.00	Mustard	13.00-222.00
Carrot	9.00-135.00	Safflower	4.00-114.00
Turnip	100.00-125.00	Sunflower	21.00-3,400
Cucumber and squashes	21.00-6,700	Sesame	24.00-40.00
Onion	353.00-9,878	Niger	17.00-45.00
Cabbage	100-300.00	Linseed	2.0-49.00

Table 3. Number of colonies required for different crops

Sr. No.	Crops	Pollination requirement
1	Apple	2-3
2	Almond	5-8
3	Citrus	2-3
4	Turnip	2-5
5	Coconut	2-3
6	Cauliflower	5
7	Lucerne	3-6
8	Grape	2-3
9	Guava	2-3
10	Mango	2-3
11	Pumpkin	2-3
12	Papaya	2-3
13	Mustard	3-5
14	Onion	2-8
15	Water melon	1-5
16	Musk melon	1-5
17	Sesame	2-3
18	Brussels sprout	5
19	Sunflower	2-4
20	Cotton	2-6
21	Sprouting Broccoli	5

Potential of beekeeping in Jammu and Kashmir

There is a great scope for growth and development of beekeeping in Jammu and Kashmir state. The state has more than 11,14,000 ha. of area under cultivation of different crops. Out of which 2,15,000 ha. is under such crops which fully or partially depend upon beekeeping for pollination. The area under pulses, fruits and vegetables, oilseeds and fodder crops is 32,000; 66,000; 71,000; and 46,000 ha., respectively. Almost all these crops require or benefit from insect pollination. On a modest rate of 3 colonies per hectare for pollination purposes, the state requires a minimum of 6, 45,000 colonies to produce field/fruit crops. But against this requirement, the state has just less than 67,000 bee colonies - a gap of more than 10 times the required number. This shows that at present there are just 0.2 colonies/ha. against the minimum requirement of 3 -5 colonies/ha.

All the above crops, provide rich sources of nectar and pollen which have unexploited so far. Besides, the state has more than 6, 58, 000 ha. area under forests. The pollination needs of wild forest plants have not been estimated but are likely to be significant which also provide ample scope for development of beekeeping by providing basic support to the beekeeping industry in the form of shelter, wood for bee hives, favourable climate and nectar pollen sources. In addition to this the state has 2, 19,000 ha. area which is still barren and uncultivated. This area could be profitably used for mass plantation of bee flora to overcome the problem of depleting floral resources and boost beekeeping potential. This activity will not only provide bee flora but also help reclamation of waste lands and degraded soils to green the environment.

There is a wide gap between the present number of bee colonies and the number that could be supported in this region. According to current estimates, the state of Jammu and Kashmir has the potential to sustain more than 6, 00,000 bee colonies to produce 9,000 tonnes of honey and provide job opportunities for 12,000 families. But despite this great potential, the desired level has not been achieved.

Problems in expanding beekeeping

There have been many problems in the further expansion of beekeeping which include:

- The honey resources and beekeeping areas have not been fully exploited.
- Non - availability of essential materials and bee colonies at the proper time.
- Lack of bee training facilities and bee nurseries.
- Lack of disease investigation facilities.
- No organized system of marketing bee products.
- Problem of migration to hilly areas due to disturbances in the state.
- Indiscriminate use of pesticides.
- Rising cost of beehives and lack of diversification of bee products.

Despite many problems, the exotic bee species have proved much superior and less problematic than the indigenous honeybee *A.cerana*. It has been found to yield several times more honey than *A.cerana* besides having many other useful traits such to disease, low swarming tendency and gentle temperament. It is expected that *A. mellifera* will serve the commercial beekeeping in most parts of the state, whereas *A.cerana* will continue to subsist in marginal beekeeping areas which are less suitable for *A. mellifera.*

Conclusions

The state has vast resources of bee flora and great scope for further expansion of beekeeping in the state. In the state where land holding is less than 0.76 ha., beekeeping can provide better food, balanced nutrition and employment to small and marginal farmers. It can also provide unemployed and under employed persons with full employment and extra income.

References

Abrol, D.P. 2009. Bees and beekeeping in India. 2nd Edition, Kalyani Publishers, Ludhiana, India 720p

Abrol, D.P1993. Insect pollination and crop production in Jammu and Kashmir, Current Science 65:265-269.

Abrol, D.P1997. Bees and beekeeping in India. Kalyani Publishers, Ludhiana, India 450p

Chahal,B.S.Brar, H.S Gatoria, G.S and Jhajj, H.S 1995. Italian honeybees and their management. Punjab agricultural university Ludhiana 116 pages.

Crane, E. 1990. Bees and Beekeeping: Science Practice and World Resources, Cornell Univ Press.

Dhaliwal, G.S, Atwal, A.S.1970. Interspecific relations between Apis cerana indica and Apis mellifera. J Apic Res 9: 53–59.

Diwan V.V and Salvi S.R. 1965. Some interesting behavioural features of Apis dorsata Fab., Indian Bee J. 27, 52

Koeniger N., Koeniger G. (1980) Observations and experiments on migration and dance. Communication of Apis dorsata in Sri Lanka, J. Apic. Res. 19, 21–34.

Koeniger N., Vorwohl G. (1979) Competition for food among four sympatric species of Apini in SriLanka (Apis dorsata, Apis cerana, and Trigona irridipennis), J. Apic. Res. 18, 95–109.

Mraz, C. 1982. Bee venom for arthritis - an update, in American Bee Journal Vol.122, No. 2, pp.121-123.

Ruttner F. 1988. Biogeography and Taxonomy of Honey bees, Springer-Verlag, Berlin.

Schmidt, J and Buchmann, S. 1992 (Research Entomologists, Carl Hayden Bee Research Center). (1992). Other Products of the Hive. The Hive and the honey bee. J. Graham (ed). Hamilton, IL: Dadant.

CHAPTER 18

Mushroom Cultivation – A Mode of Livelihood Security for Dryland Farmers

Moni Gupta[1], Sachin Gupta[2] and C.S. Kalha[3]

[1]*Division of Biochemistry and Plant Physiology, Faculty of Agriculture, Sher-e-Kashmir University of Agricultural Sciences and Technology of Jammu, Chatha, Jammu-180009.*

[2]*Division of Plant Pathology, Faculty of Agriculture, Sher-e-Kashmir University of Agricultural Sciences and Technology of Jammu, Chatha, Jammu-180009.*

[3]*Associate Dean, Faculty of Agriculture, Sher-e-Kashmir University of Agricultural Sciences and Technology of Jammu, Chatha, Jammu-180009.*

Population explosion at the world level in the recent past has mainly resulted in development of three basic problems; shortage of food, pollution of the environment and diminishing quality of human health due to malnutrition. According to the latest demographic studies the world population will be nearly 9.0 billion by 2050. It has been observed that more than 70% of the agricultural and forest products have not been put to total productivity or wasted during processing. Macrofungi (mushrooms) not only can convert these lignocellulosic biomass wastes into human food but also can produce notable bio-medicinal products which have many health benefits. Another significant aspect of mushroom cultivation is its role in decreasing pollution. Mushrooms and there medicinal products have a great potential for supplying healthy food and dietary supplements for domestic consumption as well as for export, provided that the international quality standards and timely supply schedules can be met.

The world has an immense amount of lignocellulosic biomass resource, which like solar energy is sustainable. This resource has been considered insignificant, or of no commercial and certainly no food value at least in the original form. If they are properly managed and utilized, this would lead to economic growth. The by-products produced in processing the agriculture products should be used and treated as raw materials for the production of mushrooms. A number of enzymes are produced by the mushrooms which degrade lignocellulosic materials for their own growth and fruiting. This demonstrates the impressive capacity of mushrooms for biosynthesis, which is different from the photosynthesis of green plants. Cultivated mushrooms have now become popular all over the world. Mushroom cultivation and the processing of mushroom products have been beneficial to millions of people in India, China and other developing countries in terms of financial, social and health improvement. In addition, cultivation and development of mushroom industries positively generate economic growth and already have an impact at national and regional levels. This impact is expected to continue to increase and to expand in future with sustainable research and development activities on mushroom production technology and mushroom products processing (Chang, 1999).

Mushrooms have always fascinated man owing to their sudden appearance in numbers, groups, rings, bunches and also as a single, attractive and fascinating structure. References about the mushrooms are available in most ancient literatures like Vedas and Bible. The early civilizations of Greeks, Egyptians, Romans, Chinese and Mexican appreciated mushroom as delicacy, knew something about their therapeutic value and often used them in religious ceremonies. However, mushroom cultivation did not come into existence until A.D. 600 when *Auricularia auricula* was first cultivated in China on wood logs. Other wood rotting mushrooms such as *Flammulina velutipes* (A.D. 800) and *Lentinula edodes* (A.D. 1000) were grown in a similar manner (Chang and Miles, 1987) but the biggest advance in mushroom cultivation came in France when *Agaricus bisporus* was cultivated on a composted substrate.

Historically, mushrooms were classified among the so called lower plants in the division Thallophyta by Linnaeus but modern studies have established the mushrooms in a separate fungal kingdom, Myceteae. The fungi differ from the plant and animal kingdom by their possession of a cell wall that is different in composition from that of plants and mode of nutrition that is heterotrophic but, unlike animals is absorptive (osmotrophic) rather than digestive.

In 1999, the world production of cultivated edible mushrooms was estimated to be greater than 7 million tones. The combined total market value for edible and medicinal mushrooms for that year was conservatively estimated at more than US $30 billion.

Currently, 14,000 mushroom species are known to exist. Of these 31 genera are regarded as prime edible mushrooms. To date, only 200 of them are experimentally grown, 100 of them economically cultivated, approximately 60 commercially cultivated and about 10 have reached to industrial scale production in many countries. The most acceptable varieties among the cultivated types are *Agaricus bisporus* (button mushroom), *Lentinus edodes* (Shittake), *Volvariella spp.* (Paddy straw mushroom).

By definition "Mushroom is a macrofungus with a distinctive fruiting body which can either be epigeous (above ground) or hypogeous (underground) and large enough to be seen by naked eye and to be picked up by hands" (Chang and Miles, 1992). The structure that we call a mushroom is in reality only the fruiting body of the fungus. The vegetative part of the fungus, called the mycelium, comprises a system of branching threads and cordlike strands that branch out through the soil, compost, wood log or other lignocellulosic material on which the fungus is growing. The most common type of mushroom is umbrella shaped with pileus (cap) and stipe (stem), e.g., *Lentinula edodes* and some species additionally have an annulus (ring) e.g., *Agaricus bisporus* or a volva (cup), e.g., *Volvariella volvacea*, or have both e.g., *Amanita phalloides*. They lack chlorophyll and grow on dead and decaying organic materials, as phytoparasite or in mycorrhizal association with plants.

Ainsworth and Bisby's Dictionary of the fungi (8th Ed.) divided mushrooms into four categories: **1)** Those that are fleshy and edible into the edible mushroom category, e.g., *Agaricus bisporus;* **2)** mushrooms that are considered to have medicinal applications are referred to as medicinal mushrooms e.g., *Ganoderma lucidium;* **3)** Those that are proved to be or suspected of being poisonous are named poisonous mushrooms e.g., *Amanita phalloides;* **4)** those in a miscellaneous category, which includes a large number of mushrooms whose properties remain less defined. These may tentatively be grouped as "other mushrooms" (Hawksworth *et al.*, 1995).

Mushrooms are now getting significant importance due to their nutritive and medicinal values and income generating venture in about 100 countries. Mushrooms are rich in proteins, carbohydrates and vitamins. These are low in calorific value and hence are recommended for heart and diabetic patients. They are rich in proteins as compared to cereals, pulses, fruits and vegetables on dry weight basis. In addition to protein, they also contain carbohydrates, fibers and minerals. They also contain vitamin B and C which include thiamine, riboflavin, niacin, biotin and pantothenic acid. Folic acid and vitamin B12 which are normally absent in the vegetable foods are present in mushrooms. Besides, vitamins they also contain minerals like potassium, sodium and phosphorus in high quantities. They generally do not contain calcium and iron. Since mushrooms posses low calorific value, high protein, high fiber content, high K: Na ratio and low fat (rich in linoleic acid and devoid of cholesterol) make mushroom the choice of the dietician for those suffering with or prone to, obesity, hypertension etc.

Edible mushroom production also represents an attractive method of improving the nutritional quality of lignocellulosic wastes for use as an animal feed stock. The feed value of these lignocellulosics is limited by the low polysachharides degradation achieved during the digestion within the rumen. This restricted digestibility is due to the presence of lignin which acts as a barrier depriving the cellulolytic and hemicellulolytic enzymes access to the polysaccharide components. Among the various physical, chemical and biological methods used for upgrading the digestibility and nutritive value of agricultural wastes, biodegradation by using white rot fungi including mushrooms have been found promising (Sharma, 1997).

Besides their nutritional value, the way they are cultivated today is more important. Their indoor cultivation utilizing the vertical space and their cultivation on a wide variety of cheap and waste materials including agricultural wastes gives them more importance. The success to isolate pure culture through tissues and spores was the turning point in the process of commercial mushroom production in the world. In India mushroom production started in 60's but it was during 90's that there was sudden jump in mushroom production due to Hi tech projects set up in collaboration with the foreign companies. This resulted in significant increase in mushroom production from 4,000 tonnes (1985) to 30,000 tonnes (1995) and 38000 tonnes (1997) and at present it is estimated around 50,000 tonnes/annum.

At present world mushroom production is estimated to be around 5 million tons/annum and is increasing @ 7% per annum. Mushrooms consumption in western countries is almost 85% of the total world production and remaining 15% by rest of the world. China is major supplier of both button and as well as specialty mushrooms. Today, mushroom farming is being practiced in more than 100 countries and the production of mushrooms was more than 7 million MT in 1997 while mushroom production in India is presently 70,000 tonnes. In developed countries like Europe and America, mushroom production is a Hitech industry with complete computer controlled growing systems (Tewari, 2005).

Mushroom science is the discipline that is concerned with the principles and practices of mushroom cultivation and the practical experience can only be obtained through personal participation. A basic knowledge of three basic sciences *viz.* microbiology, fermentation and environmental engineering is helpful to understand the science of mushroom cultivation. The concept of mushroom cultivation that promises to produce a highly nutritious food of excellent taste from waste materials without making extensive demands on land or having requirements for expensive equipment seems quite fascinating and interesting. This is the reason that many people have moved to undertake mushroom growing on a commercial basis. Unfortunately, although simple in concept, mushroom growing is a complicated business. When entered into by untrained individuals who do not have a basic understanding of the mushroom cultivation science, there are chances of failure.

In this chapter, the various phases of mushroom technology are underlined so as to encourage the rural youth especially women folk of dryland and hills to take up mushroom cultivation so as to enhance their socio-economic status.

Phases of Mushroom Cultivation

The main objective of mushroom growers and researchers of this field is to increase the yield of mushrooms from a given surface area per unit of time by use of high-yielding strains, by shortening the cropping period or by increasing the number of high yielding flushes. The successful grower requires the scientific knowledge that is provided through research and the practical experience obtained by personal participation and training in the practice of mushroom cultivation. There are different operations in mushroom farming, each of which must be performed properly for the success of enterprise. Failure in any phase will result from decreased harvest to total loss of the yield.

The practices of mushroom cultivation consist of six major phases. The sequence of phases is: **(1)** selection of a mushroom species, **(2)** selection of a fruiting culture, **(3)** development of spawn, **(4)** preparation of compost, **(5)** spawn running, **(6)** mushroom development. The different phases are discussed here under:

a) Selection of a mushroom species

The market, region or the area where the mushroom after production is to be sold needs to be kept in mind before going for mushroom cultivation. If the mushroom is to be marketed in fresh condition, then it must be a species that is acceptable on the basis of its taste appeal to the people in the area where it is cultivated. This can be determined by market surveys of previously cultivated species or by testing for market acceptability with fresh mushrooms brought from other places for this purpose.

After determining the species that is acceptable to the local consumer, the other specific considerations are **a)** availability of raw materials for substrates and **b)** prevailing environmental conditions so that it will not be excessively costly to maintain the necessary temperatures for mycelial run and for mushroom development. There is a considerable variation among edible mushroom species in the temperatures suitable for vegetative growth and fruit development and it is very important that both these temperatures must be considered in selection of an acceptable mushroom.

Selection of a fruiting culture

After determining and selecting the mushroom species on the basis of acceptability among customers, availability of plentiful substrates and the environmental conditions, the next requirement of the grower is to arrange a suitable fruiting culture of the mushroom. Fruiting culture is defined as a culture that has the genetic capacity to form fruiting bodies under suitable growing conditions. When grown under the proper conditions, this culture produces fruiting bodies. Tissue culture derived from the stipe or pileus of the mushroom of either homothallic or heterothallic species can be used to establish fruiting cultures. Establishment of the tissue culture is the method used to isolate and propagate spore less strains. Spore less strains of *Pleurotus* spp. are of great commercial interest, because normal spore producing strains shed spores early in the development of fruit bodies and continue this up to harvesting resulting in heavy spore density in air in mushroom houses. Unfortunately this results in respiratory track problems and allergic reactions in mushroom workers. Consequently, there is an interest in developing spore less mutants that will produce fruiting bodies equivalent to those of accepted commercial spore-forming stocks in yield, flavour, texture, fruiting time etc. Obviously, a strain or stock that does not fruit or fruits poorly cannot be considered a fruiting culture.

Development of Spawn

Mushroom spawn serves as the inoculum or seed for the substrate in mushroom cultivation. It is a medium through which the mycelium of a fruiting culture grows. In spawn making the entire process of preparation should be performed under aseptic conditions and the pure culture spawn should be free from contaminating organisms. Unsatisfactory spawn owing to genetically unsuitable fruiting culture, degenerated or old age may lead to failure in achieving a satisfactory harvest. Although, ideal environmental conditions and management are absolute necessary for quality spawn, a stable strain or stock possessing the genetic characteristics required by the growers can't be overruled.

Although, the potential of spawn is set by the genetic constitution of the fruiting culture used in its manufacture, the substrate material is also very important. The substrates used in spawn manufacture may be different from the materials used in the cultivation of the mushroom, or they may be the same. Substrates may be used singly or in combination. Some of the substrates used in spawn making include various grains (rye, wheat, sorghum), rice straw cuttings, cotton waste, rice hulls, cotton seed hulls etc. Spawns are frequently referred to as grain spawns or straw spawns.

The spawn substrate serves mainly as the vehicle that carries the vegetative mycelium of the mushroom that is used to inoculate the growing beds. An important point to be kept in consideration is the availability of spawn substrate and the related cost.

Preparation of compost

Compost generally referes to a mixture consisting largely of decayed organic matter that is used for fertilizing and conditioning land for horticultural purposes. In mushroom sciences, a substrate rich in available nutrients does not necessarily constitute a satisfactory medium for growing mushrooms. The reason behind is that the material first must be sterilized or else bacteria and molds will grow in abundance and curb the growth of mycelium and the development of mushrooms. That is, when spawn is inoculated into a raw, unsterilized substrate, the naturally occurring microorganisms quickly gain dominance and retard or even prevent the development of mushroom mycelium. The basic function of composting is to reverse that situation, i.e., to prepare a medium with characteristics that promote the growth of mushroom mycelium to the practical exclusion of other organisms. Certain chemical and physical properties must be built in the substrate for accomplishing this.

The nutrients in the compost must be in a form that is readily available to the mushrooms. For example, most mushroom species cannot use nitrates so it is essential that nitrogen should be present in the compost materials in the form of protein. The compost must also be free of toxic substances, which inhibit the growth of spawn. Thus, conditions non favourable to the growth of these toxins producing organisms must be employed during composting. In regard to the physical qualities of the substrate, good aeration, ability to hold water without becoming waterlogged, a proper pH and good drainage are desirable features. Presence of microorganisms that bring about the breakdown and decay of the substrate materials during the composting processes is also essential in compost.

In practice, composting is accomplished by piling up the substrates for a period of time during which various changes take place so that the composted substrate is quite different from the starting material. A substrate consisting of agricultural and chemical materials other than horse manure, when composted is called synthetic compost. Synthetic composts have been devised with numerous formulations of just about every type of agricultural waste product and residue.

Principles of the complicated process of composting are known and have been worked out in great detail, but in practice modifications are necessary to meet various situations that may be encountered. Such modifications might be necessary keeping in view the availability of the raw materials, the micro flora in the composting area, the facilities in the growing area and the most important is the species of mushroom to be cultivated.

Spawn running

Compost prepared is put into properly sterilized beds or bags and thereafter the spawn is introduced. The amount of spawn used as inoculum varies from species to species. In general, larger amounts of spawn results in beds becoming filled out with mycelium more rapidly, however, the use of greater amounts of spawn increases production costs.

During spawning the spawn is removed from the container (glass bottle or polypropylene bags), it is broken into small pieces by crushing and crumbling with fingers. Pieces of spawn may then be broadcast over the bed surface and then pressed down firmly against the substrate to assure good contact or they may be inserted 2 to 2.5 cm deep into the substrate. Spawn running (mycelial running) is the phase during which mycelium grows out from the spawn and colonizes the substrate. Good mycelial growth is essential for mushroom production and at the same time it is essential to maintain proper conditions of the beds, proper temperature and humidity for the species in the growing chamber. The beds should be watered lightly with a fine sprinkler so that bed surfaces do not dry up. Improper environmental control is a common cause of poor spawn run. Mushroom house management is very important in this stage of mushroom growth.

Mushroom development

Under suitable environmental conditions, which are commonly different from those for spawn running, primodia formation occurs, followed by production of fruiting bodies. Improper aeration leading to increase in CO_2 in the vicinity of mushroom beds may inhibit primodia formation or later stages of mushroom development (Wood, 1984). Conditions of pH and temperature required for fruiting are also commonly different than from mycelial running (Kurtzman, 1979).

The appearance of mushrooms is commonly in rhythmic cycles called flushes. Mushrooms are picked at different maturation stages depending on the species and on consumer preference and market value. The method of harvesting also varies with the species. Suitable temperature, humidity and ventilation conditions must be maintained during the cropping period, because these factors affect the number of flushes and total yield of the mushrooms.

In J&K state, mushroom industry made its beginning in 1960's when button mushroom was grown during 1964 on experimental basis by the Department of Agriculture, J&K Govt. under natural growing conditions. Concomitantly, work on edible fungi was taken up by R.R.L at Srinagar. Mushroom growing was at its best in seventies and eighties in the state but owing to the disturbed conditions in valley most of the production units were closed. At present about 500-600 tons of button mushroom is grown in J & K state annually and most of the produce is sold locally. No large scale commercial unit with round the year cultivation is located in the state (Kalha *et al.*, 2007). Being seasonal activity in Jammu the commonly cultivated mushroom *Agaricus bisporus* shares the major part (more than 90%) of production followed by *Pleurotus* spp. (10%) Efforts are needed to diversify the mushroom production scenario and include a few mushrooms in the calendar so that instead of being seasonal occupation, seasonal activity, mushroom growing should become a round the year activity. For the purpose a calendar (Table1) for round the year mushroom cultivation under Jammu conditions has been developed by Division of Plant Pathology, SKUAST-Jammu (Gupta and Kalha, 2009). As is evident from the table that in total six crops of different mushrooms can be taken in Jammu region.

A number of substrates like wheat straw and paddy straw required for the cultivation for different mushrooms are amply available in the region. This substrate is either given as feed to the animals or gets wasted in the fields. As already discussed spent substrate of mushroom cultivation can serve as a better feed than raw substrate. Therefore, for proper utilization of these agricultural wastes mushroom cultivation is the best option.

Year round mushroom cultivation under Jammu conditions (Jammu subtropics)

Type of mushroom	Species/strain	No. of crops	Cropping period
Agaricus bisporus	S-11, U3, S-310, S-791	Two crops	September - February
Pleurotus spp (Dhingri)	*P. florida*, *P. flabellatus*, *P. sajor caju*, *P. eous*, *P. membranus*	Two crops	March - May
Calocybe indica (Milky mushroom)	OE-46, APK	Two crops	June- August

After harvesting mushrooms, the spent compost can be used as either feeding material for animals and for growing earthworms in vermicomposting. Finally, the residues can then be used as organic manure. In the whole scheme of things, all wastes would be utilized. This is the concept of zero emissions or total productivity (Pauli, 1996). Mushrooms not only can become nutritious protein rich food, but they also can provide nutriceutical and pharmaceutical products. The most significant aspect in mushroom cultivation is the recycling of waste by products during each stage of mushroom production and the creation of pollution free environment. The significant impact of mushroom cultivation, mushroom derivatives and products on human welfare in the 21st century can be considered globally as a Non green revolution (Chang, 1999).

References

Chang, S.T., 1999. Global impact of edible and medicinal mushrooms on human welfare in the 21st century: non green revolution. Int. J. Med. Mush. 1:1-7.

Chang, S.T and Miles, P.G. 1987. Historical record of the early cultivation of Lentinus in China. Mushroom J. Tropics 7 : 31-37.

Chang, S.T. and Miles, P.G. 1992. Mushroom Biology: a new discipline. Mycologist 6 : 64-65.

Gupta, S. and Kalha, C.S. 2009. Mushroom cultivation. In "Proceedings of the training programme on Conservation farming through efficient use of resources to sustain livelihood of dryland farmers of North-West Himalyas" held at SKUAST-J from 16th Jan. to 5th Feb. 2009. pp. 281-286.

Hawksworth, D.L., Kirk, P.M., Sutton, B.C. and Pegler, D.N. 1995. Ainsworth & Bisby's Dictionary of fungi, 8th Ed. CAB International, Wallingford, U.K.

Kalha, C.S., Vaid, A., Tandon, G., Raina, P.K. and Gupta, A. 2007. Year round cultivation of mushroom. Technical Bulletin of Division of Plant Pathology, SKUAST-J.

Kurtzman, R.H., 1979. Mushrooms: single cell protein from cellulose. In "Annual report on fermentation processes". Vol. 3 (Ed Perlman, D.), Academic Press, New York.

Pauli, G. 1996. Breakthroughs: What business can offer society, Epsilon Press, Surrey, U.K.

Sharma, S.R. 1997. Scope of speciality mushrooms in India. In "Advances in Mushroom Biology and Production (Eds. Rai, Dhar and Verma). Mushroom Society of India, Solan. 193-203.

Tewari, R.P. 2005. Mushrooms, their role in nature and society. In "Frontiers in mushroom biotechnology (Eds. Rai et al.). N.R.C.M. Solan pp. 1-8.

Wood, D.A. 1984. Microbial processes in mushroom cultivation; a large scale solid substrate fermentation. J. Chem. Tech. Biotechnol. 34: 232-240.

CHAPTER 19 Strategies for the Management of Dry Temperate Pastures and Grasslands of North Western Indian Himalayas

Naveen Kumar and B. R. Sood

[1] *Professor, Department of Agronomy, College of Agriculture, CSKHPKV-Palampur, H. P. - 176062.*

[2] *Retd.Professor, Department of Agronomy, College of Agriculture, CSKHPKV-Palampur, H. P. - 176062.*

Himalaya is the highest and youngest mountain ecosystem in the world. It is spread across eight countries of Asia including India, China, Nepal, Afghanistan, Pakistan, Bangladesh, Bhutan and Myanmar. Indian Himalayan region covers 14 states (fully or partly) and 90 districts in two distinct geographical flanks of the western and eastern regions. The Western Himalayan ranges extend form Jammu & Kashmir, Himachal Pradesh, Shiwaliks of Punjab and Haryana, Uttrakhand and up to the western border of Nepal.

The Indian Himalaya in the eastern and western flanks exhibits a great diversity in climate, land forms, ethnicity, physiographic features, land use resources availability, biodiversity and livelihood production systems and socio economic conditions. Climatically the Himalayan region encompasses areas having the highest tropical rainfall of around 5000 mm on eastern side to arid and semi-arid cold trans-Himalayan receiving very scanty annual rainfall e.g. Leh with 37mm only. Temperature varies drastically with altitude as well as latitude. The moisture distributions vary with aspect, the southern slopes appearing drier and are less vegetated than the northern slopes. In addition, the distribution of snow, ice, glaciers and the location of isolated peaks, lakes and valleys modify the climate. Vegetation wise, the region abounds with tropical and temperate forests, sub-temperate to alpine pastures, dry scrub and steppe vegetation to almost desert-like scrub and forages in the far west.

Temperate Indian Himalaya

The temperate Himalayan zone is spread over the states of Himachal Pradesh, Jammu and Kashmir and Uttrakhand. Grazing, agro-wastes and tree fodder are the main source of forage supply in the entire region. The alpine zone is completely devoid of trees and is under heavy grazing pressure for 2-3 months from July to mid September. In the region due to heavy snow it is not possible to get winter crops and cultivation during summer (April-September) is a common feature. The land holdings are very small and it is food and cash crop which are mostly cultivated by the farmers. The dry temperate region of Indian Himalaya lies mainly in the rain shadow zone of the main Himalayas and is characterized by extremely cool winter and heavy snowfall and temperature plunge down even up to -30^0 C to -40^0 C. Winters are from November to March and during this period the area remains cut off from rest of the world. The crop season is from March to September. The mean monthly temperature during crop season ranged between 3^0 to 26^0C, average annual rainfall and snowfall are 200mm and 200cm, respectively.

Livelihood of farmers

In the region a large share of economy of hill people is mainly dependent on livestock and its products and the huge population relies on grasslands and the forests for their forage needs. Livestock rearing is the most important component invariably in different production system being followed in the Himalayan Region. Besides contributing to a balanced diet, livestock supports crop production system by providing much needed farm power and manure: so vital for hill agriculture. Natural pastures/grasslands, forest lands, crop residue, fodder trees, wasteland or interspaces of orchard, cultivable fodder crops, field bunds, weed growth of field crops are some of the fodder resources in the region. Among all these resources pasture and grasslands are the main fodder resources. But owing to low productivity these resources are hardly enough to meet the forage demand of ever 50% of existing livestock population.

Present status of the pastures

The dry temperate and alpine pastures are unique features of the Himalayan ecosystem. These areas spread between 1500-4500m altitudes are known as *Margs* in Kashmir, *Thach* in Himachal Pradesh, *Buggyal* in Kumaon and *Payar* in Garhwal in Uttrakhand. Most of the pastures are spread over long, flat, undulated or sloppy terrain. The botanical composition of these pastures contain up to 95 per cent of grass component, while the alpine pastures have 41 per cent of grasses. The legumes contribute only 1.37 per cent of the entire floristic composition. The herbage availability and the nutritive values of various pastures indicate that the overall availability of herbage is very poor and declines at higher altitudes. On the contrary, the altitude plays a significant role in the enhancement of the nutritive value. In Himachal Pradesh, the protein content of herbage from the pastures at 500 m altitude is only 4.5 per cent, while the same gets enhanced to 9.8 per cent at an altitude of 3500 m. Similarly, in Uttrakhand, 6.5 per cent of crude protein has been reported in the herbage at 1550 m, while it increased to 9.8 per cent at 3600 m altitude. Besides low biomass availability these pastures have been infested with abnoxious and poisonous weeds like *Stipa, Sambucus, Aconitum, Calcifuges, Adonis, Artimesia* etc. (Misri, 1998).

A survey of the pastures and grasslands has revealed that besides over grazing, absence of improved grass and legume species and presence of inferior and unproductive species are some of the factors responsible for low productivity. The present level of production of these grasslands and pastures is 2.5 to 5.5t/ha which is about 20% of their actual potential productivity. Melkania and Singh (1989) found that net above ground biomass varied from 279 to 1568 g/m^2 in lower Himalaya, 219 to 285 g/m^2 in mid Himalaya and 233-372 g/m^2 in higher Himalaya. Green forage availability from pastures of Kashmir Himalaya varied from 4.7 to 29.1 t/ha (Misri, 1988), from Uttrakhand Himalaya 1.62 to 3.96 (Ram and Singh 1994) and for temperate pastures of Himachal Pradesh Himalaya it varied from 1.5-1.79 t/ha and in alpine and sub-alpine pastures from 0.5 to 1.0 t/ha (Singh, 1995). In cold and temperate grasslands of semi-natural grasslands the stocking capacity varied from 0.6 to 1.9 sheep/ha/annum (Tincheng and Yuangang, 1989).

In dry temperate region of Indian Himalaya, arid climate (annual rainfall 12 cm) accounts for sparse vegetation cover. Livestock rearing is an integral part of rural economy but acute shortage of fodder appears a major bottleneck in any livestock improvement and development activities. In Indian northern western Himalayan hills of Himachal Pradesh, Jammu Kashmir and Uttrakhand, more than 90% of the population depends directly on agriculture and animal husbandry. The rearing of livestock is an integral part of people's life. In high hills of this region very little area is put under cultivated fodder crops and grasslands and pastures are the primary source of fodder for the livestock.

Constraints

In the nutshell the major constraints which has a significant influence on the negative forage productivity in the region are:

- negligible rainfall,
- harsh climate,
- mono-cropping season (April to September),
- fragile ecosystem and absence of adequate vegetative cover-hence more soil erosion,
- shallow depth of soil with low water percolation rate,
- small land holdings, very little area under cultivation (0.6% of the total area-that too near water resources),
- lack of farmers awareness regarding fodder resources improvement approaches.

Opportunities and options

Owing to small land holdings use of non cropped land into productive use with judicious use of irrigation water appears a viable preposition to mitigate the fodder scarcity in the region, simultaneously fruits like apple, apricot and other trees like salix, poplar and robinia holds very good potential in the region. Preliminary studies have indicated successful establishment of fruits and other multiple purpose trees with under story covers of grasses and forage legumes. The under story vegetative cover has better effects on soil and water conservation which in turn helps in successful establishment and production of improved horti-pastoral and silvi-pastoral systems in the region. Studies have identified adaptable and suitable grasses and forage legumes for the region.

Most important methods for the improvement of these pastures and grasslands in the region is to introduce ecologically adapted superior grass/legume spp. adopting suitable planting/seeding techniques with proper fertilization under controlled grazing conditions. In order to improve the productivity of grasslands under dry temperate Himalayan region, emphasis needs to be given on several aspects described in the following text.

Introduction of improved grass and legume species

The successful establishment of a pasture requires more skill and care, as compared to other crops. The method of introduction of improved grasses and legumes in the natural grasslands should be cost effective with minimum soil working. Introduction of improved grasses and legumes species in the native vegetation not only help in increasing total biomass production but also improve the quality and seasonal distribution of the produce.

Improved Grass Species for different dry temperate climatic conditions

Species	Variety	Time of sowing /plantation	Mode of propagation	Green fodder yield (q/ha)
Grasses				
Tall fescue	Hima-1 Hima-4	October- November	Seed Root slips	300-400
Orchard grass	Comet	October- November	Seed Root slips	300-400
Legumes				
White clover	PLP Comp. PWC-3	April	Seed	300-400
Red clover	PRC-3	April	Seed	250-300
Lucerne	Anand-3	April	Seed	400-500

Introduction of superior grass and legume species can bring a breakthrough in the forage production programme. Studies have conclusively indicated that improved grasses/ legumes either sole or in mixture increased the forge production of local grassland by 100 to 165 per cent (Table 1) under dry temperate conditions of Himachal Pradesh. In addition to these species, Pangola grass (*Digitarea decumbans*) and Rhodes grass (*Chloris gayana*) also holds very good promise for the hills of Uttrakhand.

Table 1. Effect of different species on the herbage yield

Improved species	Forage yields (q/ha)	
	Green	Dry
Local grasses	61.0	22.6
Orchard grass+ fescue grass	129.2	50.0
Red clover + Lucerne	122.2	40.0
Orchard grass+ fescue grass + Red clover + Lucerne	162.2	54.0

(Anon, 2001)

Fertilization

Except the excreta of grazing animals, no fertilizer is added to pastures or grasslands in the region. Judicious use of fertilizer for pastures is imperative to boost the vegetative growth of the species which is the ultimate objective of the fodder production programmes. Depending upon the ecological conditions, the pasture grasses responds to 40-120 kg N/ha and 30-80 kg P_2O_5/ha (Dogra *et al.* 1997, Sood and Sharma 1996, Singh 1995, Sood and Bhandari, 1992)

In Himachal Pradesh natural grasslands dominated by local species of the region fertilized with 40 kg N and 60kg P_2O_5 ha^{-1} increased the herbage yield by two to three times. Similarly, grassland introduced with improved grasses + legumes fertilized with 80 kg N and 60kg P_2O_5/ha can be cut three times in a season (Anonymous, 2001). Nitrogen application in two equal splits (first dose in March -April and second dose after the first cut) is the best preposition to harvest highest herbage yield.

Cutting Management

Frequency of cutting also influences the yield and quality of herbage. The areas with high temperatures may require larger interval and low intensity of cutting. The response to cutting of a forage plant depends upon the seasonal yield of carbohydrates and their storage. Tall fescue (*Festuca arundinacea*) produces higher dry matter when it was cut at 30 days interval during second year (Singh *et al.* 1993_b). Cutting grasses twice in a season under natural grasslands recorded higher fresh forage yield than one cut or three cuts. The crude protein content was higher with two cuts compared to one cut (Kaul and Sood, 1986). Singh (1995) also suggested double cutting during July - October. In Uttrakhand hills Para grass provides highest yield with four cuttings than the single cut (Melkania, 1995).

The grasslands experience intense grazing pressure. Controlling the time, duration and intensity of grazing appears to be the key factors in grazing management. Periods of rest allow grazed perennials to replenish leaf area, seed set and store food reserves in their roots (Merrill, 1983; Adams *et al.*, 1991). Grazing contributes more than 50% of the herbage requirement for sedentary and semi-migratory flocks, while for migratory flocks 100% herbage is provided by grazing. Continuous grazing in the same range impedes new growth. The frequent grazing leads to the disappearance of nutritive species and infestation by less palatable species and weeds. Rotational grazing system was found superior as compared to continuous grazing. Rotational grazing has steadily gained the popularity in last two decades, because it offers better control over livestock distribution and feeding pattern with goals of periodically resting vegetation. Though, rotational system of grazing gives better results but it is not an easy task to adopt this strategy on large scale in hilly situations. The most desirable system of grazing would be that of periodic or deferred grazing by limiting the number of livestock on the basis of carrying capacity of grasslands. This system would give a better chance for self seeding and stand recovery. Frequent cutting and grazing leads to reduction in forage yield along with the dominance of unproductive grass sp. Delayed cuttings produce poor quality herbage.

Silvi-pastoral and horti-pastoral systems

Most of the grasslands and pastures under heavy grazing pressure suffer extensive soil erosion coupled with have low production potential and loss of soil fertility. Under the situation adoption of improved ecosystems of silvi-pastoral appears to be the most viable alternative. Under trees a bonus productivity of improved grass species can be harvested. Horticultural systems also offer tremendous scope for the introduction of improved forage species to realize higher herbage productivity in view of the availability of shade tolerant improved grass species.

Korane and Singh (1990) have suggested planting of some range grasses under pine and Cedrus trees for supplementing the availability of fodder (Table 2).

Table 2. Yield of different grasses under pine and Cedrus trees.

Grass species	Green forage yield (t/ha)	
	Under pine tree	Under Cedrus tree
Pangola	14.0	13.6
Rhodes	4.7	3.0
Spear	0.5	-
Para	0.4	2.0
Guinea	0.4	2.5
Kikuyu	-	6.4

For Himachal Pradesh and Jammu & Kashmir hills planting of *Robinia*, Poplar (*Populus indica*), salix (*Salix denticulata)* with improved grass and legume species is advisable.

In these systems, forages are grown in the wide inter row space of fruit trees for economic utilization of orchard lands. Introduction of *Fescue* grass in apple orchard gave 83.5% higher green forage yield over local grasses in high hills of HP. (Table 3).

Table 3. Effect on various grass species under apple orchard

Grass species	Green forage yield (t/ha)
Fescue	6.7
Orchard	3.1
Rye	2.1
Local	1.1

(Sharma and Jindal, 1989)

Forage from the horti-pasture is consumed fresh and is also conserved as hay for winter. Some best fodder plant combinations for horti-pasture in high hills are given in (Table 4). There are ample scope to exploit the shaded environment of forestlands and other tree covers for harnessing the productivity in terms of herbage yields in the region.

Table 4. Fodder production under horti-pasture

Forages	Seed rate (kg/ha)	Green forage yield (t/ha)
Orchard grass + Red clover	5+4	42.0
Rye grass + Red clover	5+4	48.0
Brome grass + Red clover	15+4	39.0

(Misri, 1988)

Forage production situation in hills of dry temperate region of north western Himalaya is very alarming. On the basis of long term research on various aspects for the improvement of natural grasslands in the region, it has been found that the productivity of these grassland can be enhanced significantly by planting improved grass and legume species using appropriate techniques of planting either in sole stand or with fodder/fruit trees, following judicious use of fertilizers and adopting suitable cutting and grazing schedules. This all has to be supplemented with extension efforts to induce the farmers to adopt improved production strategies.

References

Anonymous. 2001. Annual Progress report of NATP: Pasture improvement and legume introduction : soil- plant - animal relationship. pp. 14

Dogra K.K., Katoch B.S., Sood B.R. and Singh G. 1997. Production potential and quality of silviherbage systems vis-a-vis natural grasslands in the humid sub tropics of H.P. *Range Management and Agroforestry* 16(2): 165 –168.

Kaul, A.K. and B.R. Sood. 1986. Production and quality of a natural grassland as influenced by cutting frequency and nitrogen application. *Indian J. Range Mgmt.* 7 (1) : 5-9

Koranne, K.D. and J.P. Singh. 1990. Forage production strategy for Kumaon Himalayas. *Paper presented at International Symp. on water erosion sedimentation and resource conservation held at Dehradun, 9-13 Oct.*

Melkania N.P. and Singh J.S 1989. Ecology of Indian grasslands. In: *Perspective in Ecology* (Ed. J.S Singh and B. Gopal), Jagmonder Book Agency, New Delhi.

Melkania, N.P. 1995 Forage Resources and forage production systems in central Himalaya In: *New vistas in forage production. Edt. by* C.R. Hazra and Bimal Misri. pp. 197-202

Misri B.K. 1988. Forage production in alpine and sub-alpine regions on northwestern Himalaya. In: Pasture and Forage Crop Research, A State of Knowledge Report (ed. Punjab Singh). RMSI, Jhansi. 43-45

Misri, B. 1998. Forage Production in alpine and sub-alpine region of north western Himalaya . In: *Pasture and forage crop research. A state of knowledge report* edt. Panjab Singh. RMS1, Jhanshi: 43-45

Ram Jeet and Singh S.P. 1994. Ecology and conservation of alpine meadows in central Himalaya, India. In: *High altitudes of the Himalaya* (Ed. Y.P.S. Rangtey and R.S Rawal). Gyanodya Prakashan, Nainital.

Sharma, J.R. and K.K. Jindal 1989. Introduction of superior varieties of grasses in orchards. *Paper presented at the workshop on pasture and grassland improvement at HPKVV, Palampur 12-13 Oct.*

Singh K.A., Rai R.N., Gupta H.K. and Singh L.N. 1993_b. Cutting management of some temperate grasses in Sikkim. *Range Management and Agroforestry* 14 (1): 23-27.

Singh V. 1995. Technology for forage production in Hills of Kumaon. In : New Vistas in Forage Production (ed. Harzra, C.R and Misri Bimal). AICRPF (IGFRI). Publication Information Directorate, New Delhi. pp. 197 -202.

Singh, L.N 1995. Temperate Pastures and their Management, IGFRI, Jhansi.

Sood B.R. and Sharma V.K. 1996. Effect of nitrogen and phosphorus levels on the productivity of natural grassland. *Haryana Journal of Agronomy* 12 (1) : 68-74.

Sood, B.R. and J.C. Bhandari 1992. Response of *Setaria anceps* cv. Narok to nitrogen and phosphours in natural grasslands of Kangra vallley. *Range Mgmt. and Agroforestry* 13(2): 139-41

Sood, B.R. and V.K. Sharma 1996. Effect of nitrogen and phosphorus levels on the productivity of a natural grasslands. *Haryana J. Agronomy* 12 (16): 68-74.

Tingcheng Z. and Yuangag Z. 1989. Past, present and future use of pastoral zones of eastern and central Asia. In : proceeding XVI International Grassland Congress, Vol. III : 1725-1729.

CHAPTER 20

Rain Water Management for Efficient Crop Planning in Rainfed Areas

Mahender Singh, Parshotam K. Sharma and Brij Nandan

Division of Agronomy, Faculty of Agriculture, Sher-e-Kashmir University of Agricultural Sciences and Technology of Jammu, Chatha, Jammu-180009.

Agricultural production at any place is largely determined by weather agronomic practices, soil water management and crop species. Among them, weather which is most variable plays a vital role. Rainfall is a prime factor in crop production. Due to variations of weather systems and orographic influences, variety of rainfall patterns are produced (Anonymous, 1995). The crop growing period, which usually coincides with the length of rainy season, is subject to year to year variation. Identification of suitable moisture availability periods and adaptation of appropriate crops to match these periods are basic criteria to achieve optimum production. Many agricultural operations revolve around the probability of receiving definite amount of rainfall. The probability of rainfall can be used for various purposes such as land use planning, identification of crop growing periods, choice of cropping pattern, resource allocation etc. Biswas and Khambete (1979) and Biswas and Basarkar (1982) used the same methodology to delineate the dry farming tract of Maharashtra and Gujarat respectively in to different zone using taluka wise rainfall data. Crop planning requires a comprehensive understanding of amount and distribution of rainfall over a long period. The success of rainfed crop depends on proper distribution of rainfall during its growth period.

Significance of Rainfed Areas

Rainfed agriculture has tremendous role in boosting the grain production in India. Rainfed agriculture supports 40 percent of human population and two thirds of livestocks. Ninety one percent of coarse cereal's, 90 percent pulses, 80 percent oilseeds & 65 percent cotton are grown in dry lands. Rainfed agriculture in India has been practiced since time immemorial. Besides arable crops, farmer's dependence of livestock farming as a source of income is very high in rainfed areas emphasizing the importance of not only raising the productivity of food crops but also augment the feed and fodder needs. Unlike the develop countries such as USA & Australia where agriculture is mechanized. Farmer in India has developed innovative methods to practice agriculture under rainfed situations. Rainfed agriculture has been one of the priorities of Government of India since independence. Development of rainfed areas is now being taken up on watershed basis involving the participation of beneficiaries. A number of technologies were evolved which primarily aimed at resource conservation & management critical for the long term sustainability of natural resources of soil, rain water vegetation.

Constraints in rainfed farming

Agricultural production at any place is largely determined by weather, agronomic practices, soil water management and crop species. Rainfed areas suffer from a number of constraints. Farmers are resource poor & soils are highly degraded with low water retentive capacity and multiple nutrient deficiencies. Aberrant behavior of monsoon causes frequent drought leading to shortage of food & fodder. The high risk nature of dry land agriculture lead to a vicious circle of low investments.

- low yields
- Poor technology adoption
- Poor status of living

A number of technologies were evolved which were primarily aimed at resource conservation and management critically for the long term sustainability of natural resources of soil, rain water and vegetation. It is very important that the natural resources in rainfed areas are not only conserved, but also efficiently utilized with sustainable rainfed farming as ultimate goal. However weather related uncertainties and poor socio economic resource have constraint the adoption of improved dry farming technology on wider scale.

Climate as Determinant in Cropping System Management in rainfed areas

The climate or weather is the natural resource and is an important basic input in agricultural planning and strategies. The Indian economy is mostly agrarian based and depend on onset of monsoon and its further behavior. Agriculture develops harmony with the normal weather or climate over a place. Utilization of climatic information, current weather conditions, and predictions at various time scales is extremely important in rainfed agriculture. However, there is still a lack of awareness about the enormous potential and economic value for weather/climate information. Every plant process, related with growth, development and yield of a crop and every in season and off season farm operations such as a ploughing, harrowing, land preparation, sowing, weeding, irrigating, manuring, spraying, dusting, harvesting, threshing, storage and transport of farm produce are influenced by weather. The high yielding varieties suitable to a particular agro climate is the most important factors in realizing their yield potential. Sowing time of the crop can be adjusted for utmost advantages of environmental factors best suited to various growth stages of the crop. The characteristics of crop genotypes play an important role ignoring optimum yield with available soil moisture as well as optimum weather conditions.

In rainfed agriculture planting and crop selection are the functions of the climate. Use of climate in planning could reduce, but not eliminate the risk and uncertainty of crop production resulting from unfavorable weather. Optimal use of climatic resources requires inventories of the climatic elements and the adaptabilities of the plants and animals to a given agricultural production system. Major application includes selection and adaption of production systems, with the associated crop variety, level of inputs and cultural practices. Further, cost benefit analysis of microclimate modifications will enhance use of climatic information.

Weather forecast for rainfed farming

Prediction at seasonal levels in the tropics has significant potential to improve weather based planning in farm management. In rainfed areas, seasonal/annual prediction is of far greater importance than daily weather forecasting. The seasonal forecasts primarily give information on departure of rainfall from normal for the country and season, as a whole. For example, the parametric and power regression models (Gowariker et al., 1991) demonstrated the predictability of south west monsoon rainfall over the Indian subcontinent using large scale atmospheric circulation features. This information is useful at national and regional levels of crop planning decisions such as crop selection and scheduling of inputs. However, development of monthly and possibly lower period sub-divisional rainfall forecasts are essential to meet the requirements at farm-level strategic decision making.

In rainfed agriculture, distribution of rainfall from week to week is very important rather than total amount, for optimum growth and development of different phenophases of crops. The planting of crops is mainly dependent on the arrival and subsequent progress of monsoonal rains. Inter-annual variability in onset and progress is well known fact. Because of this variability, considerable year to year fluctuations are experienced in the length of the growing season in different part of the country. The intra-seasonal variability may lead to prolonged dry and wet spells. Dry spells induce moisture stress and have a major impact on the productivity particularly when they overlap the critical phenophases of the crop under plough.

Stimulus of rainfall characteristics in rainfed agriculture

The optimization of agricultural production with a given quantity of water and cropping pattern depends upon the rainfall, soil characteristics and atmospheric demand of water. The success of rainfed agriculture depends upon rainfall and the paramount need is to plan the optimal utilization of rain water. The rainfall is major limiting factors in success of rainfed agriculture. The rain water should be manage in such a way that crop plants get the supply of water as and when required for their optimum growth. The important features of rainfall in relation with plant growth are amount, distribution, variability and intensity with space and time. These factors determine the effect of rainfall on establishing the crop stand, population, vegetative growth, seed development, crop maturity etc. in crop production the amount, periodic distribution of rainfall rapidly and intensity of individual rains received and the runoff plus seasonal evaporation are some of significant rainfall features. The amount of rainfall that falls during any given short time is more significant than that the average amount of a long period (Visher, 1967) as rainfall is highly variable over short period both in time and space. Wide fluctuations in rainfall from year to year and from place to place cause a stressed environment in form of drought and floods. The impact of extreme values of temperature, sunshine or commencement of rain week, frequency of wet and dry spells, end of the growing season, probability of rainfall provide useful guidance for planning cropping practices to optimize the production, at micro-regional level.

a) Importance of rainfall probability

Rainfall probability may bring out the influence of water availability on crop growth. The probability of rainfall and availability of water are very useful for agricultural climatological approaches for possible adjustments in existing cropping patterns. The study of rainfall probability is of great importance in evaluating the rainfall charactertics more sensitively. The dry farming tract in India is defined as the region with mean annual rainfall between 40cm and 100cm. Sarkar et al. (1982) identified seven homogenous zones in the dry farming tract based on the assured weekly rainfall distribution. Rainfall probability may help in crop planning, choice of the crop and its variety, sowing time, intercropping, double cropping and fertilizer application etc. thus, it will indicate the periods during which the suitable agronomic practices have to be followed because of the stable rainfall features. Khambete and Kanade (1985) suggested that the weekly assured rainfall < 16mm at 30 percent probability level or <4 mm at 50 percent probability level is the criteria for drought proneness of a station. The moisture supply through total rainfall is not generally a good indicator for a good harvest under rainfed conditions. This is because a few heavy rains spells account for most of the seasonal rainfall (Ananthakrishnan, 1976). It is thus necessary to evaluate short term of probabilities of success and failure of a given crop at a place (Virmani, 1980: Sarkar and Biswas, 1988).

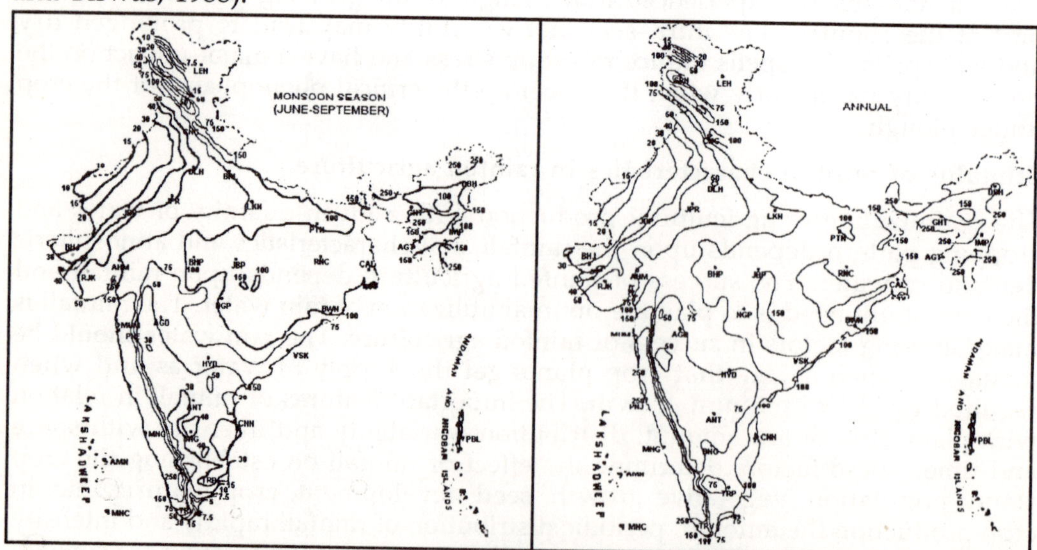

Fig.1. Normal Seasonal and Annual Rainfall (cm) in India

b) Onset, cessation and length of rainy season

The knowledge of the probable dates of commencement and ends of rainy season is very useful in deciding the crop planning under rainfed agriculture. The time of onset of monsoon is necessarily the time suitable for sowing of dry land crops. Sowing is favoured and can commence only when soil moisture build up take place at least upto the rooting depth, after meeting evaporation, percolation and runoff losses. The forecast of sowing rains are crucial in rainfed as well as dry farming regions. The delayed in land preparation and sowing may lead to complete crop failure, sometimes due to reduction in crop growing season.

Table 1. Cropping systems for crops having different lengths of growing period.

Length of growing period	Cropping sytem that can be adopted
<75 days	Perennial vegetation monocropping of short duration pulses
75 – 140 days	Monocropping
140 -180 days	Intercropping
>180 days	Double cropping

The length of the crop growing periods becomes more critical for crop growth in years of early cessation of monsoon rains. The duration of rainfall period plays an important role for the selection of such crops and varieties in crop planning which may complete their life cycle during that period (Table 1). Frequency of irrigation is determined by the dates of onset and withdrawal which are not easy in arid climate due to intermittent and patchy nature of rainfall. Such types of information are very useful for agronomic practices like preparation of seed bed, manuring, sowing, weeding harvesting, threshing etc.

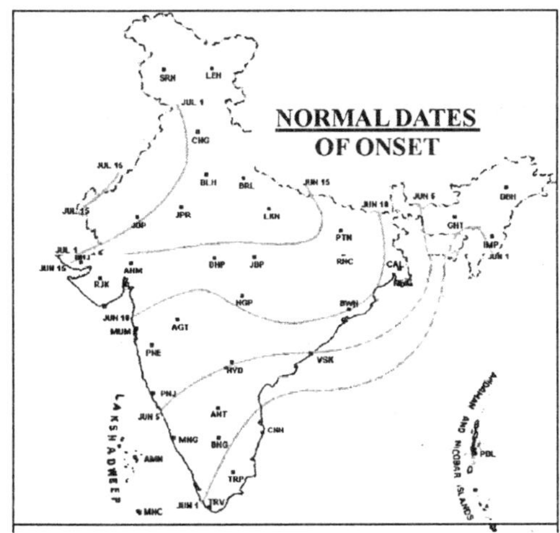

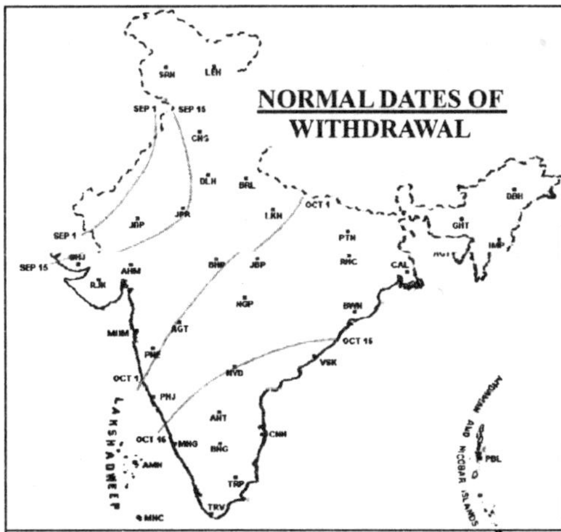

Fig. 2. Normal Dates of Onset and Cessation of monsoon Season in India

c) Dry spell/break in monsoon

The knowledge of the period of occurrence of dry spells in rainfed agriculture is valuable in selecting crops and their varieties to obtain the required level of drought tolerance or to provide life saving irrigation at appropriate times. Thus, with knowledge of dry spells occurrence a farmer can adjust sowing period in such a way that moisture sensitive stages do not fall during dry spells. Dry spells after sowing further aggravate stress on crops.

Rainwater harvesting in rainfed areas

Water is main limiting factors in this region. Distribution of rain both in space and time and the great variation in intensity of individual rain showers, results in considerable runoff (10-25%). Estimates show that about 10 percent of rainfall is lost as runoff from rainy season cropped in deep vertisols and about 25 percent from alfisols. Most of runoff occurs in a few storms of high intensity (Kanwar, 1982). An attempt should be made first to retain as much rainwater in the soil on which it falls, so as to provide a favourable moisture regime to the crop. The excess rain water-one that exceeds the infiltration and storage capacity of the soil- may be harvested nearby in the same field or at another convenient point in the same watershed. The farm ponds once formed the most appropriate devices for harvesting rainwater *in situ*. The system has lost ground because of several reasons. Improvement of tank irrigation systems need be given high priority in many states, particularly Andhra Pradesh, Karnataka, Tamil Nadu, Maharashtra, Madhya Pradesh, some part of Uttar Pradesh, Rajasthan, Gujarat, Orissa and Jammu & Kashmir. Benefits from runoff collection and reuse are quite substantial for *rabi* crops even in medium to low rainfall areas and even greater advantage realized in high rainfall regions. Harvesting runoff water and storage in farm ponds is a distinct possibility in red soil areas of Karnatka. Ponds may be 200m^3 for 0.6 ha catchment and 2000 m^3 for 6-7 ha catchment areas. Nearly 50 percent of the stored water can be used for protective irrigation. About 25-30 percent of catchment area can be given protective irrigation 2-3 times (Havangi, 1982).

Inter row and Inter plot Harvesting System

In situ water harvesting studies were initiated at CAZARI in 1969. Singh *et al* 1973 found that ridge furrow (60cm: 40 cm) and micro-catchment system of 4 percent slope gave 203 percent and 121 percent higher yield than the regular flat planting. It was further revealed (Singh, 1976) that micro catchment having a catchment to cropped area ratio of 0.5 resulted in high soil moisture availability in the system. Besides micro- catchment runoff farming, Mann and Singh (1977) reported that ridge –furrow configuration (inter-row water harvesting) is also an appropriate technique for increasing soil moisture availability and has greater adaptability to farm situation in rainfed areas. For taking arable crops in inter-row water harvesting system, it is advisable to save energy input by practising tillage in the furrows and ridges kept untilled on a permanent basis. This will also give a set- row cultivation effect. Further, for reducing wind action in furrows, Singh and Bhati (1988) suggested the laying out of the system against the prevalent wind direction of south west. The ridge exerts a partial shading effect on the adjoining furrows for 6 to 7 hours a day under high sunlight and reduces evaporation in first stage by 25-35 percent.

At times, sandy soil texture in the rainfed areas presents a limitation in realizing the full benefits from ridge -furrow technique. To overcome this limitation, construction of basin -like structures at the lower base of catchment were suggested. The basins will store water when the incoming water from rainfall exceeds the soil moisture storage capacity. This water could be recycled as supplemental irrigation during dry spells to meet the plant evapotranspiration. Singh, 1997 found that a catchment area of 31.5 m^2 with 5 percent slope is adequate for optimum water harvesting. In normal and high rainfall years receiving 300-500 mm rainfall, the 2-m soil substratum profile of plot having runoff catchments contains twice as much moisture as in that of flat plots. In low rainfall years (200-250 mm rainfall) the 1-m soil profile carried 40-60 percent higher moisture regime than flat beds (Singh and Gupta, 1989).

Traditional Methods of Water harvesting

Water is scarce commodity in rainfed areas and its availability for domestic and crop production use has always been a challenging job. Different systems were used by rain water harvesting, the dominant among these systems are tanka, nadi, khadin and roof water harvesting. These traditional systems which are slowly getting abandoned in favor of some recently developed technology.

a) Tanka

Construction of tanka on individual family basis has caused psychological impact of pride of ownership to the beneficiaries. Plantation of suitable species of trees around the structures at many places has helped in the environmental improvement.

It is underground tank constructed for collection and storage of runoff water from a natural catchment, artificial catchment, or from a roof top. To overcome the problem of leakage, pollution, and high evaporation losses in traditional tanka, the improved design by providing vertical surfaces with stone masonry or cement concrete structure, the base having 10-15 cm thick concrete layer. Silt entry in the structure is restricted by providing silt traps at inlet points. Improved deigns with capacities ranging from 1000 L to 60000 L have been developed.

b) Nadi

Nadi (village pond) is constructed for storing water available during rainy season. Nadi system of water harvesting is one of the oldest practices and is still the common sources of water for conjunctive use. The capacity of nadis generally ranges from 1200 m^3 to 15000 m^3, depending upon physiographic conditions and rainfall pattern. Traditional nadi has the limitation of high evaporation losses from free water surface, seepage losses through porous sides and bottom, and heavy sedimentation due to degradation in the catchment.

c) Khadins

Khadin is a unique land use system developed by farmers of western Rajasthan, particularly in Jaisalmer district, where rocky catchment and valley plains occur in proximity. The harvested water stands in the khadins throughout the monsoon season and crops such as wheat, chickpea and rapeseed mustard are sown by first week of November after the water recedes. Any surplus water is discharged through the slice before sowing (Kolarkar et al., 1983). Soil moisture and fertility are well maintained in the khadin by the deposition of fine sediment as a result of an earthen bund and crops mature without irrigation. The depth of water trapped in a khadin varies from 50 cm to 125 cm, depending on the catchment characteristics, rainfall pattern, and storage capacity of khadin bed. During past three decades khadin farms have been developed on a large scale under different programmes of different State Government. The Improved deigns of khadin to optimize the process of infilteration, runoff generation and routing, as well as soil and water storage capacity (Khan, 1997).

Crop -Water Relations

Water stress in plants, particularly in post rainy season, is common phenomenon. It is characterized by decrease in osmotic and total water potential, accompanied by loss of turgor, reduced diffusion of carbon dioxide into plant leaves, and therefore, reduction in photosynthesis and decrease in growth. However the plant response to water stress is governed by soil, plant and environmental factors. The degree and duration of water stress and the growth stages, at which it occurs, considerably modify the crop response.

Available Soil water and Plant Water Use

The conventional available water capacity is defined as difference in volumetric water content between field capacity (-0.33 bar) and permanent wilting point (-15.0 bars of soil potential). But even below -15.0 bars, matric potential crops are known to deplete soil water. In vertisols the crops such as Dolochos lablab survived even at 0.4kg/ kg of of soil moisture level, while permanent wilting point of these soils is 0.22-0.28 kg /kg. Water content at field capacity of different soils of rainfed regions of India is given in Table 2.

Table 2. Water holding capacity of some soils of rainfed region of India

Soil/ Region	Depth (cm)	Upper limit in moisture storage (0.33bar) Kg water per kg dry soil	Texture
Inceptisols	120	0.25 - 0.27	Silty loam to silty clay
Aridisols	200	0.07 - 0.14	Loamy sand
Entisols	130	0.24 - 0.35	Silty loam to sandy loam
Vertisols	150	0.32 - 0.35	Clay to clay loam
Alfisol	90	0.11 -0.15	Sandy loam

Technologies for Improving Water -Use Efficiency

Success of dry land cropping lies in improving land use, water use efficiency (WUE) and nutrient use efficiency which eventually should enhance biomass productivity on sustainable basis.

I) Land Configurations

Appropriate land configurations such as ridge and furrow, broad beds, graded strips, raised bed sunken bed, terracing, and inter-row and inter-plot water harvesting systems hold great promise for in situ conservation of soil, water, and plant nutrients. They are most effective in moderate to high rainfall regions where such losses are significant.

II) Ridge and furrow system

Different dry land crops like soybean, maize and sorghum recorded significantly high yields when planted on ridges developed on a Vertisol, Planting of sorghum on ridges in Kovilpati in Tamil Nadu also led to significant yield advantage. Furrows not only provide safe disposal of runoff, but also facilitate its retention by soil. This enables the crop to withstand dry periods better.

III) Raised bed and sunken bed

In regions with assured rainfall of more than 1000 mm and having relatively flat topography. The upland kharif crops suffer due to poor drainage during the periods of continuous and intense rainfall. On slopy lands on the other hand, runoff causes severe soil, water and nutrient losses. Raised and sunken bed system developed at the Jawaharlal Nehru Krishi Vishwa Vidyalaya (JNKVV), Jabalpur provides surface drainage, encourages in situ rainwater conservation, and retards soil erosion. The system consists of an array of alternating raised and suken beds. The dimensions of raised and suken beds would depend on many factors namely, amount of rainfall, its intensity, runoff yield, water intake rate and soil, and soil surface conditions. The WUE of crops under raised and sunken bed system was higher (3.95 kg ha-1 cm-1) as compared to flat (3.3 kg $ha^{-1}cm^{-1}$) system. The system has shown promise in managing and partially reclaiming calcareous sodic soils.

IV) Broad-bed and furrow system (BBF)

The system developed at the International Crops Research Institute for the Semi-Arid Tropics (ICRISAT), Patancheru consists of 90-150 m wide beds created along the contour to a 0.6% slope and separated by furrows that drain into grassed waterways. This system permits safe disposal of runoff and conserves soil and water in situ and enhances grain yield.

Rainwater management in rainfed areas

Various soil and water conservation measures relevant for rainfed agriculture include:

In situ measures for rainwater management in rainfed areas

- Off season land treatment
- Conservation furrows
- Ridge and furrow system in cotton
- Cover cropping
- Micro catchment for tree systems

Medium term measures rainwater management in rainfed areas

- Stone and vegetative field bunds for soil and water conservation
- Graded line bund helps in efficient drainage
- Trench cum bund for soil and water conservation

Long term measures rainwater management in rainfed areas

Water harvesting

- Contour trenching for runoff collection
- On-farm reservoirs
- Ground water recharge structures (percolation tanks)
- Recharge through defunct wells

These rainwater management strategies have good scope for adoption in regions receiving more than 700 mm rainfall. Based on the rainfall regime, the above interventions can be prioritized. Before planning for long term measures, availability of runoff may need to be assessed for ensuring judicious investment. On other hand, in-situ conservation measures have large scale applicability in every region as they ensure well distributed moisture regime and may provide succor against short term dry spells.

While there should be continuous endeavour to ensure the rainfed areas to overcome from fluctuations of monsoon, it is also necessary to ensure the sustainability of the limited irrigated facilities available (through well, tank, small scale lift irrigation schemes) by following the improved methods of water management (furrow, sprinkler and drip). Crop diversification is to be promoted towards low water consuming crops which may enhance the overall water productivity of rainfed areas. In order to bring awareness among farming community on water availability and utilization pattern in watersheds, ground water monitoring on watershed/hydrological unit basis through farmer participatory mechanism need to be promoted with proper water budgeting approach encompassing water needs for all sectors within the watershed.

With the emphasis of watershed programme shifting from purely technical interventions to improve the livelihoods, the watershed programs need to be designed for more resilience to overcome the short term climate variability.

Efficient Use of Rain water

Increasing the stored water in the root profile and increasing its efficiency, remains the crux of soil and moisture conservation strategy. According to Kanwar (1982), the most rewarding practices that result in greater efficiency of rain water use are

- Use of crops and their varieties that require less water
- The tailoring of varieties to fit with the period of moisture availability
- Improve fertility management and rain water management
- Synergisistic effect of soil and water management and crop management systems

Response to supplemental irrigation varies with crops, time of irrigation, methods of water application, and fertilizer management.

Crop -Water Relations

Water stress in plants, particularly in post rainy season, is common phenomenon. It is characterized by decrease in osmotic and total water potential, accompanied by loss of turgor, reduced diffusion of carbon dioxide into plant leaves, and therefore, reduction in photosynthesis and decrease in growth. However the plant response to water stress is governed by soil, plant and environmental factors. The degree and duration of water stress and the growth stages, at which it occurs, considerably modify the crop response.

Moisture and effect on crop productivity

The major moisture perturbations affecting water use by the plants are:

- Change in ambient air humidity and
- Changes in availability of soil water for plant roots. Dry air or soil drought both cause stomatal closure and some reduction in leaf transpiration.

a) Changes in moisture availability

Change in moisture in the plants availability bring about changes in epidermal turgor pressure in the leaf tissue and hormones, such as cytokinin and abcisic acid sent by the root tips to the stomatal guard cell. Other hormones either transported by roots or synthesized in the leaves such as auxins may also be involved stomata close due to soil dryness and turgor is lost. Recovery is slow.

b) Increased evaporational demand

Due to increased evaporation demand which induces stomatal closure. However the stress caused by low atmospheric humidity is reversible and plant may recover on the resumption of water supply.

c) Growth and development is also affected by shortage of water

Shortage of water results in a partial closure of stomatal pores and loss of leaf water potential which results in slowing of cell division and expansion. A growing cell must maintain high turgor potential to force immature cell wall to stretch and expand. Under water stress, therefore leaf area development is slow down. About a critical limit of deficit soil moisture defined by soil characteristics and rooting depth of the plant, plant growth ceases and green leaves wilt. A prolonged stress finally results in a senescence and dropping of leaves.

d) Crop yield and productivity

Despite technological development, water availability is climatic factor affecting crop productivity and yields in arid and semi arid regions.

The water stress affects all physiological, morphological, biochemical, all phonological, groth and development processes as well as harmone relations, nutrient uptake, metabolism, photosynthesis, turgor etc. All the processes inside and outside the plant are affected. This results finally into the reduction in crop yields.

Drought management strategies in rainfed agriculture

In absence of any assured irrigation facility with ever growing changing patterns of temperature and precipitation, rain water management technology would play a greater role in rainfed areas. Renewed focused with incentive measures for in situ especially in low to medium rainfall regions with farmers themselves taking the leadership in conservation should be the local theme. The predominant interventions to overcome climate related impacts in rainfed areas includes

- Soil and water conservation practices
- Agronomic interventions
- Nutrient management practices
- Livestock based interventions
- Development of alternate land use plans

These interventions have a role to play in all agro-eco systems except their order of priority changes, which basically depends on rainfall, status of natural resources like soil and water.

Crop based interventions need to be planned based on amount and distribution of rainfall, availability and further augmentation of water resources with watershed programme. Based on the resources availability, broad guidelines for watershed programme are as follows:

Crop strategies based on amount of rainfall in rainfed condition

Regions with rainfall <500mm

Priority should be given for ensuring drinking water facility even during lean season. In situ conservation coupled with farm/ field boundaries should be given emphasis. Deep soils only should be encouraged for cultivation. Livestock based farming system should be encouraged. Fodder needs can be met by growing grasses in soils with low to medium soil depth. Runoff harvesting can be possible only if watershed receives runoff from upstream areas.

- Based on rainfall suitable crops during kharif season are pearlmillet, guar, mungbean, Cowpea, castor and Til. For Rabi season crop suitable are gram, Mustard, lentil and taramira.
- The cropping sequences are suggested are Pearmillet- Mustard, Pearmillet-Gram in case of late onset & normal withdrwal. Cowpea-Taramira, Guar-Barley, Til- Gram, when normal onset and early withdrawal.

Regions with rainfall of 500-700 mm

Crops can be grown in medium to deep soils with high available water content. Runoff harvesting could be possible in few cases for critical/supplemental irrigation. Horticulture can be promoted to large scale. Land capability based land use planning with emphasis on alternate land use need to be promoted.

Under this situation

a) Maize –chickpea as sequential system

b) Soybean –chickpea as sequential system

c) Maize intercrop with pigeon pea as intercrop system

d) Soybean intercrop with pigeon pea as intercrop system

In high rainfall (700mm) year's soybean –chickpea gave maximum economic return, followed by soybean /pigeonpea, maize-chickpea and maize/chickpea. In normal rainfall (600 mm) years Maize –Chickpea gave maximum economic return and maize/pigeonpea was not much different. These were followed by soybean -chickpea and soybean /pigeonpea cropping system. Mungbean- Wheat, Jowar -lentil under normal condition.

Regions with rainfall of 700-1100 mm

Farming system can be promoted medium to deep soils with medium to high available water content can be promoted for cultivation. Cropping and livestock based systems can be promoted. Runoff harvesting for supplemental irrigation can be planned within water shed. In few cases, residual moisture within fields or pre sowing irrigation for rabi crop is also possible and there is a need to explore the possibilities based on location specificity. In areas where the rainfall of 1000-1100 mm is received through south west monsoon, integrated aqua based farming system with fisheries in medium to low lands of rainfed rice (rice- fish-duck) system also be encouraged.

References

Anonymous (1995). Weekly Rainfall Probability. In "Weekly Rainfall Probability for selected stations of India". Vol. 1, pub. By Additional Director General of Meteorology, IMD-5. pp. i-viii.

Ananthakrishnan, R. 1976. Impact of Seasonal rainfall in rainfed areas. In proceeding of Symposium on Monsoon, Indian Institute of Tropical Mteorology, Pune, India. Pp. 254-324

Biswas, B.C. and Basarkar, S.S. (1982). Weekly rainfall probability over dry farming tract of Gujarat. Annals of Arid Zone, Vol. 21(3).

Gowariker, V, Thapaliyal, V. Kulshrestha, S.M., Mandal, G.S., Sen Roy, N., Sikka, D.R. 1991. Apower regression model for long range forecast of south west monsoon rainfall over India. Mausam, 42:125-130.

Havangi, G.V.1982. Water harvesting and life saving irrigation for increasing crop production in red soil areas of Karnatka. Pages 159-170 in proceedings of the Symposium on Rainwater and dry land agriculture, Indian National Science Academy, New Delhi, India, 3rd Oct, 1980 New Delhi, India.

Kanwar, J.S. 1982 Rainwater and dryland agriculture- an overview in proceedings of the Symposium on Rainwater and dry land agriculture, Indian National Science Academy, New Delhi, India, 3rd Oct, 1980 New Delhi, India. Pp. 1-9

Khambete, N.N. and Biswas, B.C. (1978). Characteristics of short period rainfall in Gujarat. Indian J. Met. Hydrol. Geophy. Vol.29 (3), pp. 521-527.

Khambete, N.N. and Kanade, V.V. 1985. A weekly rainfall analysis of dry farming tract in Karnataka. Mysore J. Agric. Sci. 19:1-10.

Mann, H.S and Singh, R.P. 1977. Crop production in the Indian arid zone. In: Desertification and its control. New Delhi, India: ICAR. Pp215-224.

Sarkar, R.P. and Biswas, B.C. 1988. A new approach to agro climatic classification to find out crop potential. Mausam 39: 343-358.

Sarkar, R.P., Biswas, B.C. and Khambete, N.N.1982. Probability analysis of short period rainfall in dry farming tract of India. Mausam 33: 263-284.

Virmani, S.M. 1980. The agricultural climate of the Hyderbad region. ICRISAT, Patacheru, Andhra Pradesh, India: ICRISAT.

Singh, H.P. and Bhati, T.K.1988 Ridge –Furrow system of planting dryland crops for soil and rainwater management in the Indian arid zone. In Challenges in Dryland Agriculture-A Global Prospective. Proceeding of International Conference on Dryland Farming, Aug.15-19, Amarillo/Bushland, Texas, USA. Pp. 575-577

Singh, H.P. and Gupta, J.P. 1989. Scope of water harvesting and control of water losses in deep percolation. Presented at International Symposium on Managing Sandy Soils, Central Arid Zone Research Institute, Jodhpur, India, Feb.6-10, 1989.

Singh, S.D. 1976. Trans. Ind.Soc. of Desert Technology 1(1):83.

Singh, S.D. (ed.) 1997. Water Harvesting in Desert. New Delhi, India: Manak Publications Pvt. Ltd. 170 pp.

Singh, S.D. ,Dauley, H.S. and Muthana, K.D. 1973. Runoff farming - making the best of available water. Ind. Farming 23: 7.

CHAPTER 21

Constraints and Strategies for Increasing Pulse Production in Dry Farming Situations

Bikram Singh[1], Anil Kumar[2] and B. C. Sharma[2]

[1]*Division of Plant Breeding and Genetics, Faculty of Agriculture, Sher-e-Kashmir University of Agricultural Sciences and Technology of Jammu, Chatha, Jammu-180009.*

[2]*Division of Agronomy, Faculty of Agriculture, Sher-e-Kashmir University of Agricultural Sciences and Technology of Jammu, Chatha, Jammu-180009.*

Pulse crops are wonderful gift of the nature. They provide nutritious food, feed and fodder. They also have ability to thrive well in fragile eco-systems where other crops often fail. Being rich in quality protein, minerals and Vitamins, they are inseparable ingredients in diet of vast majority of Indian population. The protein from pulses is easily digestible, relatively cheaper and with high biological value. The lysine rich protein of pulses is considered to supplement the deficiency of amino acid in cereal dietaries and comes at par with milk protein in terms of biological efficiency. Pulses have the ability to trap atmosphere nitrogen in their root nodules in association with Rhizobium bringing qualitative improvement in physical, chemical and biological properties of soil. Have ability to fix atmospheric nitrogen 58-109 kg/ha in mungbean and 55-140 kg/ha urdbean in symbiotic association with Rhizobium bacteria, which enables these to meet their own nitrogen requirement and also benefit the succeeding crops (Ali,1992). On account of these virtues, pulses remain on integral component of subsistence farming system of semi-arid tropics since time immemorial.

In recent years, the role of pulses in sustaining soil productivity has also been recognized under intensive cropping system of irrigated areas. Thereby, arresting deleterious effects on natural resources, due to continuous cultivation of high water and input intensive cereal crops. Pulse crops fit well in mixed/intercropping systems, crop rotations and as dry farming short duration crops. On account of short duration and photo thermo insensitivity, they are considered excellent for crop intensification and diversification.

Globally, India has the largest area under pulse crops (22-24m ha) with annual production of 13-15 million tonnes. However, their productivity is quite low (635 kg/ha). Among various constraints, poor crop management practices assume prime position. The on-farm demonstration shows that the improved agro-techniques evolved in years are capable of doubling the pulse productivity. Therefore, aggressive efforts are required to popularize the improved production technologies.

While gradual transformation of Indian agriculture from subsistence to magnificence has made country food sufficient, daily per capita availability of pulses has substantially declined from a comfortable position of 69 g in 1961 to a mere 32 g at present. Availability of pulses per capita/year has declined. In developed world dependence on animal protein is 56%. But in developing countries only 12% protein comes from animal source whereas 80% protein requirement comes from plant sources mainly pulses. This makes the country nutritionally vulnerable despite increased availability of food grains. To alleviate various problems associated with protein energy malnutrition of more than 53% of the malnourished in the country, the availability of pulse should be at least of 50g/capita/day. 1 gm of protein/capita/day for each kg weight is recommended by WHO. If, average body weight of 60 kg is there, one requires 60 gm/day of proteins.

Increased production of animal and fish products in most of developing countries is encouraging sign. However, increase supply of plant protein is needed especially in India where majority of the people are vegetarian and prefer pulses in their regular diet. Pulses are also major source of vitamins like riboflavin, thiamine, niacin and iron. There is now renewed interest in consumption of vegetable protein due to its fiber-rich nature in comparison to animal protein which is more in the form of concentrate (Srivastava and Ali 2004).

Table 1: Nutritive value of various Pulses (in per centage)

Crop	Protein	Carbohydrate	Fat	Minerals
Chickpea	18-22	52-70	4-10	Ca,P,Fe
Fieldpea	20-24	60-65	1.5-2	Ca,Fe
Lentil	24-26	55-60	1-1.5	Ca,P,Fe
Rajmash	20-24	58-62	1.15	Ca,P,Fe
Arhar	21-25	52-68	0.50-0.85	Ca, P, Fe
Mungbean & Urdbean	24-28	59-65	1-1.5	Ca, P, Fe

Major Pulse Crops of India

In India major pulses grown can be classified into two groups

i. Rabi pulses: Gram (*Cicer arietinum*), Lentil (*Lens culinaris* Medik), Fieldpea (*Pisum sativum*); Lathyrus (*Lathyrus sativus* L.) or grasspea called Khesari and Teora in Hindi, Rajmash (*Phaseolus vulgaris* L.).

ii. Kharif pulses, Urdbean or black gram (*Vigna mungo* L Hopper) Mungbean (*Vigna radiata* L. Wilczek), Pigeonpea or Redgram (*Cajanus cajan* L. Millsp),

Cowpea (*Vigna sinensis*), Horsegram or kulthi (*Macrotyloma uniflorum*), Kidneybean (*Vagina acontifolia*).

Production Scenario

In India, both production and productivity are lower than rest of the world. In order to ensure "better household nutritional security" as per recommendations of the international conference on nutrition (ICN) concerted efforts are needed at this stage to improve further the productivity of pulse crops in most of the pulses growing states in India (Ali *et al* 2003. These crops are likely to play in the future for sustainable agriculture and rural development (SARD).

Table 2: All India Area, Production and Yield of Pulses

S No	Year	Area (m ha)	Production(m tonnes)	Yield (Kg/h)
1	1949-50	20.17	8.16	405
2	1959-60	24.83	11.80	475
3	1969-70	22.02	11.69	531
4	1979-80	22.26	8.57	385
5	1989-90	23.41	12.86	549
6	1999-2000	21.19	13.35	630
7	2005-06	22.40	13.38	598
8	2006-07	23.76	14.20	498
9	2007-08	23.81	15.11	635

Annual import has increased from 0.50 million tonnes to 1.80 million tonnes during 1997-98 to 2001-02 and contribution of pulses in the national food basket reduced from 17% to 7%. To make India self reliant in pulse production, at least 16.0 million tonnes production target was fixed to be achieved by terminal year (2006-2007) of the Xth plan.

Compared to cereals, progress for both the production and productivity of pulses had been rather slow during the last three decades. Growth rate for pulses (1.4%) had been slow, resulting in wide gap between the ratios of pulses to cereals from 1:14 in 1960 to 1:32 in 1990. Availability of pulses per capita per annum has gone down from 10.3 kg in 1971 to 8.5 kg in 1991 whereas that of cereals has gone up from 235.4 kg in 1971 to 286.3 kg in 1991. On the contrary availability of pulses in other regions has increased. Hence urgency for increased productivity of pulses is quite evident.

Existing constraints

The virtue of pulses to give more yields even under adverse conditions has pushed them to more and more neglected areas. The major constraints, in general, are as fallows:

Agro-ecological constraints

The intrinsic characteristic of pulse crop to make the more efficient use of the scare farm-resources has gone against them as over 92 per cent of pulses are grown in rainfed areas. This has become the main barrier for their timely planting and getting desired plant stand in field. The marginal and sub-marginal rainfed lands occupied by pulses are generally poor in soil fertility and moisture retention capacity and therefore, yields are poor. Crops with high yielding varieties of cereals, oilseeds and cash crops, have often pushed pulses to more and more marginal areas.

Biological constraints

Pulses have remained neglected crops ever since they were brought into cultivation. The primitive characters like perennial habit, indeterminate growth, prolonged flowering over extended period of time, deep root system, compound leaves, considerable flower drop and shattering pods associated with pulses are the testimony of the fact that their domestication was aimed more for their survival under adverse conditions rather than high yields. Such criteria of domestication led to several weak links in productivity which are the major inherent constraints.

The pulses are highly prone to a large number of insect pests and diseases as compared to cereals and other crops. Several fungal, bacterial and viral diseases often ruin these crops. Similarly, a large number of insect pests such as pod borers, jassids, thrips, bugs, etc., feed on these crops and reduce their yields drastically. Stored grain pests cause considerable loss to pulses, compelling farmers to sell their produce immediately after harvest. These havocs discourage farmers to take up pulses cultivation on large scale.

Management/Institutional constraints.

Lack of Improved varieties: New varieties in pulses have only marginal yield advantage but possess better tolerance to biotic and abiotic stresses than new cereals. It is likely that selection pressure under subsistence agriculture had been responsible for erosion of high yielding genes in most of the pulses (Singh and Satyanarayana 1997). Hence, concerted efforts to improve productivity of pulses are required. There is also an urgent need to improve the harvest index (HI) by adopting the concept of new plant type or by exploiting heterosis in some of the crops as pigeonpea. Harvest index in pulses is only 15-20% compared to 45-50% in case of high yielding varieties of cereals (wheat & rice).

Seed production: Lack of good quality seed and poor replacement ratio is another major constraint with the result farmers are not able to derive the benefit of technological advancement. The inadequate availability of seeds of improved varieties of pulses continues to be a stumbling block of increased pulses production. As a result of this, the improved varieties could not reach the farmers and traditional land races which are poor yielder and susceptible to a large number of diseases are grown. Certified seeds of improved varieties of all pulses must be produced in sufficient quantity by State Seed Corporations, National Seed Corporation as well as cooperative and private seed producers.

Research: Since pulse are only 3 per cent of total food grains in Asia-Pacific region and hence did not get required policy support in many countries for research and development activities. Pulses being many in numbers do not find concerted attention to the extent other crops have received. India is the chief pulses producing country in the world and the possibility of inflow of research information from other countries parallel to rice and wheat is practically negligible. The fundamental research base on pulses as a whole is miserable. A voluminous backlog exists in research of basic nature to elucidate the factors which inhibit the productivity and to find out their appropriate remedies for realizing the full yield potential in a given agro-ecological condition.

Extension: The technology gap between research institutes and farmers is very wide. The available technology has the potential to double the present level of productivity but there is urgent need for sound and sustained extension approach. Pulses being grown on marginal and risk-prone areas, farmers are not willing to use costly inputs as fertilizer, pesticide, herbicides, especially the micro-nutrients as Zn, S, Mg, Ca etc. Starting dose of 20-25 kg N/ha and 40-50 kg P/ha could result in significant advancement in yield. Similarly, just one supplemental irrigation could substantially improve productivity.

Training: Facilities for training of scientists, developmental and extension workers and farmers are quite meager. This has resulted in communication gap about latest technology and feed back. The farmers need to be educated regarding production potential of improved varieties of pulses.

Plant protection: Pulses are prone to insect pests and diseases. Losses due to disease and insect pest also take heavy loss in case of pulses. Cutworm, defoliation, pod borers, bruchids, podfly, fungal, bacteria and viral diseases are quite common. Breeding for disease and pest resistance is, therefore, an area of high priority in case of pulse crops. It is ironical that even after raising a successful crop of pulses, the farmer has to undergo heavy losses because he fails to control insect pests as he does not have enough resources to buy sprayers/dusters and insecticides. To make a dent in elevating pulses production, the government must provide service units for cluster of villages where from the farmers may get necessary equipments and insecticides, timely on subsidized rate.

Table 3: Major diseases of pulses

Mungbean/urdbean	Chickepa	Lentil	Fieldpea
Yellow mosaic virus	Fusarium wilt and root rot	Rust	Powdry mildew
Leaf crinkle virus	Ascochyta blight	Stem rot	Fusarium wilt
Macrophomina blight	Botrytis gray mould	Ascochyta blight	

Table 4: Major insect pests attacking pulses

Mungbean/urdbean	Chickepa	Lentil	Fieldpea
Thrips, caliothrips	Aphid, Aphis craccivora		Leaf minor
Jassids, E.Keri	Gram pod borer	Lentil pod borer	Stem borer
Whitefly, B.tabaci	*H.armigera*	Black aphid	Green aphids
	Termites, *Odontotermes spp.*		

a) **Microbial culture and quality control**: It has been established that yield of pulses can be stepped up substantially by use of efficient *Rhizobium* culture, but unfortunately the facilities available for its production are not sufficient. The use of the culture has not picked up. There is no institutional mechanism to guarantee quality control also.

b) **Processing technology:** The conventional milling of pulses is laborious, time consuming, completely dependent on weather and subject to high losses of the edible portion. It is estimated that about 1.5 million tones of pulses valued at Rs. 750 crores per annum are lost directly or indirectly due to inefficient milling technology. There is definite need for modernization of processing machines and establishment of small scale dal milling units in the villages.

c) **Socio-economic constraints**: Pulses have subsidiary status in the total farming systems of peasants, perhaps because of several prevalent notions and belief which determine farmers response like (i) cereals are the staple food, (ii) pulses are not major cash crops being mainly consumed at the farmer's family level, (iii) pulses may not be able to exploit a good resource base, (iv) low stability of production of pulses due to number of diseases and insect pests, and (v) high damage in storage and highly fluctuating post-harvest prices and other associated problems of local level marketing.

Trends in Area, Production and Productivity of Pulses in J & K from 1980 onwards:

The area under pulses was continuously on decrease from 48.60 th ha in 1980-81 to 27.54 th. ha in 2003-04 but from 2004-05 onwards it has again increased to 30.90 th.ha. With the decrease in area the productivity levels that was hovering around

600 kg/ha also has also slashed down to around 500 kg/ha. The reasons for decline in acreage and productivity could be attributed to following main reasons:

Table 5: Area, Production and productivity of Pulses in J&K State

S. No.	Year	Area (000'ha)	Production (000'q)	Productivity (q/ha)
1.	1980-81	48.60	337	6.93
2.	1990-91	41.32	268	6.49
3.	1995-96	32.57	152	4.67
4.	1999-2000	29.27	145	4.95
5.	2000-01	27.25	128	4.70
6.	2002-03	28.86	142	4.92
7.	2003-04	27.54	132	4.79
8.	2004-05	30.90	152	4.91
9.	2005-06	29.27	135	4.61

- The maximum area under pulses during 1980-81 in rabi and kharif was coming from now called Ravi- Tawi command area which was once rainfed belt with no source of assured irrigation. With the availability of canal irrigation water the farmers have opted for better remunerative crops as wheat and oilseeds in rabi and rice in kharif. Now again SKUAST-J & Directorate of Command Area are motivating the farmers to bring more area under pulses by adopting improved technology.
- The area for kharif pulses of Jammu province was coming from rainfed dryland areas of zone I (Kathua and Jammu districts) Zone II (Rajouri) Zone III (Udhampur) where Urdbean was the main crop. The Yellow Mosaic Virus have been the major problem of these areas and due to non availability of sufficient quality seed of resistant cultivars, the farmers opted for maize and other crops. Now with recommendation of new resistant varieties of Urdbean by SKUAST-J the area under pulses is on increase.
- Due to continuous increase in population and fragmentation of land holding farmers have opted for other means of livelihood as government service and industrial works thereby putting little efforts on their fragmented holding. This has resulted in decrease in productivity levels since last two decades.
- The pulses have become a crop of marginal, less productive and rainfed areas with no inputs from farmers at all. Now with field demonstrations by SKUAST-J farmers are being motivated to adopt improved production technology.
- During the past two decades there have been aberrant climatic changes in this region, and pulses being the most sensitive and least stable in productivity have high risk of failure thereby forcing farmers to opt for more stable crop. There is still need for evolving stable variety of various pulses.

The maximum area of pulses in Jammu region is under Urdbean followed by Chickpea and Fieldpea. Urdbean and mungbean are major pulses during Kharif and Chickpea, Fieldpea and Lentil during Rabi.

Strategies for Improving Pulse Production

1) **Choose appropriate high-yielding varieties** combining resistance to prevalent major diseases and pests

Recommended mungbean varieties for North West Plain Zone

Variety	Pedigree	Yr. of release	Released by	Salient features
PDM 54	Bahraich selection	1987	GBPUA&T Pantnagar	For sub tropical and intermed. zones (upto 750m) both for spring & Main season
ML 131	ML 1 x ML 23	1980	PAU, Ludhiana	For sub tropical and intermediate zones, YMV & PM resistant
Pant M 3	LM 294-1 x L 80	1985	GBPUA & T Pantnagar	Maturity 72-77 days, tol. to pod borer attack, YMV resistant, Av. yield 10.5 q/ha
ML 5	No54xHy645	1976	PAU,	Maturity 71-82 days, tol. to pod borer
ML 818	5145/87xML 267		PAU, Ludhiana	Maturity 67-73 days, low whitefly incidence, tol. to pod borer
Pusa 105	Sel. M 178	1987	IARI, Delhi	Matures in 70-73 days
Pusa vishal		2000	IARI, New Delhi	

Recommended chickpea varieties for North West Plain Zone

Variety	Pedigree	Yr. of release	Released By	Salient features
C - 235	JP 85 x C 1234	1975	PAU, Ludhiana	Tol. to pod borer, *Ascochyta blight*, suitable for rainfed areas
Gaurav	_	1985	CCS HAU Hisar	Res. to *Ascochyta blight* and grey mold
Avrodhi	_	1987	U.P.	_
PBG - 1	GG 578 x NEX 206	1988	PAU, Ludhiana	Late sown conditions, Res. to *Ascochyta blight*, grey mold & drought, maturity 150-160 days
Pusa/BG 372	P 1231 x P 1265	1993	IARI, New	Tol. to pod borer
GNG 469 (Samrat)	Annegeri 1 x H75-35	1996	Sriganga-nagar, Raj.	Drought res., tol. to frost & low temp., maturity 160 days, for rainfed areas, bold seeded

DCP 92-3	Germ. sel.	1997	IIPR, Kanpur	Maturity 155 - 165 days
BG1053(K) (Chamatkar)	Sel. from ICCV 3	1999	IARI, New Delhi	Tol. to pod borer & drought, maturity 165 - 175 days
RSG 888	RSG 44 x E 100 ym	2002	RAU, Durgapura	Tol. to pod borer, for rainfed conditions
RSG 931	RSG 44xRSG 524	2003	-Do-	For rainfed conditions
SCS-3 (Shivani)	**Sel. from IPC - 97 67**	**2005**	**SKUAST-J, J&K**	**Mod. tol. to pod borer, drought res., maturity 165 days**
RSG963 (Aadhar)	RSG 524 x PDG 84-16	2004	RAU, Durgapura	Late sown, reddish brown, medium bold seed, maturity 135 days
H 82-2	F 61 x L 550	_	CCS HAU	Late sown, Mod. res. to wilt

Recommended lentil varieties for North West Plain Zone

Variety	Pedigree	Year of release	Released from	Salient features
PL - 406	P 495 Sel.	1980	GBPUAT	Small seeded
LH 84-8 (Sapna)	L 9/12 x JLS 2	1991	CCSHAU Hisar,	Bold, Res. to rust, Maturity 135 days,Yield15 q/ha, semi-spreading
L - 4076	PL 234 x PL 639	1993	IARI, Delhi	Bold, Res. to rust, Maturity 135 days, Yield14q/ha, semi-spreading.
Pant L - 4	VPL 175 x (PL 184 x P 288)	1993	GBPUAT Pantnagar	Small,Res. to rust & wilt, Maturity 140days,Yield16q/ha,semi-spreading
DPL 15 (Priya)	(PL 406 x L 4076)	1995	IIPR, Kanpur	Bold, Res. to rust, Tol. to wilt, Maturity 135 days, Yield 15.5 q/ha
L-4147(PVaibhav)	(L 3875 x P4) x PKVL	1996	IARI, New Delhi	Small seeded, Res. to rust, Maturity 134 days, Av. Yield - 17.8 q/ha
DPL 62 (Sheri)	JLS 1 x LG171	1997	IIPR, Kanpur	Bold, Res. to rust, Tol. to wilt, Maturity 135 days, Av.Yield16 q/ha

Recommended fieldpea varieties for North West Plain Zone

Variety	Pedigree	Yr of release	Released by	Salient features
HFP 4 (Aparna)	T 163 x EC 10916	1988	CCS HAU Hisar	Dwarf, Maturity 145 days, Res. PM, leafless, yield:26 q/ha
DMR 7 (Alankar)	6587 x L 116	1996	IARI, New Delhi	Tall, Maturity 115-135 days, Res. to PM, bold seeded, Av. yield - 24 q/ha
HFP 8909 (Uttra)	EC 109185 x HFP 4	1996	CCS HAU Hisar	Tall, Maturity 125 - 130 days, Res. to P. M., Av. yield - 26 q/ha
KPMR 522	KPMR 156 x HFP 4	2001	CSAU, Kanpur	Dwarf, Maturity:134 days, Res. to PM, Av. Yield: 22.9 q/ha
DDR 27 (P panna)	HFP 4 x Pea 1542	2001	IARI, Delhi	Very early, Maturity109 days, Res. to PM, Av. yield - 17.6 q/ha
Rachna	T 163 x T10	1982	CSAU, Kanpur	Early maturing, PM resistant

2) **Use certified seeds:** Non-availability of quality seeds in adequate quantity is one of the major constraints in pulse production. The seed replacement rate is very low (2 to 5%). Increasing seed replacement rate from 4% to 10%. Production of quality seed from 40,000 mt to 100,000 mt.

3) **Integrated pest/disease management:** On an average, 20-40% crop is annually lost due to damage caused by pod borers in pigeonpea and chickpea. Wilt and root rots cause heavy loss to pigeonpea and chickpea crops. Effective IPM module is available for management of targeted pest and diseases. Popularization of seed treatment with *Trichoderma sp.* against wilt and root rot. Conducting IPM demonstrations on chickpea and pigeonpea against key insect pests and diseases.

4) **Integrated Nutrient Management**: Adhere to adequate and balanced use of fertilizers. Apply 15-20 kg N, 30-40 kg P_2O_5 and 20 kg S per ha to meet the nutrient requirement of the crops. Promoting use of single super phosphate through appropriate incentive package. Creating awareness among farmers about importance of sulphur and zinc through demonstration and providing incentive to farmers for their use. Inoculate the seeds with rhizobium culture before sowing. Promoting use of quality bio-fertilizers. Laying front line demonstrations on use of bio-fertilizers.

5) **Protective Irrigation**: Avoid moisture stress at flowering and pod development stages. Popularization of sprinkler sets with package incentive in the rainfed areas.

6) **Post Harvest Management:** Pulses suffer heavy losses due to stored grain pests. The quality of seeds stored in the traditional storage structures also deteriorates. Further, there are no small processing units to convert pulse grains into Dal and other byproducts. This compels the growers to dispose off their produce immediately after harvest at low price. Educating farmers for Dal Mills and metal storage bins. Establishing model pulse processing centers at district levels in major pulse producing areas. Involving KVKs and NGOs to impart training about processing and value addition to entrepreneurs.

7) **Technology Transfer:** Transfer of improved pulse production technologies have been the most neglected component since, and consequently the benefit of improved varieties and production technology could not be harnessed. Conducting of front line demonstrations on improved varieties and production technology to disseminate seed of new varieties and popularize improved production technology involving KVKs, NGOs, SAUs and private sectors. Massive programme on mass media, printed literature in vernacular languages and Kisan Gosthies Special training programme for (i) farmers, (ii) seed growers, (iii) extension workers (iv) development officers of the concerned central nodal Development Directorate (DPD).

8) **Farm Implements:** Ridge planting has been found very effective in ensuring optimal plant stand and consequently higher yield. Distribution of simple ridger with package incentive would be quite useful.

9) **Improve drainage facility:** Poor drainage/water stagnation during rainy season causes heavy losses to pulses account of low plant stand and increased incidence of diseases as Phytophthora.

10) **Timely sowing**, preferably by mid July for urdbean and mungbean and first fortnight of November for chickpea, pea and lentil.

11) **In blight affected** areas delay the sowing of chickpea upto mid November.

12) **Use recommended seed rate** to maintain optimum plant population of 1.5-2.0 lakhs for early pigeonpea, 3.33 lakhs for kharif urdbean/mungbean, 7-8 lakhs for summer mungbean, 2.5-3.6 lakhs for chickpea and pea.

13) **Plant the crop at proper spacing**; mungbean, urdbean, chickpea at 30-35 cm, and pigeon at 45 cm in case of early duration varieties and 60 cm in case of late duration varieties.

14) **Popularize summer mungbean** and spring urdbean as catch crops.

15) **Adhere to timely weed control**. Keep the field weed free for the initial 6 weeks. Apply 1.0 kg pendimethalin or 1.0 kg basalin/ha(soil incorporation) before sowing.

16) **Grow mungbean as catch crop** after the harvesting of wheat crop in the first week of April.

Strategy for future research

Suitable strategy on pulses is necessary to overcome the existing constraints.

1. **Development of multiple stress tolerant varieties:** Varieties developed in past with resistance to single stress needs to be modified by developing varieties having resistance to more than one stress for greater insurance. In India, there is need to incorporate resistance/ tolerance against web blight, anthracnose, leaf crinkle all virus and pre-harvest sprouting for kharif crops.

Table 6. Promising accession for major biotic stresses in mungbean germplasm

Resistance/ tolerance to	Accessions evaluated	Promising accessions
MYMV	29	ML 337,ML 423,ML 428
	39	ML-370,ML 428,ML 459,Ml 506,ML 508,ML 537
CLS	60	MB 55,Sitakundu
MYMV and CLS	37	E 321,NCM 69,NCM 68
	150	PBM 49,ML 820,Pusa 9472
Powdery mildew	170	ML 223,ML 395
	36	JRUM 1

Macrophomina blight	102	PLU 241,PLU 137, Sel 1
Leaf blight	609	Line 112,BR 10
Cyst nematode	14	CO 282/1,VB 4
Root knot nematode	300	PLU 648

Table 7 Promising accession for major biotic stresses in urdbean germplasm

Resistance/ tolerance to	Accessions evaluated	Promising accessions
MYMV	415	NP 21,340/1,Line 100,7382
	975	
	670	IC 27026,IC 106088,IPU 99-127
Cercospora leaf spot	100	TEU 95-1, Pusa 3, UPU 91-7, UPU 95-1, P 38, UPBU 51, UG 407
Powdery mildew	87	-
	48	P 115
Macrophomina blight	102	PLU 241,PLU 137, Sel 1
Leaf blight	609	Line 112,BR 10
Cyst nematode	14	CO 282/1,VB 4
Root knot nematode	300	PLU 648

2. **Tailoring plant type for target environment**. Lack of efficient plant types for different situation which can provide more than 1.5 tonnes yield, is still major constraint in breaking the yield plateau in mungbean and urdbean. Presently, available plant are photo-thermo sensitive with indeterminate growth habit and low harvest index. For mungbean high yielding with 85-90 days for Kharif season and 65-70 days for spring season combining determinate growth habit, high harvest index and reduced photoperiod sensitivity are required. For summer/spring extra early varieties of 55-60 days with synchronous maturity are desirable. Vegetative growth should get terminated with flowering and assimilates should be transported to developing pods. The major yield limiting barriers are lack of seedling vigor, excessive flower production, flower drops, poor pod seeding, poor harvest index, low response to inputs, narrow adaptation, indeterminate growth habit, staggered maturity and sensitivity to photoperiods and temperature (Saini and Das,1997). The phenomenon of compensation among yield components is still considered to be main yield limiting factor (Kumar 2005).

3. **Broadening the genetic base:** Limited variability has been exploited in varietal development programmes. Pedigree analysis of released cultivations indicated that a small number of parents with high degree of relatedness were repeatedly used in crossing programme (Tickoo *et al* 2006). In mungbean 49 varieties developed through hybridization involved only 71 ancestors (Kumar *et al*.2004). The top ten ancestors contributed 79% of genetic base with T 1 appearing in 35% of varieties, followed by T 49, BR2, G65 and

Madira. In urdbean ancestors involved are only 26 with relative genetic contributions of top ten ancestors as high as 69%. T 9 is the most frequent used ancestor appearing in 64% of varieties followed by D6-7, G 31, Netiminumum and AB 1-33 (Gupta and Kumar 2006). It indicate very narrow genetic base of the released varieties of these legumes.

Use wild & cultivated vigna spps to incorporate normal characters & broaden genetic base *V. radiata var.sublobatusx V. mugo, V. ttrilobus , umbellate, glabosha, trilobata, V. mungo var. sylbestris* for CLS, MYMV.

Table 8: Variability ranges for different quantitative traits in mungbean/urdbean

Character	Moongbean	Urdbean
Days to flowering	34.0-69.9	29.0-80.7
Days to maturity	52.0-104.0	-
Plant height (cm)	22.6-130.0	16.6-280.0
Primary branches	1.0-8.6	-
Branches pe rplant	--	1.0-15.7
Pods per plant	2.0-95.6	3.0-259.0
Pod length (cm)	4.4-9.7	3.0-5.3
Clusters per plant	2.0-29.8	-
Seeds per pod	3.4-14.5	3.6-9.0
100 seed weight (g)	1.0-3.9	1.9-5.6
Yield per plant (g)	0.3-26.3	0.5-8.3

4. **Molecular breeding**: DNA marker techniques are useful innovations and have tremendous potential in increasing the efficiency of conventional breeding programmes. DNA marker technique can be used for tagging important genes which can subsequently be used for early, easy and precise selection of promising genotypes. Molecular maps of mungbean and urdbean with 255 RFLP markers and 148 markers have already been developed and many useful traits like bruchid resistance, PM resistance, seed size, etc., have been tagged (Chaitieng *et al.* 2006). Efforts are underway to develop well saturated maps of legumes so as to map important genes/QTLs responsible for trait of economic importance.
5. **Biotic stresses:** Disease and insect pest cause considerable yield loss in pulses. Mungbean yellow mosaic virus (MYMV), Cercospora leaf spot and powdery mildew are of considerable economic importance. During vegetative stage defoliators like hairy caterpillars, semilooper and caterpillar are common pests. Activity of thrips starts at bud stage and poses serious problem when the crops attains peak flowering, resulting in heavy flower drop. There is no resistant variety against these pests.
6. **Abiotic stress:** Erratic weather conditions as heavy rains and intermittent/terminal drought lead to significant yield losses in Pulses. Pre-harvest sprouting especially in mungbean pose a serious thereat to timely sown crop dur-

ing rainy season. Intense heat and hot winds during May-June lead to flower drop and poor pod set in spring / summer crop.

7. **Production Technology**: Agronomic practices for pulses under relay cropping in rice fallows and intercropping with cereals, cilseeds, and commercial crops needs to be standardized. Population management, fertilizer use and weed management need attention for popularization of these crops.

Conclusions

Pulses would continue to be the major source of protein in India. National policy concerning thrust to be given on pulses in view of their role in nutritional security as well as sustainability would determine the extent to which their production and productivity are improved. Unfortunately pulses have not received the required policy research developmental support in past being grown mostly in the rainfed and marginal areas, the productivity of pulse crops has often remained low in the region. However, considerable intra-regional variation exists, thereby indicating the potential for increasing both the production and productivity of pulse crops provided concerted and systematic efforts are made to improved the existing scenario.

However, the present technology available if passed on to the farmers will certainly increase the production substantially. For this purpose, several programmes through technology mission of oilseeds & pulses have been initiated.

It is encouraging to note that with a little push to improve further the productivity for which required technology is in place, considerable progress is possible to meet our future demands of protein from grain legume. Pulses also have to play a significant role in future towards sustainability of various cropping systems. Hence scope exist for horizontal expansions in the region, besides vertical growth.

References

Ali, M., Kumar, S. and Singh N. B. 2003. *Chickpea Research in India* (Edited) Indian Institute of Pulses Research Kanpur India 344p.

Ali, M.1992. Effect of summer legumes on productivity and nitrogen economy of succeeding rice (*Oryza sativa*) in sequential cropping. *Indian Journal of Agricultural Sciences* 62:466-467

Chaiteng, B., Kaga, A., Tomooka, N., Isemura, T., Kuroda, Y. and Vaughan, D. A. 2006. Development of black gram (*Vigna mungo* L. Hepper) linkage map and its comparison with azaki bean (*Vigna angularis* (willd) ohwi and ohashi) linkage map. *Theoretical and Applied Genetics*. 113:1261-1269

Gupta, S. and Kumar, S. 2006. Urdbean breeding. In *Advances in Mungbean and Urdbean. Eds.* Masood Ali and Shiv Kumar. IIPR, Kanpur. Pp-149-168

Kumar, S. 2005. New Plant type in urdbean. Pulses Newsletter 16(3): 4-5

Kumar, S., Gupta, S., Chandra, S. and Singh B.B.2004. How wide is the genetic base of pulse crops. In *Pulses in New Perspective* (Eds. Maqsood Ali, B.B. Singh, Shiv Kumar and Vishwa Dhar). Indian Society of pulse Research and Development. Kanpur, India. Pp 211-221.

Saini, A.D. & Das, K. 1979. Studies on growth & yield of three mungbean (*Vigna radiata* L.) cultivars. *Indian Journal of Plant Physiology* 22:147-154.

Singh, D.P. and Sayanaragara, S. 1997. Creating higher genetic yield potential in urdbean. In *Recent Advances in Pulses Research* (Eds. A. N. Asthna and Maqsood Ali) Kanpur, India. Indian Society of Pulses Research & Development. Pp. 157-171.

Srivastava, R.P. and Ali, M. 2004. *Nutritional Quality of common Pulses,* Indian Institute of Pulses Research. Kanpur :Pp. 65

Tickoo, J. L., Lal, S.K., Chandra N. and Dikshit, H. K. 2006. Mungbean breeding. *In Advances in Mungbean and Urdbean.* (Eds. Masood Ali and Shiv Kumar). IIPR, Kanpur pp. Pp. 110-148

CHAPTER 22 Biotechnological Intervention for Crop Improvement Under Moisture Stressed Conditions

Gyanendra Kumar Rai and Anil Kumar Singh

Division of plant breeding and Genetics, Faculty of Agriculture, Sher-e-Kashmir University of Agricultural Sciences and Technology of Jammu, Chatha, Jammu-180009.

Dehydration stress is one of the most serious yield-reducing stresses in agriculture. Drought stress is especially important in countries where crop agriculture is essentially rainfed. In India drought have a devastating effect on food security. While irrigation is the method of choice in averting drought stress in many areas of the world, alternative low-input approaches are being explored, and biotechnology offers a promising array of tools that may be useful in achieving drought tolerance in plants. One such tool is the low input approach to crop production by which crops are modified to suit the environment in which they are growing, rather than modifying the environment to meet the needs of the crop. This approach is advantageous in areas where water supplementation by irrigation is either difficult or unaffordable.

What happens on plants during drought?

Drought stress causes an increase in solute concentration in the environment, leading to an osmotic flow of water out of plant cells. This in turn causes the solute concentration inside plant cells to increase, thus lowering water potential and disrupting membranes along with essential processes like photosynthesis. These drought-stressed plants consequently exhibit poor growth and yield. In worst case scenarios, the plants completely die. Certain plants have devised mechanisms to survive under low water conditions. These mechanisms have been classified as tolerance, avoidance, or escape.

ROSes may be bad

Central to signal transduction pathways related to drought and other stresses are reactive oxygen species (ROS), which are molecules, formed by the incomplete one-electron reduction of oxygen. Under stress, ROS formation is usually exacerbated. Drought stress leads to the disruption of electron transport systems; therefore, under water deficit conditions, the main sites of ROS production in the plant cell are organelles with highly oxidizing metabolic activities or with sustained electron flows: chloroplasts, mitochondria, and microbodies. ROS are generally damaging to essential cellular components, and plants have evolved various ROS scavenging mechanisms. These include the enzymes superoxide dismutase (SOD), catalase, and peroxidases, as well as oxidized and reduced glutathione.**Fig.1** show the relationship between ROS-scavenging mechanisms, stress, and the damage to cellular membranes and macromolecules.

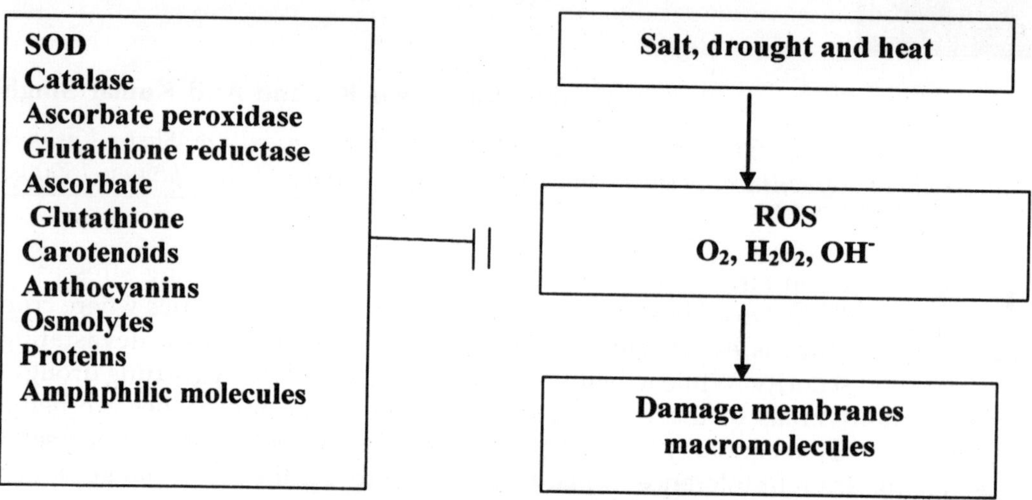

Fig. 1. Reactive oxygen species pathway.

Some researchers have focused on expressing genes for enzymes involved in ROS scavenging to enhance plant protection against oxidative stress. Transgenic alfalfa (*Medicago sativa*) expressing Mn-super oxidedismutase cDNA tended to have reduced injury from water-deficit stress, and this improvement was also seen in field trials in yield and survival.

Secrets of resurrection

What does it take to rise from the dead? This is a question scientists working on resurrection plants have been exploring recently. Resurrection plants can tolerate almost complete water loss in their vegetative parts. Researchers are trying to unlock the secrets of the resurrection plant *Xerophyta viscosa* in an attempt to achieve drought tolerance in crops. These plants can be a source of drought tolerance genes for transgenic crop improvement. To withstand periods of drought, resurrection plants practically "die" (by losing all their vegetative parts) and then "rise again" when water becomes available. Their vegetative tissues lose all free water and then rehydrate once water becomes available again. Resurrection plants minimize ROS formation and also upregulate various antioxidant protectants during drying and rehydration. The group has identified a novel stress-inducible antioxidant enzyme, *XvPer1*, by differential screening of a cDNA library of *X.viscose*.

Osmoprotectants

Osmolytes are involved in signaling/regulating plant responses to multiple stresses, including reduced growth that may be part of the plant's adaptation against stress. In plants, the common osmolytes are proline, trehalose, fructan, mannitol and glycinebetaine. The protection mechanisms are not yet fully understood, but they are thought to work via osmotic adjustment, stabilizing macromolecules, and scavenging ROS. One proposed transgenic strategy has been to overproduce osmolytes. However, transgenic plants overproducing osmolytes often exhibit impaired growth.

Trehalose, a non-reducing disaccharide, protects biological molecules in response to different stress conditions in many microorganisms. Plant biologists are interested in channeling trehalose metabolism (Fig.2) to enhance stress tolerance in plants. Trehalose is made from UDP-glucose and glucose-6-phosphate via a two step process. 6-phosphate is converted to trehalose by a phosphatase encoded by the bacterial *otsB* gene. Tobacco plants transformed with bacterial *ots A* have greater ability to retain water and greater ability to photosynthesize under water stress.

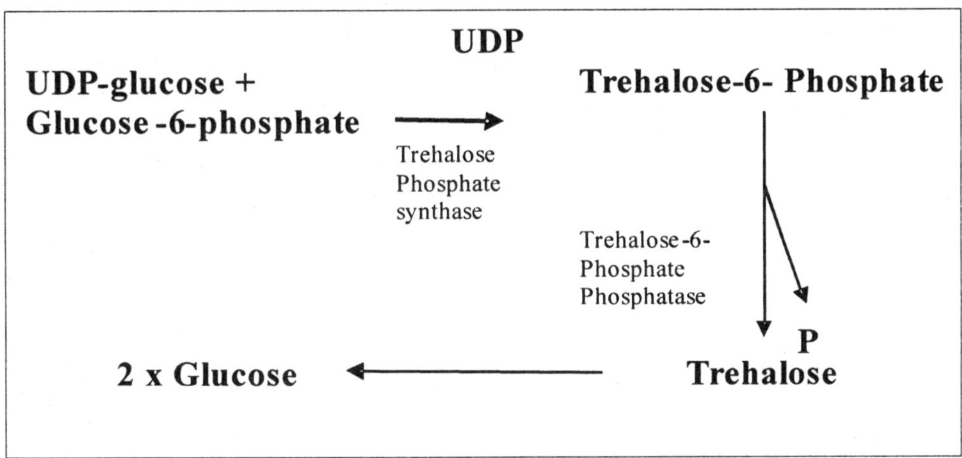

Fig. 2. Trehalose, Synthesis and metabolism

Protection only when needed

Genes imparting protection from drought stress can be expressed in plants in two ways: they can be expressed all the time, whether or not the plant is under stress; or they can be engineered to express only when there is drought stress. The second method is more favored, as it limits the side effects of the manipulations. One of the challenges biologists face in trying to engineer for drought tolerance is that drought tolerance and/or resistance traits are often negatively correlated with productivity. To achieve protection only when needed, scientists use promoters that are stress-inducible, typically abscisic acid (ABA) inducible promoters.

Comparative maps for the gramineae

Comparative maps allow transfer of information about genetic control of traits from species with small diploid genomes, such as rice (*Oryza sativa* L.), to species with more complex genomic structures (increased repetitive DNA, polyploidy) and less economic support. Because of the size and complexity of the genomes,it may not be appropriate to sequence the entire genomes of wheat (*Triticum* ssp.), rye (*Secale cereale* L.), oat (*Avena sativa* L.), or barley (*Hordeum vulgare* L.). However, alternative strategies involving identification of gene-rich regions of the Triticeae genome and comparison of the genome structure and genetic colinearity with rice, maize (*Zea mays* L.),sorghum (*Sorghum vulgare* L.), and other species provide Triticeae researchers with the knowledge and tools necessary for genetic parity with simpler genomes. The Gramineae family encompasses a diverse group of species that have been classified into two major clades based on molecular phylogenetic studies (Soreng and Davis, 1998). The Panicoideae subfamily including maize, sugarcane (*Saccharum*), sorghum and millet (*Pennisetum*) make up one clade while the other clade contains the Pooideae subfamily wheat, barley, rye and oat. Rice and wild rice, belong to the subfamily Oryzoideae. The genome of cultivated rice is considered to resemble an ancestral grass genome with a high base chromosome number (x=12) and relatively small genome 191. size of 430MB (Argumuganathan and Earle,1991). Molecular markers have been used to develop comparative chromosome maps for several members of the Gramineae (Moore *et al.*, 1995; Devos and Gale, 1997) and these have been used to study genes of agronomic importance across species (Snape and Laurie,1998). Crop species of the Poaceae display a remarkable level of genetic similarity despite their evolutionary divergence 65 million years ago (Paterson *et al.*,1995). Large segments of the genomes of maize, sorghum, rice, wheat, and barley conserve gene content and order (Gale and Devos,1998), although the correspondence has been modified by chromosome duplications, inversions, and translocations. For the domesticated grasses, the conserved linkage blocks and their relationships with rice linkage groups provides the insight into the basic organization of the ancestral grass genome (Wilson *et al.*,1999).

To date, most comparative mapping among the grasses has relied on RFLP probes (cDNAs or genomic clones) to establish gross gene orders and distance in specific chromosome segments. Only to a limited extent have researchers employed cloned genes, ESTs, mutant phenotype loci or QTLs in comparative genomics.

Despite the progress in comparative mapping, the application of this technology, especially for wheat, rye, oat and barley will not be realized unless scientifically sound strategies for studying drought tolerance are devised that allow researchers to utilize genetic tools and information developed for model species. This will require more detailed comparative genetic analysis from the DNA sequence of genes all the way to comparative analysis of QTL.

Genetics of abiotic stress tolerance

Phenotype is the product of genotype and environment, assessment of the desired genotype is highly dependent on the proper environmental conditions. Biotic stresses such as drought, temperature, salinity, and others generally reduce crop productivity. It has been estimated that crops attain only about 25% of their potential yield because of the detrimental effects of environmental stress (Boyer, 1982). The abiotic stresses are location-specific, exhibiting variation in frequency, intensity and duration. Stresses can occur at any stage of plant growth and development, thus illustrating the dynamic nature of crop plants and their productivity.

There are several definitions of drought which include precipitation, evapotranspiration, potential evapotranspiration, temperature, humidity and other factors individually or in combination (Renu and Suresh, 1998). Drought is the primary abiotic stress causing not only differences between the mean yield and the potential yield but also causing variation from year to year (yield instability). Although selection for genotypes with increased productivity in drought environments has been an important aspect of many plant breeding programs, the biological basis for drought tolerance is still poorly understood. Also, drought stress is highly heterogeneous in time (over the seasons and years) and space (between and within sites), and is unpredictable. This makes it difficult to identify or simulate a representative drought stress condition.

It has been predicted that in the coming years rainfall patterns might shift due to an increase of the global temperature caused by burning of fossil fuels and the corresponding increase in atmospheric dioxides (Guido and Paul, 1994). Consequently, farming communities in the Northern Hemisphere could become increasingly dependent on drought tolerant varieties. Crop productivity in a water-limited environment derives from mechanisms that either permit tolerance of episodes of cellular dehydration or that minimize water loss and thereby maintain a favorable water status for leaf development.

Different mechanisms may render a plant drought tolerant: (i)the ability of a plant to escape periods of drought, especially during the most sensitive periods of its development; (ii)the ability of a plant to recover from a dry period by producing new leaves from buds that were able to survive the dry spell; commonly considered less interesting from the breeder 's point of view; and (iii) the ability of a plant to endure or withstand a dry period by maintaining a favorable internal water balance under drought.

Selection for drought tolerance while maintaining maximum productivity under optimal conditions has been difficult (Zavala-Garcia *et al.*,1992). It has been reported that photosynthesis and several other related physiological traits differed significantly between drought. tolerant and drought susceptible genotypes

(Gummuluru *et al.*,1989). Several characteristics have been considered important in adaptation to stress. For example, osmotic adjustment, in which the plant increases the concentration of organic molecules in the cell water solution to "bind" water is one example of a mechanism that alleviates some of the detrimental effects of water stress by promoting both avoidance and tolerance (Blum,1989). Instantaneous leaf water efficiency, defined as the ratio of leaf photosynthesis to transpiration measured simultaneously, has also received considerable scrutiny with respect to its postulated adaptive significance for plants growing under drought stress (Morgan and Le Cain, 1991). A thicker layer of waxy material at the plant surface and more extensive and deeper rooting are others. Physiological and biochemical traits that might enhance drought tolerance have been proposed, but only a few of these mechanisms have been demonstrated to be causally related to the expression of tolerance under field conditions (Ludlow and Muchow,1990).

There is a lack of knowledge about the processes between the DNA sequence of a gene, and a trait (the "phenotypic gap"). The analysis and manipulation of complex traits such as drought tolerance and plants grown in stressful and dynamic environments is a challenge. There are several ways to reduce the phenotypic gap. These ways gradually reveal the function(s) of the genes and their connection(s) with the phenotypes. There are many metabolic changes in response to drought stress. One of the most notable changes is the synthesis and accumulation of low-molecular weight, osmotically active compounds such as sugar alcohols, amino acids, organic acids and glycine betaine (Good and Zaplachinski,1994). The accumulation of these compounds leads to osmotic adjustment as indicated by an increase in the intracellular osmotic potential of the cell (Morgan, 1984).

Genetics of drought tolerance in the Crops

Mapping quantitative trait loci associated with drought tolerance Numerous QTL mapping studies examining drought tolerance and related traits in maize, rice, barley and wheat have demonstrated that this trait is affected by several loci, each of which have relatively small effects (Quarrie,1996).

Several studies have mapped loci associated with morphological traits under drought conditions. In maize a reduced anthesis-silking interval (ASI) is one of the traits most commonly associated with drought tolerance (Ribaut *et al.*,1997). Four of the 5 QTL for ASI from Agrama and Moussa (1996) appear to map in the same chromosomal regions (chromosomes 1,5, 6,8) as those in Ribaut *et al.*(1996) who identified 6 QTL for this trait. In addition, Ribaut *et al.* (1997) identified two "stable" QTL for grain yield that coincided with QTL for kernel number per plot. Lebreton *et al.*(1995) mapped QTL for physiological traits associated with drought tolerance in maize. They measured stomatal conductance, ABA of different tissues, leaf water relations parameters, fluorescence, root pulling force, and nodal root number. Xylem ABA content and stomatal conductance were associated with root characteristics. They found that xylem and leaf ABA content were positively correlated with nodal root number and root pulling force and negatively correlated with stomatal conductance. This was supported by coincident QTL on chromosome 3 for xylem ABA content and nodal root number; whereas, QTL for stomatal conductance and root pulling force were linked but not overlapping on the same chromosome.

Champoux *et al.*(1995) an early QTL study in rice and found more than 45 QTL associated with leaf-rolling under field drought stress and root-morphology traits. Twelve of the 14 QTL associated with leaf rolling were also associated with root thickness, root/shoot ratio, or root dry weight per tiller. Using the same mapping population, Ray *et al.* (1996) evaluated root penetration. They found that some of these QTL corresponded to QTL for root morphology. Later, Lilley *et al.* (1996) extended the results of those two studies by evaluating osmotic adjustment and relative leaf water content. A single locus on chromosome 8 near RG1 and RZ66 was found to be associated with osmotic adjustment at 70%water potential. Teulat *et al.* (1998) also mapped genes for osmoregulation in barley and one of the QTL on barley chromosome 7H matched the homoelogous chromosome location reported for rice by Lilley *et al.* (1996). For this same chromosome region, Teulat *et al.*(1997) mapped QTL controlling relative water content and number of leaves under water stress. Champoux *et al.* (1995) mapped QTL for root morphology and leaf rolling in the homoeologous rice chromosome region. However a major gene mapped in wheat for osmoregulation (Morgan and Tan,1996) appears to be distal to this region based .on rice/wheat comparative maps (Van Deynze *et al.*, 1995c). Teulat *et al.* (1998) mapped other osmoregulation genes in barley on 6H near WG286 and on 2H near E9-4 and mwg 720. The E9-4 locus also corresponded to a homoeologous rice chromosome segment reported to be associated with lethal osmotic potential in rice (Lilley *et al.*, 1996). Price *et al.*(1997) evaluated several root growth characteristics in rice and in a companion paper (Price and Tomos, 1997) identified several QTL for maximum root length, volume and thickness. In a comparison with Champoux *et al.* (1995) using a different mapping population, there were 3 QTL in common for maximum root length (chromosomes 2,11,9 or 5), 2 for root thickness (chromosomes 2,3), and 1 for root volume (chromosome 12). The root penetration study by Ray *et al.* (1996) found that the QTL for root penetration on the long arm of chromosome 6 corresponded to root length and the one in the central region of chromosome 11 corresponded to root growth. In a third rice mapping population, Redona and Mackill (1996), identified a root length QTL that corresponded to a QTL in the Ray *et al.* (1996) study for root length at 14 days. Li *et al.*(1999) developed NILs for 4 different rice chromosome regions associated with total root weight, deep root weight and shallow root weight. Most of the NILs for the target regions did not exhibit the change in the root trait predicted and several of the NILs for the selected allele had increased height and reduced tiller number suggesting linkage drag was a problem.

The stay-green trait has been associated with drought tolerance in sorghum (Borrell *et al.*,1999). It was found that 3 loci (linkage groups B, G, and I) accounted for 34%of the total variance of stay green under drought stress. These chromosome regions correspond to parts of maize chromosomes 10, 8, and 3, respectively. Yadav *et al.* (1999) reported QTL for drought tolerance and yield components in two mapping populations of pearl millet (*Pennisetumm glaucum* (L.) R.Br.). Grain yield QTL on linkage groups 2, 5, and 6 were identified and QTL for components of yield corresponded to each of them. Homoeology to maize chromosomal regions has not been published.

A more comprehensive approach to studying drought tolerance has been advanced by INRA researchers using proteomics (de Vienne *et al.*,1999). Using large-scale 2-D gel electrophoresis, they quantify protein spot intensities and these are mapped as protein quantity loci (PQL). This approach can aid in the detection of regulatory genes and in identifying candidate genes. This approach was used to evaluate a maize RIL population under mild drought stress. Differentially expressed proteins from leaf tissue were sequenced for identification. One of the proteins was an ABA/water stress/ripening induced protein located on chromosome 10 that had previously been found to be induced by water stress in other species (de Vienne *et al.*,1999). The location of this candidate gene corresponded with a QTL for xylem sap ABA content, leaf senescence, and anthesis/silking interval. Other PQL reported by Prioul *et al.* (1999) to correspond to QTL for drought responsive traits included those on chromosomes 1 (Sh2 -ADPgulcose pyrophosphorylase), 2 (invertase), 5 (invertase), 6, 8 (sucrose-phospahte synthase), 9 and 10 (Prioul *et al.*, 1999).

Abscisic acid has been demonstrated to play an important role in plant response to water stress.Recent studies have mapped QTL for maize-leaf ABA content under drought stress (Sanguineti *et al.*,1999). Sixteen QTL for ABA content corresponded with QTL for at least one of the following traits: stomatal conductance, drought sensitivity index,leaf temperature, leaf relative water content, anthesis-silking interval and grain yield. An increase in ABA content was generally associated with decreased stomatal conductance and grain yield but increased leaf temperature. However, the opposite effect was observed for a QTL on chromosome 7 that aligned with a QTL from a previous study for root pulling resistance suggesting that elevated ABA stimulated the development of a more extensive root system (Lebreton *et al.*,1995). In a study of 140 wheat genotypes, associations of yield components, carbon isotope discrimination (CID), ash content in flag leaf and kernels,and canopy temperature, revealed that CID explained approximately 30%of the total variability of dryland grain yield (Nachit, 1998).

Genes with up-regulated expression in response to drought

Plants respond to changing environmental stimuli with the expression of specific sets of genes that allow the plants to adapt to the altered environmental conditions.

One of the most common environmental stresses to which plants are exposed is drought stress. The two most productive approaches to establishing the basic responses of plants to drought involve studying candidate genes and differential screening. Comparing the expression of genes thought to be important for drought tolerance, such as the enzymes in drought-induced metabolic pathways under drought versus non-drought conditions can provide useful information. A second approach uses differential screening to isolate up-regulated genes. These experiments have been successful in describing many genes encoding proteins of known function associated with desiccation (Table 1). While most of these genes are induced by the plant hormone abscisic acid (ABA), several have been shown to be unresponsive to ABA (Yamaguchi-Shinozaki *et al.*, 1992).These findings suggest the existence of two separate signal transduction pathways responding to intracellular dehydration, an ABA-responsive and an ABA-independent pathway (Nordin *et al.*,1991). ABA is synthesized through the carotenoid biosynthesis pathway. ABA concentration is altered when there are changes in cellular dehydration. Reduction of turgor results in rapid synthesis of this phytohormone. The synthesis itself requires nuclear gene expression and translation (Guerrero and Mullet, 1986).Increased levels of ABA can, in turn, induce changes in gene expression resulting in stomatal closure in leaves, inhibition of photosynthesis and the growth of leaves, stems and hairy roots. When subjected to osmotic stress or abscisic acid, some vascular plants such as barley respond with an increased accumulation of the osmoprotectant, glycine betaine (betaine), being the last step of betaine synthesis catalyzed by betaine aldehyde dehydrogenase (BADH). Manabu *et al.* (1995) have cloned and characterized a BADH cDNA from barley and described the expression pattern of BADH transcript. Two cDNA clones,BADH1 and BADH15, putatively encoding betaine aldehyde dehydrogenase have also been isolated from sorghum and characterized (Andrew *et al.*, 1996).

Table 1. Genes up-regulated by drought stress and encoding polypeptides of known function

cDNA	Plant	Feature	Reference
pSS1;pSS2	*C.plantagineum*	Sucrose synthases	
PPPC1	*Mesembryanthemum crystallinum*	Phosphoenolpyruvate carboxylase	
PBAD	*Hordeum vulgare*	Betaine aldehyde dehydrogenase	Ishitani *et al.*,1995
CAtP5CS	*Arabidopsis thaliana*	Pyrroline-5-carboxylate synthetase	Yoshiba *et al.*,1995
RD28	*A.thaliana*	Water channel	Yamaguchi-Shinozaki *et al.*,1992
SAM1;SAM3	*Lycopersicon esculentum*	*S* -adenosyl-L-methionine synthetases	Espartero *et al.*,1994

rd19A;rd21A	*A.thaliana*	Cysteine proteases	
UBQ1	*A.thaliana*	Ubiquitin extension protein	Kiyosue *et al.*,1994b
PMBM1	*Triticum aestivum*	L-isoaspartyl methyltransferase	Mudgett and Clarke,1994
SC514	*Glycine max*	Lipoxygenase	Bell and Mullet,1991
cATCDPK1 and cATCDPK2	*A.thaliana*	Ca 2+ -dependent, calmodulin-independent protein kinases	Urao *et al.*,1994
PKABA1	*T.aestivum*	Protein kinase	Anderberg and Walker-Simmons,1992
CAtPLC1	*A.thaliana*	Phosphatidylinositol-specific phospholipase C	Hirayama *et al.*,1995
Apx1 gene	*Pisum sativum*	Cytosolic ascorbate peroxidase	Mittler and Zilinskas,1994
Sod 2 gene	*P.sativum*	Cytosolic copper/zinc superoxide dismutase	White and Zilinskas,1991
P31	*L.esculentum*	Cytosolic copper/zinc superoxide dismutase	Perl-Treves and Galun,1991
Pcht28	*L.chilense*	Acidic endochitinase	Chen *et al.*,1994
Atmyb2	*A.thaliana*	MYB-protein-related transcription factor	Urao *et al.*,1993
ERD11;ERD13	*A.thaliana*	Glutathione *S* -transferases	Kiyosue *et al.*,1993
CAtsEH	*A.thaliana*	Soluble epoxide hydrolase	Kiyosue *et al.*,1994a

Dehydrins [late embryogenesis abundant (LEA) D11 family] are also produced in a wide variety of plant species in response to dehydration, low temperature, osmotic stress, seed drying and exposure to abscisic acid.

Inheritance studies, including QTL analysis, in several crop plants have revealed apparent co- segregation of *Dhn* genes with phenotypes associated with dehydrative stress, such as drought and freezing (Close,1996). Despite their widespread occurrence and abundance in cells under dehydrative conditions, the biochemical role of dehydrins remains elusive. In some species, dehydrin loci are located within quantitative trait loci (QTL) intervals for important phenotypic traits including winter hardiness in barley and anthesis-silking interval in maize. Lang *et al.*(1998) studied the variation in the dehydrin gene family of barley using 3'fragments of dehydrin cDNA clones (DHN1-4) as hybridization probes. The results of this work indicated that there are two clones (DHN1 and DHN7)that represent allelic alternatives at the *Dhn1* locus. Wheat *Dhn* genes were mapped to

4DS, 5Bl, and 6AL, also six maize *Dhn* probes identified eight maize *Dhn* loci on chromosomes 1, 3, 4, 5, 6, and 9 (Close,1996). Some genes are induced by drought, others by low temperature. This variation, together with cross-hybridization between *Dhn* genes highlights the necessity of gene-specific methods to study the *Dhn* multigene family.

Genes that protect plants from drought

There has been substantial progress in identifying genes for resistance to various abiotic stresses such as temperature, salinity and drought. Table 2 shows some transgenic plants for improved drought tolerance using genes that have been isolated and tested as drought resistance genes. Among these genes, alanine aminotransferease that has been isolated from barley roots (Muench and Good,1994) and D-myo-inositol methyltransferase from *M.crystallinum* that has been transformed to *Nicotiana tabacum* L.

Table 2.Transgenics produced for improved drought tolerance

Transgenic over-expression	Plant	Reference
alanine aminotransferease	Tobacco	Muench and Good, 1994
Dmyo inositol ethyltransferase	Tobacco	Elena *et al.*,1997
Fructan	Tobacco	Pilon-Smits *et al.*,1995
HVA1	Rice, wheat	Xu *et al.*, 1996

Plant transformation resulting in stress-inducible, stable solute accumulation appears to provide protection under drought and salt-stress (Elena *et al.*,1997). Fructans are polyfructose that are produced in only 15%of all flowering plant species,including wheat and barley (Renu and Suresh, 1998). It functions mainly as a storage carbohydrate but being soluble may help plants survive periods of osmotic stress induced by drought or cold (Bieleski,1993). To investigate the possible functional significance of fructans in drought stress, Pilon-Smits *et al.* (1995) have introduced the bacterial gene and *SacB* from *Bacillus subtilis* encoding levan sucrose into tobacco,a non fructan-accumulating plant. The transgenic tobacco produced bacterial fructans and was examined for growth performance under PEG-mediated drought stress.The growth of the transgenic plants was higher under drought stress compared to the wild type tobacco.

Xu *et al.*(1996), has adopted the transgenic approach to investigate the function of the *HVA1* protein in stress protection of rice. *HVA1*, is a group 3 LEA protein that is expressed in barley aleurone and embryo during late seed development correlating with the seed desiccation stage (Hong *et al.*, 1988). The transgenic rice plants exhibited constitutive high expression of *HVA1* protein in leaves and roots. The R1 progeny of three transgenic plants was used for evaluation of the growth performance under water deficit and salt stress treatment. The appearance and development of the major damage symptoms such as wilting, dying of old leaves and necrosis of young leaves caused by the stress conditions were delayed in the transgenic plants. The better performance of R1 transgenic lines under stress conditions was correlated with higher level of *HVA1* protein accumulated in their R0 plants.

Regulation of gene expression under drought stress

It has been expected from molecular studies that the basic tools will be provided to modulate stress tolerance. One important component in this tool kit is regulatory elements that are responsive to environmental signals and lead to specific gene expression. The expression of a number of genes from various species has been shown to be induced by drought stress. Responsive to drought or ABA was demonstrated by monitoring steady state transcript levels or by protein analysis. Compared to the number of genes expressed in response to drought stress and/or ABA the number of corresponding promoters analyzed is small. Most of the promoters have been isolated from LEA genes that are abundantly expressed in dehydrated seeds. Table 3 shows activities of some promoters in transgenic plants.

Table 3. Characterization of promoters in transgenic plants

Gene	Isolated from	Reporter gene activity found in	Reference
Rab 16B	Rice embryos	Tobacco embryos	Yamaguchi *et al.*,1990
Em	Wheat embryos	Tobacco embryos	Marcotte *et al.*,1989
Rab 17	*Nauze* embryos	Arabidopsis embryos	Vilardell *et al.*,1994
Rd 22,	*A.thaliana* dehydrated plants	Tobacco dehydration ABA	Iwasaki *et al.*,1995
Rd 29	*A.thaliana* dehydrated plants	Inducible in almost all vegetative tissue in dehydrated Arabidopsis plants, also cold, ABA, salt inducible tobacco	Yamaguchi-Shinozaki and Shinozaki,1993,
CdeT27-45	*Craterostigma* dehydrated	Tobacco and Arabidopsis plants embryos, mature pollen	Michel *et al.*,1993

Conclusions

Physiological events during drought stress and their genetic control is essential to define which genes are regulatory, which are primary gene products positively contributing to stress tolerance, which genes may serve as markers for the physiological stage of the plant and which gene products can be considered as secondary stress-induced metabolites. However, comparative genetic analyses can greatly facilitate the discovery of genes that contribute to this complex trait by allowing scientists to transfer information between species. Furthermore, genetic variation for components of drought tolerance may differ widely among species and this genetic variation is crucial to understanding the underlying mechanisms.

References

Agrama, H.A.S. and Moussa, M.E. 1996. Mapping QTLs in breeding for drought tolerance in maize (*Zea mays* L.). *Euphytica*, 91: 89-97.

Anderberg, R.J. and Walker-Simmons, M.K. 1992. Isolation of a wheat cDNA clone for an abscisicacid-inducible transcript with homology to protein kinases. Proc. Natl. Acad. Sci. USA, 89: 10183-10187.

Andrew, J.W., Hirofumi, S., David, R., Robert, J.J. and Peter, B.G. 1996. Betaine aldehydedehydrogenase in Sorghum: Molecular cloning and expression of two related genes. *Plant Physiol.*,110: 1301-1308.

Argumuganathan, K. and Earle, E.D. 1991. Nuclear DNA content of some important plant species. PlantMol. Biol. Rep., 9: 208.

Bieleski, R.L. 1993. Fructan hydrolysis drives petal expansion in the ephemeral daylily flower. *Plant Physiol.*, 103: 213-219.

Blum, A. 1989. Osmotic adjustment and growth of barley genotypes under stress. *Crop Sci.*, 29: 230-233.

Borrell, A.K., Tao, Y. and McIntyre, C.L. 1999. Physiological basis, QTL and MAS of the stay-greendrought resistance trait in grain sorghum. In: *Workshop on Molecular Approaches for the Genetic Improvement of Cereals for Stable Production in Water-Limited*, 21-25 June 1999. CIMMYT, El Batan, Mexico. (http://198.93.240.203/Research/ABC/WS Molecular/ WorkshopMolecularcontents.htm).

Boyer, J.S. 1982. Plant productivity and environment. *Science*, 218: 443-448.

Champoux, M.C., Wang, G., Sarkarung, S., Mackill, D.J., O'Toole, J.C., Huang, N. and McCouch, S.R. 1995. Locating genes associated with root morphology and drought avoidance in rice via linkage to molecular markers. *Theor. Appl. Genet.*, 90: 969-981.

Chen, R.-D., Yu, L.-X., Greer, A.F., Cheriti, H. and Tabaeizadeh, Z. 1994. Isolation of an osmotic stress and abscisic acid-induced gene encoding an acidic endochitinase from *Lycopersicon chilense*. *Mol.Gen. Genet.*, 245: 195-202.

Close, T.J. 1996. Dehydrins: Emergence of a biochemical role of plant dehydration proteins. *Physiol.Plant.*, 97: 795-803.

De Vienne, D., Leonardi, A., Damerval, C. and Zivy, M. 1999. Genetics of proteome variation for QTL characterization: Application to drought-stress responses in maize. *J. Exp. Bot.*, 50: 303-309.

Devos, K.M. and Gale, M.D. 1997. Comparative genetics in the grasses. *Plant Mol. Biol.*, 35: 3-15.

Elena, S., Wendy, C., Hans, J.B. and Richard, G.J. 1997. Increased salt and drought tolerance byD-Ononitol production in transgenic *Nicotinan tabacum* L. *Plant Physiol.*, 115: 1211-1219.

Espartero, J., Pintor-Toro, J.A. and Pardo, J.M. 1994. Differential accumulation of S-adenosylmethionine synthetase transcripts in response to salt stress. *Plant Mol. Biol.*, 25: 217-227.

Gale, M.D. and Devos, K.M. 1998. Comparative genetics in the grasses. Proc. Natl. Acad. Sci. USA,95: 1971.

Good, A.G. and Zaplachinski, S.T. 1994. The effect of drought stress on free amino acid accumulation and protein synthesis in Brassica napus. *Physiol. Plant.*, 90: 9-14.

Guerrero, F.D. and Mullet, J.E. 1986. Increased abscisic acid biosynthesis during plant dehydrationrequires transcription. *Plant Physiol.*, 80: 588-591.

Guido, R. and Paul, R. 1994. Drought tolerance research as a social process. *Biotechnology and Development Monitor,* 18: 5.

Gummuluru, S., Hobbs, S.L.A. and Jana, S. 1989. Genotypic variability in physiological characters and its relationship to drought tolerance in Durum wheat. *Can. J. Plant Sci.*, 69: 703-711.

Hirayama, T., Ohto, C., Mizoguchi, T. and Shinozaki, K. 1995. A gene encoding a phosphatidylinositolspecific phospholipase C is induced by dehydration and salt stress in *Arabidopsis thaliana. Proc.* Natl. Acad. Sci. USA, 92: 3903-3907.

Hong, B., Uknes, S.J. and Ho, T.-H.D. 1988. Cloning and characterization of a cDNA encoding a mRNA rapidly induced by ABA in barely aleurone layers. *Plant Mol. Biol.*, 11: 495-506.

Ishitani, M., Nakamura, T., Han, S.Y. and Takabe, T. 1995. Expression of the betaine aldehydedehydrogenase gene in barley in response to osmotic stress and abscisic acid. *Plant Mol. Biol.*, 27:307-315.

Iwasaki, T., Yamaguchi-Shinozaki, K. and Shinozaki, K. 1995. Identification of cis-regulatory region of a gene in Arabidopsis thaliana whose induction by dehydration is mediated by abscisic acid andrequires protein synthesis. *Mol. Gen. Genet.*, 247: 391-398.

Kiyosue, T., Beetham, J.K., Pinot, F., Hammock, B.D., Yamaguchi-Shinozaki, K. and Shinozaki, K. 1994a. Characterization of an *Arabidopsis* cDNA for a soluble epoxide hydrolase gene that is inducible by auxin and water stress. *Plant J.*, 6(2): 259-269.

Kiyosue, T., Yamaguchi-Shinozaki, K. and Shinozaki, K. 1993. Characterization of two cDNAs (ERD11 and ERD13) for dehydration-inducible genes that encode putative glutathione *S*-transferases in *Arabidopsis thaliana* L. *FEBS Lett.*, 335(2): 189-192.

Kiyosue, T., Yamaguchi-Shinozaki, K. and Shinozaki, K. 1994b. Cloning of cDNAs for genes that are early-responsive to dehydration stress (ERDs) in *Arabidopsis thaliana* L.: Identification of three ERDs as HSP cognate genes. *Plant Mol. Biol.*, 25: 791-798.

Lang, V., Robertson, M. and Chandler, P.M. 1998. Allelic variation in the dehydrin gene family of Himalaya barley (*Hordeum vulgare* L.). *Theor. Appl. Genet.*, 96: 1193-1199.

Lebreton, C., Lazic-Jancic, V., Steed, A., Pekic, S. and Quarrie, S.A. 1995. Identification of QTL for drought responses in maize and their use in testing causal relationships between traits. *J. Exp. Bot.*, 46: 853-865.

Li, Z., Shen, L.S., Courtois, B. and Lafitte, R. 1999. Development of near isogenic introgression line (NIIL) sets for QTLs associated with drought tolerance in rice. In: *Workshop on Molecular Approaches for the Genetic Improvement of Cereals for Stable Production in Water-Limited,* 21-25. June 1999. CIMMYT, El Batan, Mexico (http://198.93.240.203/Research/ABC/WSMolecular/WorkshopMolecularcontents.htm).

Lilley, J.M., Ludlow, M.M., McCouch, S.R. and O'Toole, J.C. 1996. Locating QTL for osmotic adjustment and dehydration tolerance in rice. *J. Exp. Bot.*, 47: 1427-1436.

Ludlow, M.M. and Muchow, R.C. 1990. A critical evaluation of traits for improving crop yield in waterlimited environments. *Adv. Agron.*, 43: 107-153.

Manabu, I., Toshihide, N., Seung, Y.H. and Tetsuko, T. 1995. Expression of the betaine aldehyde dehydrogenase gene in barley in response to osmotic stress and abscisic acid. *Plant Mol.* Biol., 27:307-315.

Marcotte, W.R., Russel, S.H. and Quatrano, R.S. 1989. Abscisic acid resposive sequences from the gene of wheat. *Plant Cell,* 1: 969-976.

Michel, D., Salamini, F., Bartels, D., Dale, P., Baga, M. and Szalay, A. 1993. Analysis of desiccation and ABA-responsive promoter isolated from the resurrection plant *Craterostigma plantagineum. Plant J.*,4: 29-40.

Mittler, R. and Zilinskas, B.A. 1994. Regulation of pea cytosolic ascorbate peroxidase and other antioxidant enzymes during the progression of drought stress and following recovery from drought. *Plant J.*, 5(3): 397-405.

Morgan, J.A. and LeCain, D.R. 1991. Leaf gas exchange and related leaf traits among 15 winter wheat genotypes. *Crop Sci.*, 31: 443-448.

Morgan, J.M. 1984. Osmoregulation and water stress in higher plants. *Annu. Rev. Plant Physiol.*, 35:299-319.

Mudgett, M.B. and Clarke, S. 1994. Hormonal and environmental responsiveness of a developmentally regulated protein repair L-isoaspartyl methyltransferase in wheat. *J. Biol. Chem.*, 269(41): 25605-25612.

Muench, D.G. and Good, A.G. 1994. Molecular cloning and expression of an anaerobically induced alanine aminotransferase from barley roots. *Plant Mol. Biol.*, 24: 417-427.

Nachit, M.M. 1998. Association of grain yield in dryland and carbon isotope discrimination with molecular markers in durum (*Triticum turgidum* L. var. *durum*). In: *Proc. 9th Intl. Wheat Genetics Symp.*, Saskatoon, Saskatchewan (Canada), 2-7 August 1998, pp. 218-224.

Nordin, K., Heino, P. and Palva, E.T. 1991. Separate signal pathways regulate the expression of a low temperature-induced gene in *Arabidopsis thaliana* (L.) Heynh. *Plant Mol. Biol.*, 16: 1061-1071.

Paterson, A.H., Lin, Y.R., Li, Z.K., Schertz, K.F., Doebley, J.F., Pinson, S.R.M., Liu, S.C., Stansel, J.W. and Irvine, J.E. 1995. Convergent domestication of cereal crops by independent mutations at corresponding genetic loci. *Science*, 269: 1714-1718.

Perl-Treves, R. and Galun, E. 1991. The tomato Cu, Zn superoxide dismutase genes are developmentally regulated and respond to light and stress. *Plant Mol. Biol.*, 17: 745-760.

Pilon-Smits, E.A.H., Ebskamp, M.J.M, Paul, M.J., Jeuken, M.J.W., Weisbeek, P.J. and Smeekens, S.C.M.1995. Improved performance of transgenic fructan-accumulating tobacco under drought stress. *Plant Physiol.*, 107: 125-130.

Price, A.H. and Tomos, A.D. 1997. Genetic dissection of root growth in rice (*Oryza sativa* L.). II. Mapping quantitative trait loci using molecular markers. *Theor. Appl. Genet.*, 95: 143-152.

Price, A.H., Tomos, A.D. and Virk, D.S. 1997. Genetic dissection of root growth in rice

(*Oryza sativa* L.).I. A hydroponic screen. *Theor. Appl. Genet.*, 95: 132-142.

Prioul, J.-L., Pelleschi, S., Sene, M., Thevenot, C., Causse, M., de Vienne, D. and Leonardi, A. 1999. From QTLs for enzyme activity to candidate genes in maize. *J. Exp. Bot.*, 50: 1281-1288.

Quarrie, S.A. 1996. New molecular tools to improve the efficiency of breeding for increased drought resistance. *Plant Growth Regulation*, 20: 167-178.

Ray, J.D., Yu, L.-X., McCouch, S.R., Champoux, M.C., Wang, G. and Nguyen, H.T. 1996. Mapping quantitative trait loci associated with root penetration ability in rice (*Oryza sativa* L.). *Theor. Appl.Genet.*, 42: 627-636.

Redona, E.D. and Mackill, D.J. 1996. Mapping quantitative trait loci for seedling vigor in rice using RFLPs. *Theor. Appl. Genet.*, 92: 395-402.

Renu, K.C. and Suresh, K.S. 1998. Prospects of success of biotechnological approaches for improving tolerance to drought stress in crop plants. *Curr. Sci.*, 74: 25-34.

Ribaut, J.-M., Hoisington, D.A., Deutsch, J.A., Jiang, C. and Gonzalez-de-Leon, D. 1996. Identification of quantitative trait loci under drought conditions in tropical maize. 1. Flowering parameters and the anthesis-silking interval. *Theor. Appl. Genet.*, 92: 905-914.

Ribaut, J.-M., Jiang, C., Gonzalez-de-Leon, D., Edmeades, G.O. and Hoisington, D.A. 1997. Identification of quantitative trait loci under drought conditions in tropical maize. 2. Yield components and marker-assisted selection strategies. *Theor. Appl. Genet.*, 94: 887-896.

Sanguineti, M.C., Tuberosa, R., Landi, P., Salvi, S., Maccaferri, M., Casarini, E. and Conti, S. 1999. *J.Exp. Bot.*, 50: 1289-1297.

Snape, J.W. and Laurie, D.A. 1998. Comparative mapping of agronomic trait loci in crop species. In: Crop Productivity and Sustainability - Shaping the Future, Chopra, V.L., Singh, R.R. and Varma, A. (eds), pp. 759-771. Proc. of the Intl. Crop Sci. Congress (1996) (in press).

Soreng, R.J. and Davis, J.I. 1998. Phylogenetics and character evolution in the grass family (Poaceae): Simultaneous analysis of morphological and chloroplast DNA restriction site character sets.*Bot. Rev.*, 64: 1-84.

Teulat, B., Monneveux, P., Wery, J., Borries, C., Souyris, I., Charrier, A. and This, D. 1997. Relationships between relative water content and growth parameters under water stress in barley: A QTL study. *New Phytol.*, 137: 99-107.

Teulat, B., This, D., Khairallah, M., Borries, C., Ragot, C., Sourdille, P., Leroy, P., Monneveux, P. and Charrier, A. 1998. Several QTLs involved in osmotic-adjustment trait variation in barley (*Hordeum vulgare* L.). *Theor. Appl. Genet.*, 96: 688-698.

Urao, T., Katagiri, T., Mizoguchi, T., Yamaguchi-Shinozaki, K., Hayashida, N. and Shinozaki, K. 1994. Two genes that encode Ca2+-dependent protein kinases are induced by drought and high-salt stresses in *Arabidopsis thaliana. Mol. Gen. Genet.*, 244: 331-340.

Urao, T., Yamaguchi-Shinozaki, K., Urao, S. and Shinozaki, K. 1993. An *Arabidopsis myb* homolog is induced by dehydration stress and its gene product binds to the conserved MYB recognition sequence. *Plant Cell*, 5: 1529-1539.

Vilardell, J., Martinez-Zapater, J.M., Goday, A., Arenas, C. and Pages, M. 1994. Regulation of the rab17 gene promoter in transgenic Arabidopsis wild-type, ABA-deficient and ABA-insensitive mutant. *Plant Mol. Biol.,* 24: 561-569.

White, D.A. and Zilinskas, B.A. 1991. Nucleotide sequence of a complementary DNA encoding pea cytosolic copper/zinc superoxide dismutase. *Plant Physiol.,* 96: 1391-1392.

Wilson, W.A., Harrington, S.E., Woodman, W.L., Lee, M., Sorrells, M.E. and McCouch, S.R. 1999. Can we infer the genome structure of progenitor maize through comparative analysis of rice, maize and the domesticated panicoids? *Genetics,* 153: 453-473.

Xu, D., Duan, X., Wang, B., Hong, B., Ho- Tiam-Ho, D. and Wu, R. 1996. Expression of a late embryogenesis abundant protein gene, HVA1, from Barley confers tolerance to water deficit and salt stress in transgenic rice. *Plant Physiol.,* 110: 249-257.

Yadav, R.S., Hash, C.T., Bidinger, F.R. and Howarth, C.J. 1999. Identification and utilization of quantitative trait loci to improve terminal drought tolerance in pearl millet [*Pennisetum glaucum* (L.) R. Br.]. In: *Workshop on Molecular Approaches for the Genetic Improvement of Cereals for Stable Production in Water-Limited,* 21-25 June 1999. CIMMYT, El Batan, Mexico. (http://198.93.240.203/Research/ABC/WSMolecular/WorkshopMolecularcontents.htm).

Yamaguchi-Shinozaki, K. and Shinozaki, K. 1993. Characterization of the expression of a desiccation responsive rd29 gene of Arabidopsis thaliana and analysis of its promoter in transgenic plants. *Mol Genet.,* 236: 331-340.

Yamaguchi-Shinozaki, K., Koizumi, M., Urao, S. and Shinozaki, K. 1992. Molecular cloning and characterization of 9 cDNAs for genes that are responsive to desiccation in *Arabidopsis thaliana:* Sequence analysis of one cDNA clone that encodes a putative transmembrane channel protein. *Plant Cell Physiol.,* 33(3): 217-224.

Yamaguchi-Shinozaki, K., Mino, M., Mundy, J. and Chua, N.-H. 1990. Analysis of an ABA-responsive rice gene promoter in transgenic tobacco. *Plant Mol. Biol.,* 15: 905-912.

Yoshiba, Y., Kiyosue, T., Katagiri, T., Ueda, H. and Mizoguchi, T. 1995. Correlation between the induction of a gene for pyrroline-5-carboxylate synthetase and the accumulation of proline in *Arabidopsis thaliana* under osmotic stress. *Plant J.,* 7(5): 751-760.

Zavala-Garcia, F., Bramel-Cox, P.J., Eastin, J.D., Witt, M.D and Andrews, D.J. 1992 Increasing the efficiency of crop selection for unpredictable environments. *Crop Sci.,* 32: 51-57.

CHAPTER

23 Integrated Insect Pest Management for Dryland Farming Systems

Reena[1], B. K. Sinha[2] and Surendra Prasad[3]

[1]*Dryland research Sub-station, SKUAST-Jammu, Rakh Dhiansar, Bari Brahaman, J & K.*

[2]*Division of Biochemistry and Plant Physiology SKUAST-J, FOA, Main Campus Chatha, Jammu J&K State, 180009.*

[3]*KVK, Manjhi Saran, RAU,Samastipur, Bihar*

Rainfed areas in India constitute nearly 75%, approximately 108 million hectares of the total 143 million hectares of arable land. Crop production in such areas becomes relatively difficult owing to its dependence upon intensity and frequency of rainfall. Even protective or life saving irrigation in such areas is not possible. The annual rainfall in these areas ranges from 400 to 1000 mm and that too is unevenly distributed, highly uncertain and erratic. In certain areas, the total annual rainfall does not exceed 500 mm. Such low rain dependent crop production is technically called dry land farming. Out of the total 108 million hectares of rainfed areas, about 47 million hectares are dry lands and they contribute 42% of the total food grain production in the country. They produce about 75% of pulses and more than 90% of sorghum, millet, groundnut and pulses from arid and semi-arid regions. Water is a scarce commodity in dryland agriculture. Appropriate package of practices for dryland farming has not been evolved yet, despite all the advancements in agriculture. The fruits of improved technologies have been harvested by the irrigated belt. But now with the ever-increasing human population and no further scope of vertical expansion in the irrigated areas, researcher's attention has shifted towards the dryland areas, the potential of which is yet to be harnessed. Besides, the farmers in these dryland areas are still socio-economically backward and poor. Therefore, improved dry farming is necessary for equity and prosperity.

Dryland agriculture is characterized by uncertain, ill-distributed and limited annual rainfall, extensive climatic hazards like drought, flood, etc. undulating surface, extensive and large holdings, etc. In dry land farming selection of suitable crops becomes a major limiting factor. Only drought resistant crops namely oilseeds, pulses and coarse grains like jowar, bajra, millets, etc. can be grown in dry land areas. However, in addition to these field crops, few horticultural crops, viz., aonla, guava, mango, citrus, etc. can also be included in dry land farming systems. Under such highly risky arid and semi-arid climates, two types of agriculture are generally being practiced. One is crop production or arable farming and other is mixed farming i.e. animal husbandry together with crop production and pasture management. Such agriculture is suited only where human populations are limited, but for country like India, where the population has crossed the one billion mark, we have no option left than going for extensive cultivation in profitable manner in these dry land areas.

Thus, as a beginning, Bombay research Scheme on dry farming was started in 1934 at Sholapur and Bijapur after the establishment of Indian Council of Agricultural Research (ICAR) at New Delhi in 1929. Punjab, Madras and Hyderabad were the first to start work on dry land farming immediately after the establishment of ICAR. Dry farming work in U.P. started in Jhansi and Agra at dry farming centers established in 1943-44 and 1948-50 respectively and work has been in progress since then. Crop improvement brought out very promising strains during sixties through which the crop production can be maximized in semi-arid and dry areas of the country. Keeping this in view, ICAR launched the All India Coordinated Research Project for Dry land Agriculture in 1970 with an active collaboration with government of Canada. The project started with multidisciplinary research units at 23 coordinating research centers located in various typical agro climatic regions of India with Hyderabad as headquarter. These centers have been established with an objective of accelerating the conservation development and efficient, long term use of basic resources of soil and water for a self sustaining production. Almost all the states have some area under rainfed agriculture depending upon its topography and irrigation facilities. The major dry farming areas are – the Indo-Gangetic plains of North India including districts of Rajasthan, Punjab, Haryana, North-Western M.P. and U.P., the trapian plateau of peninsular India comprising the states of Maharashtra, Karnataka and Andhra Pradesh and Plateau of granite formation.

For a wide swath of arid and semi-arid countries holding one-fourth of the world's population, scientists of the Consultative Group on International Agricultural Research (CGIAR) are developing new breeds of crops and animals that grow faster and stronger, need less water and are genetically selected for high levels of nutrition. Approximately 1.6 billion people are currently living in developing countries and regions affected by insufficient rainfall. Out of which half of the workforce earns its living from agriculture. These huge marginal regions have remained untouched by the Green Revolution. These areas have also attracted little commercial investments in agricultural technology improvement because of small markets. The CGIAR is supported by 16 international research centres. Two of these centres are working to develop new technologies for dry land agriculture.

- The International Centre for Agricultural Research in Dry Areas (ICARDA), based in Allepo, Syria.
- The International Crops Research Institute for the Semi-Arid Tropics (ICRISAT), based in Andhra Pradesh, India.

Both these centres are active in all the developing regions, with special focus on Asia, Middle East and Africa. Their priority is to improve or enhance the productivity of staple crops of dry regions, so as to provide a minimum of food security under harsh climatic conditions and with little water. These crops constitute the main product of 800 million farmers in dry regions and the population's basic food, though these crops are little known and traded in world markets.

India is yet to have a National Agricultural Policy. At a time when Indian agriculture is facing a crisis on several fronts, and as India battles at the WTO front to protect the interests of its farmers, it is time to address equity issues and reward small and marginal farmers who follow ecological approaches, maintain productivity and contribute to production, especially in the drylands. Such a policy should look at dry land agriculture with a lens that appreciates the pre-conditions of constraints and strengths of drylands. The policy should be such that it should increase the stakes of a dryland farmer in his / her farming system. The drylands can no longer be neglected, given the enormous ecological and economic contribution that dryland farmers could make.

Dry land farming vis-à-vis pest management

Even under dry land conditions, it is proved that pest infestations at several places can be effectively managed by collective action and with local plant resources. The current pest management paradigms need a complete revamping. Such changes are more urgent in case of drylands as conventional pest management costs could eat up heavily into any viable net incomes possible. Pest management strategies adopted by farmers under dryland conditions have been studied by several workers. Farmers often select well-adapted, stable crop varieties and cropping systems are such that two or more crops are grown in the same field at the same time. These diverse traditional systems enhance natural enemy abundance and generally keep pest numbers at low levels. Thus, pest management practice is a built-in process in the overall crop production system rather than a separate well-defined activity (Abate *et al.*, 2000). In Africa, crop production takes place under extremely variable agro-ecological conditions, with annual rainfall ranging from 250 to 750 mm in the northwest and in the semi-arid east and south. There, the majority of African farmers still rely on indigenous pest management approaches to manage pest problems, although many government extension programs encourage use of pesticides. Currently the main focus of pest management research activities is on classical biological control and host plant resistance breeding. With the exception of biological control of cassava mealy bug, no other research results have been widely adopted (Abate *et al.*, 2000). Hence, there is a need to develop integrated pest management techniques including the indigenous pest management knowledge of farmers. Farmers often lack the biological and ecological information necessary to develop better pest management through experimentation. Formal research should therefore be instrumental in providing the input necessary to facilitate participatory technology development such as done by Farmer Field Schools (FFSs), an approach now emerging in different parts of Africa.

Such farmer field school approach have been very successful in imparting IPM knowledge in Bhiwani district of Haryana (Singh *et al.*, 2006), Andhra Pradesh and Karnataka (Balamatti and Hegde, 2007) and Bangladesh (Beijlmakers and Islam, 2007). The conventional IPM Farmer field school approach has been modified in the South Indian dryland agriculture context to suit the needs and problems of farmers in that area (Balamatti and Hegde, 2007).

Pest management in any cropping system comprises prevention, avoidance, monitoring and suppression. Prevention encompasses those practices that intend to prevent the introduction of pests into a field as well as avoid production practices that favor the development of pest populations. Preventive practices in dryland production systems emphasize the control of grassy weeds, which serve as important sources of a variety of arthropod pests. Avoidance occurs when pests are present in the field and crop management practices are used to avoid the development of significant population densities or crop damage. However, crop rotation is of limited use against arthropod pests, with the important exception of narrow-host-range, limited mobility soil pests such as *Diabrotica* spp. rootworms in corn (Chiang, 1973). Another important avoidance strategy is the use of resistant cultivars, which are deployed against pests like green bug (*Schizaphis graminum* Rondani) in sorghum and Hessian fly (*Mayetiola destructor* Say) in wheat. However, the occurrence of arthropod biotypes may limit this approach (Bailey *et al.*, 2003; Berzonsky *et al.*, 2003; Haley *et al.*, 2004; Porter *et al.*, 1997; Su *et al.*, 2003). Another avoidance strategy is the modification of planting dates, which minimizes the synchrony between the susceptible growth stages of the host plant and the pest. For instance, delayed planting avoids aphid virus vectors (Halbert *et al.*, 1990) and Hessian fly (Cartwright and Jones, 1953) in wheat. Reduced fallow periods may promote pest avoidance as a result of increased plant diversity (Altieri and Nicholls, 2004). Though, the influence of increased crop diversity on pest abundance is variable, sufficient positive observations have been documented suggesting potential pest management benefits. Increased abundance of natural enemies due to increased availability of alternative prey and other resources may be reason behind reduced pest activity in more diverse cropping systems (Sheehan, 1986; Smith and McShorley, 2000). The presence of a second crop could affect pest abundance on the primary crop by disrupting the pest's ability to find its host or by attracting the pest away from the primary crop to the second crop (Vandermeer, 1989).

The influence of habitat manipulation, including crop diversification, on natural enemies of cereal aphids have been reviewed by Brewer and Elliott, 2004. Parasitism of Russian wheat aphid by *Diaeretiella rapae* McIntosh and *Aphelinus albipodus* Hayat & Fatima was greater in wheat grown in rotation with sunflower (*Helianthus annuus* L.) than in a wheat-fallow rotation in southeast Wyoming and north-central Colorado (Ahern and Brewer, 2002). Another study was conducted with strip intercrops of sweet corn and dry bean varying in width from 1 to 16 rows, with the aim of assessing the potential for realizing the pest management benefits of intercropping in mechanized production systems. Intercropping influences were observed in strips as wide as eight rows, with specialist pests more affected than generalists (Capinera *et al.*, 1985). Natural enemies were not affected by crop strip width in this study although natural enemies are commonly more active in field edges (Altieri and Nicholls, 1999) and thus are expected to be more abundant in narrower-strip configurations.

Monitoring of pest activity forms the basis for suppression attempts. However, under low-value dryland production systems the cost of monitoring is problematic. Much effort has gone into cost-effective methods for pest monitoring. For example, commercially available pheromone lures contain species-specific attractants that can be used to trap adult males of such important wheat pests as army cutworm (*Euxoa auxiliaris* Grote) and pale western cutworm (*Agrotis orthogonia* Morrison) (Metcalf and Metcalf, 1993). Sequential sampling plans for a variety of wheat pests, including greenbug (Giles *et al.*, 2000) and Russian wheat aphid (Legg *et al.*, 1994) also have been developed. Regional cereal aphid monitoring has been accomplished using a network of inexpensive suction traps (Halbert *et al.*, 1990). Remote sensing technology for numerous arthropod pests, diseases, and weeds of dryland crops is also under development.

Pests are also favored by continuous cropping. For example, pale western cutworm eggs are laid in fall and hatch in early spring. Larvae are thus present in the field when spring crops are sown and will remain there until maturity. It also has a broad host range and is capable of damaging a variety of grasses and broadleaf crops (Peairs, 2004). However, the pests that specialize on particular crop or have a narrow host range are disadvantaged by continuous cropping as it increases the naturally occurring biological control due to a more diverse agro-ecosystem and less monoculture. The reduction in tillage associated with intensified dryland production systems tends to promote important groups of biological control agents, including ground beetles and spiders (Stinner and House, 1990). Specialist wheat pests expected to be reduced by continuous cropping include such limited host range pests as Hessian fly, wheat midge (*Sitodiplosis mosellana* Gehin) and wheat stem sawfly (*Cephus cinctus* Norton) (Peairs, 2004).

With the reduction in fallow period and continuous cropping in dryland agro-ecosystems, pest management practices are expected to change in response to the potential shifts in the pest. Prevention practices will have reduced emphasis, while avoidance practices will benefit from increased opportunities for crop rotation. The use of resistant cultivars will remain an important strategy, but planting dates will become less flexible as more crops are added to the production system. Monitoring too will become more complicated as the number of crops, potential pests and treatment decisions increases. This in turn, will lead to an increased demand for less expensive and more efficient sampling technology, such as that being developed in the area of precision agriculture (Allen *et al.*, 1999). Suppression practices too will become more complicated. Hence the intensification of dryland production systems will increase the complexity of pest management, due to the presence of additional crops and loss of fallow period.

Crop wise IPM strategies

Cereals

Maize (*Zea Mays* L.) :- It is attacked by more than 130 insect pests, but only about half-a-dozen cause economic losses (Atwal and Dhaliwal, 1997). Among these stem borer, *Chilo partellus* (Swinhoe) (Lepidoptera : Pyralidae) is the most destructive one, the larvae of which first feeds on leaves, making shot holes and then later bore into the stem causing dead hearts. It breeds actively from March to October and then the full grown caterpillars of last generation hibernate in stubbles, stalks, etc. Few other borers, European Corn Borer; *Ostrinia nubilalis* (Hubner) and Asian maize borer; *Ostrinia furnacalis* (Guenee) (Lepidoptera: Pyralidae) also becomes serious in different parts of the world.

IPM strategies -

- Deep field ploughing after harvest, thus destroying the stubbles, weeds and other alternate hosts.
- Selection of resistant / tolerant varieties also minimizes the losses.
- Removal and destruction of plants showing shot holes or dead hearts reduces the pest incidence.
- Setting up light traps till midnight to attract and kill the moths.
- Release of bioagents in synchrony with the stage of pest *viz.*, *Trichogramma* spp. on egg masses and *Apanteles* spp. or *Microbracon* sp. adults with the larval development.
- Mix 8 kg of quinalphos 5G or Carbaryl 10 G with sand to make up the total quantity of 20 kg and apply in the leaf whorls with the help of bottle with few holes in its cap.
- Or spray quinalphos 25 EC @ 800 ml / carbaryl 50 WP @1.2 kg / Endosulfan 35 EC @ 0.07%.

Other important insect pests are blister beetle (*Mylabris pustulata* Thunberg, *M. macilenta* Mshll., *M. phalerta* Pall., *M. liflensis* Billb.) (Coleoptera; meloidae), the adults of which feed on the silk of cob, thus interfering in the fertilization process, thereby affecting seed set and yield. Cutworms, armyworms, grasshoppers, etc. are general leaf feeders. Corn worm or ear worm, *Helicoverpa armigera* (Hubner) feeds on the silk and developing grains. Aphids and Jassids are sap feeders. White grub and gujhia weevil feed on the roots. The above mentioned insecticidal sprays take care of these pests as well.

Sorghum (*Sorghum bicolor* (L.) Moench.) :-

More than 150 pest species are recorded to attack sorghum but few are important. Sorghum shootfly, *Atherigona soccata* Rondani (Diptera : Muscidae) is a serious pest in many states. High yielding hybrids are more susceptible. Maggots bore into the stem and cut the main shoot in the early stages of crop growth thereby, showing typical dead heart symptoms. Shoot bug, *Peregrinus maidis* (Ashmead) (Hemiptera : Delphacidae) and Sorghum earhead bug, *Calocoris angustatus* Lethiery (Hemiptera : Miridae) are also serious at places. The adults and nymphs of shoot bug feed gregariously on the leaves, thus retarding plant growth, while those of ear head bug sucks the sap from ears, causing the grains to remain chaffy or shrivelled. Few other pests of minor importance are sorghum midge, *Contarinia sorghicola* (Coquillett), stem borers, root feeders, armyworms, etc.

IPM strategies –

- Early sowing within 10-15 days of onset of monsoon escapes damage by shootfly.
- Growing resistant varieties.
- Setting up of hanging type plastic fish meal bait traps @ 5 per acre is highly effective in attracting adult shoot flies.
- Seed treatment with imidacloprid 70 WS @ 10g/kg seed or thiamethoxam 70 WS @ 3g/kg or isofenphos 5G @ 30 g/kg seed protects against shootfly.
- Removal and destruction of infested plants after 3-4 weeks of sowing to prevent the emergence of shootfly adults.
- Natural enemy, lady bird beetles, *Coccinella septempunctata* Linnaeus and *Menochilus sexmaculatus* (Fabricius) attack the young nymphs of shoot bug.
- Earhead bugs can be killed mechanically by shaking the earheads over a bucket of kerosinized water.
- Application of carbofuran 3G, phorate 10G, chlorpyriphos 5G in furrows @ 3 g per metre row length is also effective.
- Spraying endosulfan 35 EC or carbaryl 50 WP has also been found effective.

Bajra :-

This is a peculiar dryland crop and is not attacked by many pests. However, few pests of importance are shootbug, earhead feeders *viz.*, midge (*Germyia penniseti*) and blister beetle, root feeders white grub and leaf roller.

IPM strategies –

- Early and synchronous sowing helps in escaping the damage.
- Pulling out seedlings showing dead heart symptoms.
- Spraying with endosulfan 35 EC or dimethoate 30 EC, if need arises.

Wheat (*Triticum spp.*):-

The crop is attacked by few insect pests of economic importance. Among these wheat aphid; *Macrosiphum miscanthi* (Takahashi) (Hemiptera : Aphididae) is serious even under dryland conditions, though the incidence is more on well-irrigated and succulent crop. Both adults and nymphs suck the cell sap and devitalize the plant. Wheat thrips, armyworms, gujhia weevil (seedling eater), stem borer, leaf feeders, root eaters, etc. are few other pests of less importance. Wheat gall and Molya nematode have reported to cause severe damage to the crop in some states.

IPM strategies -

- Deep ploughing during April-May, immediately after wheat harvest kills the pupae of gujhia weevil.
- Collection and destruction of armyworm caterpillars can suppress the pest.
- Spraying with 375 ml dimethoate 30 EC or 500 ml dichlorvos 100 EC or 1000 ml quinalphos 25 EC is effective.
- The gall nematode can be controlled by separating the galls from wheat seed by floating them on water. The galls float on the surface and so can be removed easily.
- Keeping the fields fallow for one year can also suppress these nematodes.

Oilseeds

Mustard (*Brassica juncea* (Czern and Coss.) :-

These are damaged by number of pests, of which mustard aphid (*Lipaphis erysimi* (Kaltenbach) is the most serious one followed by mustard sawfly (*Athalia lugens* (Klug) and painted bug (*Bagrada hilaris* (Burmeister). The nymphs as well as adults of both aphids and painted bug suck the cell sap, while the grubs of sawflies bite holes into the leaves, sometimes skeletonizing the young leaves completely.

IPM strategies –

- Early sowing in the season escapes peak mustard aphid incidence.
- When the aphid population reaches 50-60 aphids per 10 cm terminal shoot or when an average of 0.5-1.0 cm terminal portion of central shoot is covered by aphids or when plants infested by aphids reach 40-50 per cent, foliar sprays with insecticides like 625 ml dimethoate 30 EC or malathion 50 EC or endosulfan 35 EC, 1000 ml chlorpyriphos 20 EC should be initiated. Spraying with endosulfan should be preferred as it is safer to honeybees which frequent mustard crop during flowering season.

Groundnut (*Arachis hypogea*) :-

This is attacked by several important insect pests, groundnut aphid (*Aphis craccivora* Koch), leaf miner (*Stomopteryx nertaria* Meyrick), stem borer (*Sphenoptera perotetti* G.), red hairy caterpillar (*Amsacta moorei* (Butler) and bihar hairy caterpillar (*Spilosoma obliqua* Walker), white grub (*Holotrichia consanguinea* Blanchard), thrips, leaf hoppers, etc. Aphids, thrips and leaf hoppers suck the cell sap, while leaf miner mines into the leaves, skeletonize them and web them together. Stem borer grubs bore into the stem and roots, white grubs feed on fibrous roots and hairy caterpillars defoliate the plant.

IPM strategies –

- Deep summer ploughing helps in exposing the white grubs resting in the soil.
- Early sowing escapes damage.
- Growing thrips resistant varieties like Robut 33-1, Kadiri 3, ICGS 86031 are essential as they transmit bud necrosis disease.
- Growing cowpea as intercrop (1:4) minimizes the incidence of leaf miner and hairy caterpillars.
- Field should be maintained weed free.
- Setting up of light traps up to midnight to monitor and attract leaf miner, white grubs as well as hairy caterpillars' infestation.
- Installation of pheromone traps for leaf miner.
- Collection and destruction of egg masses and early instar larvae of hairy caterpillars.
- Cutting and destroying the infested branches prevents stem borer grubs from entering the tap root.
- Application of 500 ml of chlorpyriphos 20 EC or 200 ml imidachloprid 200SL or dimethoate @ 2 ml/l takes care of almost all the pests.

Sesame (*Sesamum indicum* L.) :-

This is attacked by few economically important insect pests such as gall fly (*Asphondylia sesame* Felt.), til hawk moth (*Acherontia styx* (Westwood), shoot webbers cum capsule borers (*Antigastra catalaunalis* (Duponchel), aphids, thrips, leaf hoppers, pod bugs, etc. Gall fly maggots feed on the buds which develop into galls and produce no fruit and seeds. Gall fly is a serious pest in dryland areas. Hawk moth's larvae are voracious defoliators while the adults' sucks honey from honey combs or apiaries. Young caterpillars of shoot webbers feed on leaves and then bore into the shoots, flowers, buds and pods.

IPM strategies –

- Growing gall fly resistant varieties.
- Field ploughing during winter suppresses shoot webber activity.
- Spraying with 400 ml of endosulfan 35 EC or 300 ml of quinalphos 25 EC or 200 dichlorvos 76 WSC or NSKE 5% or 400 g of carbaryl 50 WP is effective.

Linseed (*Linum usitatissimum* L.) :-

Damage is caused by gall midge (*Dasineura lini* Barnes) maggots which feeds on buds and flowers. Few other pests of minor importance are jassids, whiteflies, thrips and hairy caterpillars.

IPM strategies –

- The adult gall midge flies can be killed by using light traps or by attracting them to molasses or jaggery added to water.
- Normal sown crop escapes peak damage by gall midges.

Safflower (*Carthamus tinctorius* Linn.) :-

Safflower aphids (*Uroleucon carthami* (Hille Ris Lambers) cause serious damage by sucking the cell sap. It is also attacked by bud fly, caterpillars, leaf thrips, etc.

IPM strategies –

- Release of *Chrysoperla carnea* @ 20,000 /acre takes care of the aphid.
- Alternatively the crop may be sprayed with 400 ml endosulfan 35 EC or 500 ml chlorpyriphos 20 EC.

Fibre crops

Cotton (Gossypium spp.) :-

More than 1326 insect species have been recorded on cotton in the world, however only 15 species are of importance. Bollworms, aphids, hoppers, thrips, whiteflies and mealybugs, stem weevil and ash weevil, red cotton bug and dusky cotton bug are important pests of cotton. Bollworm caterpillar bore into the flower buds, panicles, bolls and even growing shoots either killing the plant or causing heavy shedding of the fruiting bodies. Aphids, hoppers, thrips, whiteflies and mealybugs suck the cell sap and cause heavy loss by devitalizing the plant. With the introduction of *Bt.* Cotton in India, mealybugs, once a miner pest has assumed importance and is becoming serious. Red cotton bug and dusky cotton bug also suck cell sap and often the lint is of poorer quality or get stained. Stem weevil grubs feed on the soft tissues of cotton stem thus killing the plant. This is a serious pest in few states. Grubs of ash weevil feed on the roots of cotton seedlings while the adults feed on leaves, buds, flowers and young bolls by cutting prominent round holes.

IPM strategies -

- Timely sowing avoids peak pest attack.
- Growing brinjal in the vicinity should be avoided in ash weevil endemic areas.
- Bhendi, pigeonpea or marigold should be grown as trap crop for bollworms as well as most of the sucking pests.
- Application of neem dust formulation on cotton, divert the pests to trap crops.
- Removal of alternate weed hosts, off-season cotton sprouts.
- Deep ploughing with a furrow-turning plough by the end of February helps in reducing the incidence of bollworms.
- Seed treatment with imidacloprid 70 WS 7g or thiamethaxam 70 WS 4g / kg or chlothiamidine 600 FS 9ml / kg seed or apply NSKE 5% or neem based formulations provides protection against most of the sucking pests.
- Setting up of yellow sticky traps for whiteflies and light or pheromone trap (@ 5 / acre) for bollworms, cutworms.
- Apply profenofos 50 EC @ 1ml / l or pongamia oil 4 ml / l or indoxacarb 200 ml / ha on cotton as well as trap crops for bollworms and traizophos 0.1% or acephate 0.2%, neem oil 0.5%, acetamiprid 20SP 16g / acre, if sucking pests are serious. Application of ovicide pongamia oil can be confined to trap crops only.
- Apply biopesticides like NPV @ 200 LE / acre for bollworms.

Pulses

Bengal Gram (*Cicer arietinum* L.) :-

Chickpea is mainly attacked by gram pod borer, *Helicoverpa armigera* (Hubner), sometimes by cutworm (*Agrotis ypsilon* (Rott.) and tobacco caterpillar (*Spodoptera litura* Fabr.) also becomes serious.

IPM strategies -

- Timely sowing i.e. upto mid October or growing early maturing cultivars which complete podding by first week of March in northern region escapes peak *H. armigera* activity.
- Intercropping with mustard or linseed has been found very effective in attracting natural enemies, thus reducing pod damage in dryland areas.
- Grow tall sorghum as companion crop to serve as biological bird perches which feed on grown up pod borer larvae.
- Setting up of light or pheromone trap @ 5 / acre for monitoring and @ 20 / ha for mass trapping of pod borer adults and change the septa once in 3 weeks.

- Apply NPV @ 250 LE / ha + half dose of endosulfan 35 EC (1.25 l / ha) along with 0.1% UV retardant such as Tinopal and 0.5% jaggery.
- Apply profenofos 50 EC @ 1ml / l or pongamia oil 4 ml / l or indoxacarb 200 ml / ha or endosulfan 0.07% or quinalphos 0.05%.

Red gram (*Cajanus cajan*) :-

Pod borer, *H. armigera* is the major pest. Few other important pests are tur pod fly (*Melanogromyza obtusa* Malloch.), pod bug (*Clavigralla gibbosa* Spinola) and plume moth (*Exelastis atomosa* Walshingham). Pod fly maggots as well as plume moth larvae bore into the pods and feed on the grains, while pod bug adults and nymphs suck the cell sap. Eriophid mites also cause havoc on this crop by transmitting sterility mosaic disease.

IPM strategies –

- Spraying with 1.75 kg carbaryl 50 WP or 1 litre endosulfan 35 EC takes care of most of the pests.
- For the management of pod borer, measures are same as mentioned in case of chickpea.

Mungbean / Urdbean :-

These are attacked by several insect pests *viz.*, whiteflies, pod bugs, aphids and thrips which suck the cell sap and pod borers which bore into the pod and feed on the grains. Whiteflies also transmit yellow mosaic virus and leaf curl virus. Aphids are vectors of leaf crinckle virus. Sometimes hairy caterpillars too become severe on these crops.

IPM strategies –

- Virus infected plants should be routinely rouged out and destroyed.
- Growing virus resistant varieties.
- Setting up of yellow sticky traps to monitor and mass trap whitefly population.
- Sucking pests like whiteflies may be managed by application of triazophos 0.1%, acetamiprid 20 SP @ 16 g/acre, fipronil 0.05%, acephate 0.2%, methyl demeton 0.025%.
- Egg masses or early instars of hairy caterpillars may be collected and destroyed during routine field monitoring. Later instars of these hairy caterpillars are difficult to control as they defoliate heavily.

Lentil (*Lens esculentus*) :-

This is attacked by very few insect pests. One pest of little importance is lentil pod borer (*Etiella zinckenella* (Treitschke), the larvae of which bores into the pod and feeds on grains inside.

IPM strategies -

- Chemical control same as mentioned for chickpea pod borer.

Field pea (*Pisum sativum* L.) :-

Pea pod borer (*Etiella zinckenella* (Treitschke) and pea leaf miner (*Phytomyza atricornis* Meigen) are the important insect pests attacking field pea.

IPM strategies -

- Chemical control same as mentioned for chickpea pod borer.

Fruit crops

Mango (*Mangifera indica* L.) :-

Good number of insect pests attack this fruit grown under wide range of climatic conditions. A comprehensive list of mango pests has been documented (Tandon and Vergese, 1985; Veeresh, 1989; Golez, 1991). These pest species cause heavy losses if not managed properly. Mango hoppers (*Amritodus atkinsoni* (Lethierry) and *Idioscopus clypealis* (Lethierry), mango mealy bugs (*Drosicha mangiferae* (Green), *Drosicha stebbingi* and *Rastrococcus iceryoides*), fruit fly (*Bactrocera dorsalis* (Hendel), stem borer (*Batocera rufomaculata* DeGeer), bark eating caterpillar (*Indarbela tetraonis* Moore) and mango stone weevil (*Stenochaetus mangifera* F.) are few important insect pests. Hopper adults and nymphs suck the cell sap and are also found responsible for causing mango inflorescence malformation. Mealy bugs too suck the cell sap and devitalize the plant. Fruit flies maggots also pose serious threat to the mango fruits by feeding on the pulp and making the fruit unfit for human consumption. Stem borer grubs bore into the mango stem, thus killing the entire tree or branch. Bark eating caterpillar makes galleries and feed on the bark.

IPM strategies -

- Removal of weeds and alternate hosts of mealy bugs reduces pest load.
- Deep summer ploughing during May-June exposes the eggs and pupae of mealy bugs and fruit flies.
- Tree banding with 20 cm wide alkathene or polythene sheets (400 guage) should be done in November - December, 50 cm above the ground level to prevent the mealy bug crawlers from climbing up the tree.
- The nymphs congregating below the lower edge of alkathene sheets may be collected and destroyed.
- Setting up of methyl eugenol pheromone traps to monitor and mass trap fruit flies.
- Bait trap with 100 ml of 0.5 methyl eugenol (1ml/l) and 0.05% malathion 50 EC (1 ml/l) taken in 250 ml capacity wide mouthed bottle may be hanged trap the males. The solution should be changed fortnightly.

- Bait spray combining any one insecticides (fenthion 80 EC @ 1 ml/l, malathion 50 EC @ 2 ml/l, carbaryl 50 WP 4 g/l) and molasses or jaggery 10g/l may be used before fruit ripening.
- Frequent collection and destruction of fruit fly infested fruits
- Stem injection with insecticides like phosphamidon 85 WSC, ethofenoprox 10 EC, imidacloprid 17.8 SL, etc. may be done to check mango hoppers.
- Carbofuran 3G @ 5g/bore hole should be used for stem borer infestation.

Citrus :-

These are attacked by several insect species *viz.*, citrus leaf miner (*Phyllocnistis citrella* Stainton), mealy bugs (*Pseudococcus filamentosus* Cockerell), citrus psylla (*Diaphorina citri* Kuwayana), aphids, whiteflies, scales, fruit sucking moths, stem borer, etc. Leaf miner mines into the leaves and is also vector of greening disease of citrus. Mealy bugs, aphids, whiteflies, scales suck the cell sap. Fruit sucking moth adults suck the juice from fruits and directly damage the economic plant part.

IPM strategies –

- Destruction of alternate hosts or weed crops like *Tinospora* for control of fruit sucking moths.
- Setting up light traps or food lures (pieces of fruits) to attract adult fruit sucking moths.
- Bagging the fruits with punctured polythene bags (300 gauge) is also found effective in case of small trees.
- Release of mealy bug predators, *Cryptolaemus montrouzieri* @ 10 beetles / tree. Make one to three releases per annum depending on the mealy bug population.
- Alternatively release *Leptomastix dactylopi* @ 5000-7000 adults / ha for mealy bugs.
- Spray dichlorvos 0.20% in combination with fish oil resin soap (25g/l) for checking mealy bugs.
- Release of *Chrysoperla carnea* grubs @ 10-15 / plant is also effective against citrus aphid and mite.
- Prune the branches containing stem borer grubs during July - September.
- Inject 10 ml of chlorpyriphos 20 EC / live bore and plug the hole with wet clay to kill the stem borer grubs inside.

Guava (*Psydium guajava*) :-

Fruit flies (*Bactrocera dorsalis* (Hendel), whiteflies (*Bemisia tabaci* Genn.) and scale insect (*Pulvinaria psidii* Maskell) are important guava pests. Fruit flies cause considerable loss to the rainy season fruit crop.

IPM strategies -

- Same as suggested for mango fruit fly.
- Release of *Cryptolaemus montrouzieri* @ 20 beetles / tree is found effective against guava scale.

Pomegranate (*Punica granatum* L.) :-

Anar butterfly (*Virachola isocrates* (Fabricius) causes heavy damage to fruits. The caterpillars bore inside the developing fruits and feed on the pulp and seeds.

IPM strategies -

- Bagging of fruits before maturity helps in checking the damage.
- Collection and destruction of infested fallen fruits.
- Spraying NSKE 5% or neem formulations @ 2ml/l commencing from flowering act as oviposition deterrent.
- Spraying young fruits with phosphamidon 40 SL prevents damage.

Aonla (*Emblica oficinalis*) :-

This fruit tree is attacked by several insect pests *viz.*, Shoot gall maker (*Betousa tylophora* Swinhoe); leaf roller (*Gracillaria acidula* Forster); bark eating caterpillar (*Indarbela tetraonis* Moore) and aphid (*Schoutedenia emblica* Patel and Kulkarni). The gall maker caterpillar makes gall in the small branches and twigs thereby hindering the proper transportation of nutrients. More than 40% of the aonla leaves have been found rolled by leaf roller larvae.

IPM strategies -

- Pruning and destruction of galled twigs and branches reduces the pest load.
- Fallen rolled leaves should be collected and destroyed.
- Cleaning the bark eating caterpillar infested portion of the tree trunk and then swabbing the trunk with dimethoate 30 EC or endosulfan 35 EC checks the infestation.

Ber (*Ziziphus mauritiana*), Phalsa (*Grewia subinaequalis*) and Loquat (*Eriobotrya japonica* Lindley)

Fruit fly (*Carpomyia vesuviana*) is a major insect pest causing damage to ber, phalsa and loquat fruits. Ber beetles (*Adoretus pallens* Arrow and *A. nitidus* Arrow) are also found defoliating the ber tree.

IPM strategies -

- Same as mentioned in case of mango fruit fly.
- Installing light traps for trapping adult beetles.
- Raking around the trees during winter months exposes the hibernating grubs of ber beetles.

Beal (*Aegle marmelos*) :-

This is a very hardy plant and is almost free from the attack of any insect pests of economic importance. Sometimes it is attacked by bark eating caterpillar and trunk borer.

IPM strategies –

- Same as suggested in aonla bark eating caterpillar.

Jamun (*Syzygium cuminii* L.) :-

Only bark eating caterpillars have been found causing serious damage to the tree.

IPM strategies –

- Same as suggested in aonla bark eating caterpillar.

Tamarind (*Tamarindus indica* L.) :-

This is attacked mainly by mealy bugs (*Nipaecoccus viridis* (Newstead), fruit borer (*Phycita orthoclina* Meyrick), castor capsule borer (*Dichocrosis punctiferalis* (Guenee) and anar butterfly (*Virachola isocrates* (Fabricius). Mealy bugs suck the cell sap and in case of severe attack the leaflets as well as immature fruits fall off. The fruit borer and capsule borer larvae feed on the pulp and make the fruit unfit for culinary purposes.

IPM strategies –

- Same as suggested for mango mealy bugs.
- Spraying with 315 ml of dichlorvos 100 EC or 1.25 kg of carbaryl 50 WP in 600 liters water per ha.

Under dryland farming system, in general insect pest incidence is less. The intensification of dryland production systems will however, increase the complexity of pest management, due primarily to the presence of additional crops and to the loss of fallow period.

References

Abate, T., Van Huis and J.K.O. Ampofo. 2000. Pest management strategies in traditional agriculture: An African perspective. Annual Review of Entomology. 45: 631-659.

Ahern, R.G., and M.J. Brewer. 2002. Effect of different wheat production systems on the presence of two parasitoids (Hymenoptera: Aphelinidae; Braconidae) of the Russian wheat aphid in the North America Great Plains. Agric. Ecosyst. Environ. 92: 201–210.

Allen, J.C., D.D. Kopp, and S.J. Fleischer. 1999. 2011: An agricultural odyssey. Am. Entomol. 45: 96–117.

Altieri, M.A., and C.I. Nicholls. 1999. Biodiversity, ecosystem function, and insect pest management in agricultural systems. p. 69–84. In W.W. Collins and C.O. Qualset (ed.) Biodiversity in agroecosystems. CRC Press, Boca Raton, FL.

Altieri, M.A., and C.I. Nicholls. 2004. Biodiversity and pest management in agroecosystems. 2nd ed. Food Products Press, New York.

Atwal, A.S. and Dhaliwal, G.S. 1997. Agriculture pests of South Asia and their Management. Kalyani Publishers, Ludhiana.

Bailey, K.L., B.D. Gossen, R. Gugel, and R.A.A. Morrall (ed.) 2003. Diseases of field crops in Canada. 3rd ed. Canadian Phytopathol. Soc., Saskatoon, SK.

Balamatti, Arun and Rajendra Hegde. 2007. Our experiences with modified Farmer Field Schools in dryland areas. Leisa India 9(4): 17-18.

Berzonsky, W.A., H. Ding, S.D. Haley, M.O. Harris, R.J. Lamb, R.I.H. McKenzie, H.W. Ohm, F.L. Patterson, F.B. Peairs, D.R. Porter, R.H. Ratcliffe, and T.G. Shanower. 2003. Breeding wheat for resistance to insects. Plant Breed. Rev. 22: 221–296.

Bijlmakers, Hein and Muhammad Ashraful Islam. 2007. Changing the strategies of Farmer Field Schools in Bangladesh. Leisa India. 9(4): 19-21.

Brewer, M.J., and N.C. Elliott. 2004. Biological control of cereal aphids in North America and mediating effects of host plant and habitat manipulations. Annual Review of Entomology 49:219–242.

Capinera, J.L., T.J. Weissling, and E.E. Schweizer. 1985. Compatibility of intercropping with mechanized agriculture: Effects of strip intercropping on pinto beans and sweet corn on insect abundance in Colorado. Journal of Economic Entomology 78: 354–357.

Cartwright, W.B., and E.T. Jones. 1953. The Hessian fly and how losses from it can be avoided. USDA Farmers' Bull. 1627. USDA, Washington, DC.

Chiang, H.C. 1973. Bionomics of the northern and western corn rootworms. Annual Review Entomology 18: 47–72.

Giles, K.L., T.A. Royer, N.C. Elliott, and S.D. Kindler. 2000. Development and validation of a binomial sequential sampling plan for greenbug (Homoptera: Aphididae) in the southern plains. Journal of Economic Entomology 93: 1522–1530.

Golez, H.G.1991. Bionomics and control of the mango seed borer, Noorda albizonalis Hampson (Pyralidae: Lepidoptera). Acta Hortic. 291: 418-424.

Halbert, S., J. Connelly, and L. Sandvol. 1990. Suction trapping of aphids in western North America. Acta Phytopathol. Entomol. Hung. 25: 411–422.

Haley, S.D., F.B. Peairs, C.B. Walker, J.B. Rudolph, and T.L. Randolph. 2004. Occurrence of a new Russian wheat aphid biotype in Colorado. Crop Sci. 44: 1589–1592.

Legg, D.E., R.M. Nowierski, M.G. Feng, F.B. Peairs, G.L. Hein, L.R. Elberson, and J.B. Johnson. 1994. Binomial sequential sampling plans and decision support algorithms for managing the Russian wheat aphid (Homoptera: Aphididae) in small grains. Journal of Economic Entomology 87: 1513–1533.

Metcalf, R.L., and R.A. Metcalf. 1993. Destructive and useful insects: Their habits and control. 5th ed. McGraw-Hill, New York.

Peairs, F.B. 2004. Wheat pests and their management. p. 2529–2545. In J.L. Capinera (ed.) Encyclopedia of entomology. Kluwer Academic Publ., Boston.

Porter, D.R., J.D. Burd, K.A. Shufran, J.A. Webster, and G.L. Teetes. 1997. Greenbug (Homoptera: Aphididae) biotypes: Selected by resistant cultivars or preadapted opportunists? Journal of Economic Entomology 90: 1055–1065.

Sheehan, W. 1986. Response by specialist and generalist natural enemies to agroecosystem diversification: A selective review. Environmental Entomology 15: 456–461.

Singh, Sube, Shehrawat, P.S., Milakh, Raj and Hasija, R.C. 2006. Social aspects of sustainable dryland agriculture as perceived by the farmers. Environment and Ecology.

Smith, H.A., and R. McSorley. 2000. Intercropping and pest management: A review of major concepts. American Entomologist 46: 154–161.

Stinner, B., and G. House. 1990. Arthropods and other invertebrates in conservation-tillage agriculture. Annual Review of Entomology 35: 299–318.

Su, H., R.L. Conner, R.J. Graph, and A.D. Kuzyk. 2003. Virulence of Puccinia striiformis f. sp. tritici, cause of stripe rust on wheat, in western Canada from 1984 to 2002. Canadian Journal of Plant Pathology 25: 312–319.

Tandon, P.L.and Verghese, A. 1985. World list of insect , mite and other pests of mango. Tech. Doc. Indian Inst. Hortic. Res. 5.

Vandermeer, J. 1989. The ecology of intercropping. University Press, Cambridge, UK.

Veeresh, G.K. 1989. Pest problem in mango- world situation. Acta Hortic. 231: 551-565.

CHAPTER 24
Integrated Disease Management of Important Crops of Dryland Agriculture

Vishal Gupta, V.K. Razdan and Ranbir Singh

Division of Plant Pathology, Faculty of Agriculture, Sher-e-Kashmir University of Agricultural Sciences and Technology of Jammu, Chatha, Jammu-180009.

Indian agriculture is predominantly a rainfed agriculture under which both dry farming and dryland agriculture are included. Out of the 143 million ha of total cultivated area in the country, 108 million ha (i.e. nearly 75% area) is rainfed. In dry land areas, variation in amount and distribution of rainfall influence the crop production as well as socio-economic conditions of farmers. The dry land areas of the country contribute about 42 per cent of the total food grain production. Most of the coarse grains like sorghum, pearl millet, and other millets are grown in dryland only. By 2010 A.D., India will have to produce 300 million tones of food grains to feed approximately 1.5 billion mouths. This target cannot be realized from irrigated area alone as we have irrigation potential for 178 million hectares only. Therefore, we will have to evolve an appropriate technology for dry land farming. On the other hand, we can say that second 'green revolution' in Indian agriculture can be achieved through rainfed/ dryland agriculture.

Dryland, besides being water deficient, is characterized by high evaporation rates, exceptionally high day temperature during summer, low humidity and high run off and soil erosion. The soil of such areas is often found to be saline and low in fertility. As water is the most important factor of crop production, inadequacy and uncertainty of rainfall often cause partial or complete failure of the crops which leads to period of scarcities and famines. Dry farming or dryland farming may be defined as: "a practice of growing profitable crops without irrigation in areas which receive an annual rainfall of 500 mm or even less". Maximum yields per unit area of land can be achieved and sustained only if there is also a provision for protection of the crop against the diseases. Plant diseases cause serious threats to the successful cultivation of the important crops of dryland farming. Correct disease diagnosis is the prime requirement for recommending preventive or curative measures for effective disease management. The present chapter describes the diagnosis and management of important diseases occurring in the crops of dryland farming. Integrated disease management systems focus on the use of biological and other natural control options. The reduced use of agrochemicals coupled with more environment friendly ways to control diseases and enhance plant growth forms an integral part of Integrated Disease Management systems.

A. PEARL MILLET (*Pennisetum glaucum* (L.) R. Br.)

It is grown widely as a rainfed crop in semi-arid regions of Africa and India for food and forage purpose. Major diseases of pearl millet have been described as under.

Downy Mildew or Green ear disease

Causal organism: *Sclerospora graminicola* (Sacc.) Schroet.

Symptoms often vary as a result of systemic infection. Leaf symptoms begin as chlorosis at the base and successively higher leaves show progressively greater chlorosis. The chlorotic leaf areas support abundant white asexual sporulation on the lower leaf surface. Severely infected plants are generally stunted and do not produce panicles. Green ear symptoms result from transformation of floral parts into leafy structures. Asexual sporangia are produced during the night with moderate temperatures and high humidity.

Ergot

Causal organism: ***Claviceps fusiformis*** **Loveless.**

Cream to pink mucilaginous droplets of "honeydew" ooze out of infected florets on pearl millet panicles. Within 10 to 15 days, the droplets dry and harden, and dark brown to black, sclerotia develop in place of seeds on the panicle. Sclerotia are larger than seed and irregularly shaped, and generally get mixed with the grain during threshing. The seed set is poor or remains completely inhibited. Sclerotia germinate to form 1 to 16 fleshy stipes, 6 to 26 mm long. Each stipe bears an apical, globular capitulum, light to dark brown in colour, with numerous perithecial projections. Sclerotia germinate following rain. Conditions favouring the disease are relative humidity greater than 80 per cent, and 20 to 30°C temperature. Honeydew production promotes secondary infection caused by asexual conidia.

Smut

Causal organism: Tolyposporium penicillariae Bref.

Smut symptoms initially appear as green sori larger than the seed developed on panicles during grain filling. As grain matures, sori change in colour from green to dark brown. Sori are filled with dark teliospores. Such spores serve as a source for secondary spread. Chestnut-brown to black-brown spore balls are composed of 200 to 1400 aggregated yellowish-brown globose to sub-globose teliospores. Teliospores germinate to produce promycelia with basidiospores and sporidia

Infection occurs when sporidia suspended in rain or dew infiltrate into boot. Aerial populations of sporidia are greatest when minimum and maximum temperature range between approximately 21 and 31°C, and maximum relative humidity is greater than 80 per cent.

Rust

Causal organism:*Puccinia penniseti*. Zimm.

On pearl millet small reddish-brown to reddish orange, round to elliptical uredinia develop mainly on foliage. As severity of infection increases, leaf tissue wilts and become necrotic from the leaf apex to base. In infection sites developing late in the season, uredinia are replaced by telia which are black, elliptical, and sub epidermal. Urediniospores are generally elliptical, measuring 35x25 μm, with four equatorial germpores. Spores have yellowish brown walls and are echinulate, but more predominantly near the apex. Teliospores are generally 2-celled.

Rhizoctonia Blight

Causal organism:*Rhizoctonia solani* Kuhn

Disease can be expressed as seed decay, pre-and post-emergence damping off, stem lesions on seedlings or stem canker on more mature plants. Invasion of sheath and blade tissue can cause banding pattern. Mid-rib is usually the last part of the leaf killed. Mature plants have considerable accumulation of dead brown leaves around the base of the plant. Root system is reduced with extensive killing and discoloration. *Rhizoctonia* species often survive in soil as melanized hyphae and sclerotia, often associated with plant debris.

INTEGRATED DISEASE MANAGEMENT

- Seed treatment with Apron 2 g/kg + foliar spray with Ridomil 0.6% were most effective in reducing the incidence of pearl millet downy mildew (Mani and Hepziba, 2009). Seed treatment with Apron SD 35 @ 2 g a.i. kg^{-1} seed, followed by a foliar application of Ridomil MZ (2 g a.i. l^{-1}) has been recommended for the complete control of the downy mildew (Singh and Shetty,1990). Singh *et al.* (1993) had suggested the treatment of seed with 0.1% $HgCl_2$ solution for 10 min followed by several rinses in distilled water, hot water treatment of the seed at 55°C for 12 min followed by drying under shade and treating the seed with metalaxyl at 2 g a.i. kg^{-1} seed to inactivate the seed borne inoculums of the disease.
- The best method of managing the ergot disease in pearl millet is through the use of resistant varieties. Several ergot resistant lines with high yield potential developed at ICRISAT are sources for resistance in breeding programme (Thakur, *et al.*, 1988). The most commonly recommended method of control is use of clean seed. Dipping the seed in 20-30 per cent salt solution makes the sclerotia to floats which can be easily removed by hand.

- The control measures suggested for smut disease are the removal of smutted ears, usage of clean seed, hot weather deep ploughing, crop rotation, field sanitation and use of resistant varieties. Certain genotypes (SSCPS-252-S-4 and ICI7517-S-1) have shown high level of resistance (Thakur, *et al.*, 1986). There have been reports that intercropping mung bean with pearl millet also reduces smut incidence.
- *P. penniseti* is parasitized by two fungi, *Darluca filum* and *Tuberculina* spp (Sundaram,1962)
- All major diseases of pearl millet can effectively be managed by the use of host resistance in combination with appropriate fungicides and cultural practices.

B. Sorghum (*Sorghum bicolor* (L.) Moench)

Sorghum green millet or jowar is an important food and fodder crop of India.

Anthracnose Leaf Blight and Stalk Rot of Sorghum

Causal organism:*Colletotrichum graminicola* (Ces.) G.W. Wils.,

The fungus produces a wide range of symptoms on corn. Leaf blight symptoms progress from lower to upper leaves and vary in size and colour with host genotype. Typical symptoms on a susceptible hybrid appear as small, water soaked spots that are semitransparent and oval to elongate in shape. Spots enlarge and become tan with a wide border that varies in shade from red or orange to purple or tan. The entire leaf may become blighted if lesions coalesce, resulting in a "fired" appearance. Dark fruiting bodies called acervuli develop on dead host tissue. Leaf infection may also appear as a midrib infection. This type of infection is characterized by elongated elliptical lesions that vary in colour from red to purple to black. Leaf and midrib infections may occur independently of each other or together in such case yield loss increases. Infected panicles are lightweight, may exhibit some degree of sterility and mature early. Infected seed is discoloured, germinates poorly and may produce plants that succumb to seedling blight. The stalk rot phase of anthracnose is very similar to the panicle infection phase. Infection may occur at anytime during the growing season but symptom development is most common on mature plants. Infection occurs when conidia from the leaf blight stage are splashed or wind blown to the stalks. The initial symptom of stalk infection is a water-soaked discolouration of rind tissue in the lower internodes. Lesions take on a reddish discolouration and infected tissue is interspersed with healthy tissue. External infections are characterized by irregular bleached areas that are surrounded by a red border (host pigmentation).

Sorghum Downy Mildew

Causal organism:Perono*sclerospora sorghi* (Weston & Uppal)

Systemically infected seedlings are chlorotic and stunted. The chlorosis may be more noticeable on the lower half of the leaf. Young plants may die prematurely. Infected plants that survive the seedling stage produce a mixture of interesting symptoms. Under cool, humid conditions a white downy growth is produced on the lower leaf surface. This growth is a combination of conidia and conidiophores. Conidia are only produced at night and require a layer of moisture on the leaf for spore production to occur. As the plant matures, leaf symptoms become more striking. Older leaves may exhibit alternating parallel stripes of green and yellowish-green to white tissue. The tissue in the lighter stripes eventually dies and leaves become shredded, resembling hail injury. Heads produced on these plants are fully or partially sterile. Symptoms of local infections are a little less dramatic. Short necrotic streaks (stipples) are produced on leaf blades. Downy conidial growth on leaves is also associated with these infections.

Charcoal Rot

Causal organism:*Macrophomina phaseolina* (Tassi) Goidanich

Initial symptoms develop on roots and appear as water-soaked lesions. The lesions turn brown or black with age. The fungus continues to invade plant tissue from the crown up and causes similar water-soaking and discoloration in the pith. The pith eventually disintegrates leaving only the vascular strands intact. Numerous, small, black sclerotia form on these strands and are easily visible when stalks are split open. The most characteristic outward symptom of the disease is lodging. This often occurs in the driest part of the field. Other yield depleting factors associated with lodging are poor grain filling and premature ripening. In addition to lodging, bleaching of outer stalk tissue may also be evident.

Ergot or Sugary disease

Causal organism:*Sphacelia soogi* McRae

Ergot can be an economically devastating problem. In seed production fields, yield losses ranging between 10 to 80 per cent has been reported in India. Although yield losses associated with ergot infection can be significant, indirect losses may be even more important. Harvesting grain from ergot-infected fields can be difficult. Ergot contamination reduces grain quality and limits its use as a feedstock. Ergot or sugary disease results from the ergot fungus infects the florets in the ear head (panicle) and prevents seed set in such florets. Honeydew exudes from florets. The first symptom is droplets of liquid honeydew exude from the glumes of infected florets. Honeydew may range from colourless to start with to yellow, brown or white with a thin to viscous consistency. Honeydew exudation is profuse in Ergot infested ear head, it may drip down and smear the whole ear head, sometimes dripping down onto the soil. If the exudate is plentiful, the ear head appears with blackened grain.

INTEGRATED DISEASE MANAGEMENT

- Reducing or eliminating stresses, especially at grain filling, minimizes the occurrence of serious epidemics of stalk rots. Host plant resistance in the form of delayed senescence can withstand drought stress and eventually stalk rots. (Pande *et al.*, 1994).
- Downy mildew of sorghum can be managed by applying potassium azide as a soil fungicide. This chemical @ 1.12kg/ha reduce the incidence of sorghum downy mildew by 23 per cent. Spraying of metalaxyl 25 WP (Ridomil) at the rate of 2g a.i. per litre of water at 10 and 40 or 20 and 50 days after emergence completely manage the disease (Anahosur and Patil ,1983). Cultural control by deep ploughing (30-35 cm) has been suggested to reduce oospore population in soil. However, crop rotation is also effective in controlling the disease. Shivana and Anahour, (1990) reported that resistance to downy mildew has been located in several genotypes, some showing stable resistance.
- Adjustment in dates of sowing, judicious use of irrigation, maintenance of adequate plant population and balanced fertilizer application along with non-senescence cultivars are the main components of integrated management of charcoal rot. Avoiding high nitrogen levels and low potassium levels also helps in disease management (Pande *et al.*, 1992).
- Sowing of ergot free seed. Soaking seeds with 5 per cent salt solution helps to remove ergot infested seeds, as ergot infested seeds float in the salt solution. Spraying of triazole fungicides 3-4 times at 5-7 day intervals starting before stigma emergence has proved effective to control ergot in the seed multiplication plots of hybrid sorghums (Bandyopadhyay *et al.*,1996).

3. Wheat (*Triticum* spp.)

In India, it is the second important crop, next to rice and is extensively grown. The important diseases of wheat crops are as following.

Rusts

Three rust diseases occur on wheat viz., stem or black rust (*Puccinia graminis* pers. f. sp. tritici Schof.), leaf or brown rust (*Puccinia recondita* Rob.ex.Desm.f. sp. *tritici*) and stripe or yellow rust (*Puccinia striiformis* West.). Rusts are responsible for huge economic losses in wheat crop. Stem rust produces reddish-brown, elongated pustules on the stems, leaves, glumes, awns and kernels. These contain masses of brown spores. As the plant matures, later in the season, the pustules produce black overwinter spores. Leaf rust appears as small, round, orange pustules on leaves and leaf sheaths. As the plant matures, the pustules turn dark gray. Stripe rust develops as elongated, yellow-orange pustules in rows of varying lengths. This gives the appearance of narrow yellow stripes mainly on the leaves and on the grain heads. These later become dark brown pustules, which produce the overwinter spores. Yield losses depend on the growth stage of the crop at the time of infection and the amount of leaf tissue destroyed. Early infection of upper leaves, stems, and heads can cause high yield losses in the form of shriveled grain, reduced baking quality and impaired germination.

Powdery Mildew

Causal organism: Blumeria graminis f. sp. tritici

Powdery mildew on wheat is recognized by small, effuse patches (colonies) of cottony mycelia (masses of fungal threads of hyphae that make up the body of the fungus). These occur on the upper and lower surfaces of the leaves. As these patches sporulate, they become dull tan in colour. Chlorotic (yellow) patches may later surround the mildew colonies. As the wheat and the mildew colonies mature, the sexual stage of the fungus or cleistothecia are produced. The mildew fungus survives the summer in the absence of wheat in infested wheat debris as cleistothecia. When the new crop develops as seedlings and fall rains occur, the cleistothecia within the infested wheat debris rupture to release spores. This process is favoured by moderate temperatures and lush wheat growth. The mildew fungus, survives the winter on the infected wheat seedlings. In the spring, with the return of moderate temperatures, the typical cottony mildew colonies develop and sporulate (asexual reproduction) to infect and colonize the newly developing wheat leaves. This stage of the disease cycle is favoured by moderate temperatures and high relative humidity. The canopy within a lush stand of wheat is an ideal environment for powdery mildew to develop.

Karnal Bunt of Wheat

Causal organism: *Neovossia indica* (Mitra) Mundkar

Karnal bunt is difficult to identify in the field. Developing wheat kernels are randomly infected and usually only partially converted to the fungus, which is why Karnal bunt is sometimes called partial bunt. Infection typically occurs in only a few seeds per head and not all heads on a single plant are infected. Infected grain shows no symptoms until near maturity. The diseased portion of kernels is dark in colour and fishy smelling due to trimethylamines. The kernel usually remains whole, with only a part of the germ end converted into a black powdery spore mass, usually along the kernel groove. In extreme cases, the entire kernel is converted into spores.

Loose Smut

Causal organism: *Ustilago tritici* Jensen

The entire inflorescence, except the rachis, is replaced by masses of smut spores. These black teliospores often are blown away by the wind, leaving only the bare rachis and other floral structures. Wind blown teliospores that land on the flowers of wheat plants can germinate and infect the developing embryo of the kernel. The mycelium of the loose smut fungus remains dormant in the embryonic tissues of the kernel until the kernel begins to germinate. The mycelium then develops along with the growing point of the plant and at flowering time replaces the floral parts of the spike with masses of black spores. Infection and disease development are favoured by cool, humid conditions, which prolong the flowering period of the host plant. Yield losses depend on the number of spikes affected by the disease.

Basal Glume Rot and Bacterial Leaf Blight

Causal organism: Pseudomonas syringae pv. syringae

The leaves, culms, and spikes of wheat can be infected. Infections begin as small, dark green, water-soaked lesions that turn dark brown to blackish in colour. On the spikelets, lesions generally start at the base of the glume and may eventually extend over the entire glume. Diseased glumes have a translucent appearance when held toward the light. Dark brown to black discolouration occurs with age. The disease may spread to the rachis, and lesions may also develop on the kernels. Under wet or humid conditions, whitish gray bacterial ooze may be present. Stem infections result in dark discoloration of the stem; leaf infections result in small, irregular, water-soaked lesions. The pathogens survive on crop debris, as well as various grass hosts. It is disseminated by splashing rain or by insects and can be seed borne.

INTEGRATED DISEASE MANAGEMENT

- **Plant resistant varieties:** Varieties resistant to several pathogens are available and offer the best and long-term control measures. Only the recommended varieties for the region should be planted because these carry resistance against major diseases of that area.
- **Balanced soil fertility:** Excessive or deficient nutrient levels will affect diseases. Excessive nitrogen rates, for example, will promote powdery mildew development. Soils deficient in phosphorus but having high rates of nitrogen usually have severe rust disease levels.
- **Scout the crop:** Since foliar diseases are most damaging at the flag leaf or the leaf directly below stage, scouting and management should be started before flag leaf emergence. The two uppermost leaves contribute heavily to grain formation and head-fill. Thus, it is important to maintain the health of these two leaves. Scouting should began at growth stage 6 (first node of stem visible) and continued at least through growth stage 9 (complete emergence of flag leaf).
- Fungicide applications:
    - For powdery mildew and rusts management, one spray of propiconazole (25 EC) @ 0.1% at ear head emergence or appearance of disease (Brahma *et al.*1991 and Kalappanavar *et al.*, 2008)

- For management of Karnal bunt, one spray of propiconazole (25EC) @ 0.1 % may be given (in seed crop only) at ear head emergence stage. Two sprays of *Trichoderma viride* at Zadoks growth stages of 31-39 and 41-49 provide a non-chemical (biological control) management of disease. One spray of *Trichoderma viride* (at growth stage 31-39), followed by one spray of propiconazole (25EC) @ 0.1% at growth stage 41-49 can be given to attain near complete control.
- For control of loose smut, seed treatment with carboxin (75 WP @ 2.5 g/kg seed) or carbendazium (50 WP @ 2.5 g/kg seed) or tebuconazole (2DS @ 1.25 g/kg seed) or a combination of a reduced dosage of carboxin (75 WP @ 1.25 g/kg seed) and a bio agent fungus *Trichoderma viride* (@ 4 g/kg seed) is recommended.
- The IPM module involves the seed treatment with *Trichoderma viride* (@4g/kg seed) + carboxin (75WP @1.25g/kg seed) for the control of loose smut. In Karnal bunt prone areas, the seed crop can be given one spray of propiconazole or two sprays of *Trichoderma viride* at tillering and ear head emergence. For the control of powdery mildew, one spray of propiconazole (25 EC @0.1%) can be given at ear head emergence or appearance of disease on flag leaf, whichever is earlier.

4. Maize (*Zea mays* L.)

Common Smut of maize

Causal organism: ***Ustilago maydis*** **(DC) Cda.**

The disease produces soft tumors or galls almost on all the parts of plant above the ground. Usually the galls appear on cobs, tassels, axillaries buds, stalks and sometimes on the leaves. The infection first becomes apparent when galls of different sizes are seen. The galls may reach the size of 10 cm or more in diameter. As the plant grows more galls appear, especially at the junction of the leaf sheath and blade and at the nodes of stem. On the appearance of the tassel, small galls appear on the male flowers. The cob is very frequently attacked, the whole ear or more often only individual flowers are affected and the large galls are formed. Galls on young seedlings may result in extreme dwarfing or death of the plants. Embryonic tissues of the plant are also affected. Even when a few flowers of the cob are attacked, the amount of grain produced may be greatly reduced, as the grains near the small galls frequently do not develop. The small gall consists of mass of smut spores. The tumor remains covered with a thin whitish membrane. When the membrane ruptures, the spores are disseminated into the air.

Downy mildew of maize

Causal organism: Pernoscleropsora maydis

The leaves of the infected plants becomes pale yellow or chlorotic. The undersurface of the leaves bear wooly white growth. The plants show stunted growth. The tassels of the plants become malformed. No cob formation takes place, if formed grain development is poor and infected plants will die soon.

Common Rust of maize

Causal organism:*Puccinia sorghi* Schw.

The leaves become covered with numerous uredosori, lose their green color drop and dry up and in severe cases; the plant does not form cobs. As the crop natures brownish black pustules containing dark thick walled two celled teliospores develop. The postures burst soon dislodging the reddish brown powdery spore mass. In severe cases infection spreads to sheaths and other plant parts. The inoculum survives on the form of uredospores on maize plants, which are generally grown round the year in different regions. By means of uredospores the disease cycle is repeated several times in the same region.

Stalk rot of maize

Causal organism: *Erwiniaa carotovora* var. *zeae* (Sabet)

The symptoms develop as soft rot of stalk tissues followed by rapid wilting and plant death. The rot occurs at the lower nodes and passes up and down the stalk to a very limited extent. Sometimes dark soft decay of the rind occurs while other times the rotting area is principally in the interior of the stalk. With the progress of stalk rot leaves start yellowing and dry up. The typical characteristic of the disease is the omission of fermenting odour from the site of infection. The vascular bundles usually remain intact. Ear shoots and cobs occasionally get infected which droop down and hang simply.

Banded leaf and sheath blight

Causal organism: *Rhizoetonia solani* f.sp. *saspii* Exner

The symptoms develop on leaves and sheaths as concentric bands extending to the basal part. Severe infection leads to the blotching of the leaf sheath as well as leaf. Symptoms are also observed on under favorable conditions which show conspicuous light brown cottony mould with small, round, black sclerotia. Selerotia in the soil and plant debris act as source of primary inoculum. The collateral host contributes the same. The secondary spread takes place through contact of infected leaves with the adjoining healthy plants. The disease appears at pre-flowering stage of the plants and spreads right up to the ears. Temperature of 30° C & high humidity is conductive for the disease.

Turcicum leaf blight

Causal organism: *Setosphaeria turcica* (Lattrell) Leon. & Suggs.

Symptom appears as a long spindle shaped grayish green spots on leaves which enlarge and become necrotic. First symptom appears on lower leaves and gradually progress upwards. Under high humidity and moderate temperature the whole leaf area becomes necrotic and plants appear dead. The pathogen overwinters on maize debris either as dormant mycelium or chlamydospores.

Fusarium stalk-rot

Causal organism: Fusarium monoliformi Sheld. and F. graminearum

The symptoms become conspicuous when the kernels are in blister stage i.e dry matter is maximum in them. The pathogen commonly affects the roots crown region and lower inters nodes. When split open the stalks show purplish discoloration. The plants wilt and dark brown lesions develop on the lower inter nodes. In the final stage of infection, the pith is shredded and surrounding tissues become discolored. The pathogen is soil borne but the fresh seeds from the infected plant when sown can develop seedling blight. The disease is more severe under high plant population and is favored by dry and warm climate conditions.

Charcoal Rot

Causal organism:*Macrophomina phaseolina* (Goid.) Tassi

Premature ripening of the plant occurs. The sides of lower inter nodes become straw colored, pith becomes disintegrated and the presence of small pin-head like black sclerotia on the rind of the stalks is a distinguishing character of the affected stalks may split longitudinally into a mass of fibers. The fungus over winters as sclerotia in the soil and infects the host at susceptible crop stage through roots and proceed towards them.

Pythium Stalk rot

Causal organism:*Pythium aphanidermatum* (Eds.) Fitz.

The lower internodes near the ground level show brown elliptic lesions. In the infected part of the internodes the pith is destroyed but not the fiber vascular bundles. The stalks as a result are weakened and breaks leading to lodging of the plant. The infected plants topple but remain green and do not die up to 2 weeks after attack. White fluffy growth of the fungus on the breaking point is very common under high humidity conditions. The pathogen is soil borne on nature. Environmental factors such as time of planting, soil moisture, organic nutrient application, temperature, pH influence disease incidence and disease development.

Seed and Seedling blight of maize

Causal organism: *Fusarium, Rhizoctonia, Aspergillus, Pythium, Cephalosporium*

Infected seeds when sown, produce thin stands of the crop with gaps, pre emergency seed rot, post emergence damping off, brown sunken lesions on mesocotyl, collar rot, wilting, toppling of collapsed seedlings. The disease is seed borne. The problem becomes severe with the use of old seeds stored under high temperature and high humidity.

Leaf blight of maize

Causal organism: ***Drechslera maydis*** **Nishikado**

Small, Yellowish, round or oval spots appear on the leaves. They extend along the leaf and coalesce into longitudinal bands, which may cover a great part of the leaf. Symptoms may be confirmed to the leaves or may develop on the sheaths, stalks, ears and cobs. The lesions longitudinally elongated typically limited to a single inter vascular region, often coalescing to form more extensive dead portions. Ultimately the leaves are dried up and have blighted appearance. The plants remain stunted and the ears poorly developed. The pathogen grows both inter and intracellularly within the mesophyll cells. The pathogen survives on plant debris in the soil.

INTEGRATED DISEASE MANAGEMENT

- Breeding of resistant varieties is most important control measure against common smut of maize. Crop rotation and field sanitation might be expected to control some disease. High application of nitrogen fertilizers should be avoided. Uprooting of smut-effected plants and seed treatment with thiram or capton @ 3g/kg of seed is recommended for management.
- Rotation of maize with non-host crops like wheat or oat reduces the downy mildew incidence. Early sowing of maize escapes the infection. As by the time significant inoculums of asexual spores are produced in collateral host and plants develop resistance. Three-four sprays of fungicide Dithane M-45 @ 0.25% at 7 days interval reduces the disease. Metalaxyl in combination with copper oxychloride is an excellent control measure. Sun drying of the seed has been found to be effective for inactivation of mycelium present in it and also reduces moistures levels.
- Carbamate group of fungicides especially Zinc ethylene bis-dithiocarbmate @ 0.2% a.i. has proved to be suitable for controlling the rust of maize.
- Drenching of basal stalk region when plants are knee high stage with bleaching powder ($CaOCl_2 . H_2O$) containing 33% chlorine @ 3.3g/10 liters of water. Avoidance of water logging and proper drainage reduces disease incidence of bacterial stalk rot. Planting of crop on ridges rather than flat soil is also advocated.
- Proper drainage, adoption of host resistance, spray of propiconazole @ 0.1% solution 2-3 times are the integrated disease management strategy for banded leaf and sheath blight of maize. Use of antibiotics Validamycin proved very effective in checking the disease.
- The effectiveness of fungicides carboxin, mancozeb and propiconazole against turcicum leaf blight has been reported by other authors (Singh and Gupta, 2000; Patil, 2000). The foliar spray with mancozeb @ 0.25 per cent for three times at an interval of 10 days was found to be more effective and significantly reduced Turcicum leaf blight severity and increased grain yield. The

antagonism of *Trichoderma harzianum* and *Trichoderma viride* was also observed (Ramachandra, 2000). However, mancozeb @ 0.25 per cent found most effective in inhibiting the growth of Turcicum leaf blight. *Propiconazole* @ 0.1 per cent, carboxin power @ 0.1 per cent and zineb @ 0.25 per cent were found equally effective which can be used as an alternative to mancozeb. Among the bioagents, *Trichoderma harzianum* was found to be more efficacious in inhibiting the mycelial growth.

- Accumulation of aluminum and iron in maize plants lead them to susceptibility to invasion by stalk rot pathogens especially if the crop is grown in acidic soils. In such cases soil amendments with phosphate and lime reduces stalk rot severity. Predisposition of stalk and root rot is initiated with the beginning of senescence of root tissues because of inadequate supply of carbohydrate.
- Sieving or winding of seed lot meant for sowing eliminates light weight chaffy injured or infected seeds. Seed treatment with fungicides such as thiram/captan @ 2g/kg of seed demonstrated good improvement in plant stand and seedling vigour.

5. Chick Pea (*Cicer arietinum* L.)

Wilt of Chick Pea

Causal organism: *Fusasium oxysporium* Schlecht: Fries emend Synd. and Hans.f.sp.*ciceri*

Under field conditions the disease appears on 20 days old seedling of susceptible cultivars. In early stages, the seedling may collapse and lie flat on the ground. Slightly old seedlings may show drooping of the leaves and dull green color in the stage. When roots are split open vertically, discoloration of xylem vessel can be seen extending towards stem and branches. The pathogen is soil and seed borne and becomes systemic in the host. Infected seedling play an important role in the long distance dispersal of pathogen to new area. The pathogen can survive as chlamydospore in the soil as well as in the infected crop residues left over the field for at least six years even in absence of host.

Rust

Causal organism: Uromyces ciceri arietini

Circular, oval, reddish brown pustules which coalesce on both the surface of leaves. The pustules commonly appear on the underside of leaves, less abundant on pods and sparingly on stems. When leaves are severely infected, both the surfaces are fully covered by rust pustules shriveling followed by defoliation resulting in yield losses. The urediospores are unable to survive the summer heat in the plains of northern India.

Anthracnose

Causal organism: Glomerella lindemuthianum

All the above ground plant parts are affected attacked by the pathogen. The most characteristic symptom of disease is spotting on the pods. Firstly water soaked lesions appear on the pods later becoming brown and enlarging to form circular spots of varying size. The spots are usually depressed with dark centers and bright red, yellow or orange margins. They may occur more often on the upper surface of the leaf than on the lower surface and also occurs on the petioles and stem. As the In severe infections, the affected parts wither off. Seedlings get blighted due to infection soon after seed germination. The fungus is carried on the seeds and cause primary infection of the seedlings. Secondary infection occurs through air borne conidia. Under favorable climate condition the pathogen early survives and becomes more active.

Powdery mildew

Causal organism: Erysihe polygoni

White powdery growth occurs on the leaves and other green parts which later become dull coloured. The white powdery growth occurs in patches spreading to cover the stem and other plant parts. When the infection is sever, both the surfaces of the leaves are completely covered by whitish powdery growth. Severely affected parts get shriveled and destroyed. The infection may take place at any stage of plant growth, but more severely when the plants are flowering, and it persists until harvest. The disease also causes forced maturity of the infected plants which results in heavy yield losses.

Ascochyta Blight

Causal organism:*Acochyta rabies* (Pass.) Labr.

Initial symptoms appear near the tip of young shoot and top most leaves. Later on the disease spreads rapidly on the aerial parts of the plants, such as, petioles, flowers and pods. The spots on leaves and pods are circular, while on the stem and branches these are elongated. On the seed coat, dark lesions are formed with pycnidia which often lead to seed infection through testa as well as the cotyledons.

Botrytis Gray Mold

Causal organism:*Botrytis cinerea* pers.ex Fr.

All the aerial plant parts are attacked by this disease. Initial symptoms are water-soaking and softening of affected plant parts, viz. leaf, flowers and tender shoots. On these plant parts, brown spots are produced which are readily covered with dense growth of the fungus in the form of sporophores and mycelium. The plant parts covered under dense foliage and under wet conditions are heavily covered with sporophores. Under congenial environmental conditions, all the flowers are attacked resulting in the complete failure of the crop. The infected pods either do not produce any seed or produce only small, shriveled seed. The disease can appear at any time during plant growth, but maximum development of the disease is observed at reproductive phase.

INTEGRATED DISEASE MANAGEMENT

- Removal of host debris and deep ploughing during summer months (May-June) can reduce the inoculum of disease. Solarization of the soil by covering soil with transparent polythene sheets for 6-8 weeks during summer months effectively reduces pathogen population in the field. Seeds should be produced from disease free plants in disease free areas to avoid transmission. Chickpea wilt is mainly a soil borne disease and for better management of disease, it is essential to exploit host plant resistance. Seed borne inoculum can be controlled by treating seeds with Benlate-T (Benomyl 30%+ thiram 30%) @ 1.5g/kg seed. The infection can be reduced by soil drenching with brassical or spraying Bavistin (0.1%) around the roots and leaves at an interval of 15 days.
- Use of resistant variety is most effective control measure against rust disease. Spray mancozeb @ 3g manages the rust disease to some extent.
- Fallow crop rotation and spray mencozeb @ 0.3% or carbendazim @ 0.5/litre for the management of anthracnose of chickpea.
- For powdery mildew management dusting with finely powdered sulphur @ 25 kg/ha, 2-3 times during the crop season checks the disease.
- Calixin-M (11% tridemorph+ 36.5% maneb) and thiabendazole (3 g/kg seed) were effective in eliminating the seed borne infection of Ascochyta blight. Secondary spread of the disease can be effectively controlled by timely application of fungicides such as zineb, maneb and captan particularly keeping in view the cloudiness and rain during crop season.

6. PAPAYA(*Carica papaya*)

Foot Rot

Causal organism: Pythium aphanidermatum

The disease usually attacks the plants of one to three years of age in the field. The symptoms are observed as early as in the middle of June. The disease is characterized by the appearance of spongy, water-soaked patches or areas on the bark, at the collar region or immediately at the soil line. The patches enlarge rapidly and girdle the stem, causing the tissues to rot. The tissues become black and the entire tree topples down under a slight wind pressure and dies. If the bark is opened, the internal tissues appear dry and give a honey comb appearance. The rotting spreads to the roots and they are destroyed. If the conditions are favorable the plants dies within three weeks, top leaves remaining quit green, usually lateral roots do not develop and the tap roots often become twisted. The typical stem rot symptoms are common in 2-3 year old plants but younger plants have also been seen dying due to early infection. A damping off papaya seedling in the nurseries is also common. Seedling grown in such nurseries carry the disease and when transplanted develop symptoms.

Leaf curl

Causal organisms: Tobacco virus or Nicotiana virus-10.

Crinkling and curling of leaves accompanied by vein-clearing and reduction in the size of leaves are marked symptoms. Leaves become leathery and brittle and the inter venial areas are raised on the upper surface due to hypertrophy which product regosity. The most prominent symptom is the downward, rolling of leaves and the veins get thickened and turn dark green in color. The affected plants become partially completely sterile depending on the stage at which infection takes place, thereby causing serious losses. Trees do not die quickly and also they seldom recover from the attack of the disease. The severely affected plants fail to flower or bear fruit. In the advanced stage of the disease, defoliation takes place and plant growth is stunted.

Anthracnose

Causal organism: *Colletotrichum gloeosporiodes*

In the initial stages small, light yellowish brown lesions appear on the skin of fruits which enlarge and become dark brown in color. The spots are water soaked and sunken on the fruit. The centers of these spots later turn black and then pink when the fungus produces spores. The flesh beneath the spots becomes soft and watery which spreads to the entire fruits. The lesions remain covered with the abundant black dots indicating the aecrvuli of the fungus. On cutting the fruits through diseased portions, the pith and rind both were found to be rotten up to about 1 cm in depth. There is another symptoms of the disease in which lesions on the fruits enlarge and become very irregular. A pink cottony growth appears in the centre of such lesions which is due to abundant development of the conidia of the fungus.

Powdery mildew

Causal organism:*Oidium indicum* and *Oidium caricae*

The fungus grows superficially mostly on the undersurface of the leaves withdrawing nutrients from the cell of the leaf surface by specialized absorbing structures called haustoria. On the underside of diseased leaves are found patches of whitish powdery material. Early less conspicuous symptoms are pale yellow spots near the veins. The fungal mycelium grows on the undersurface of papaya leaf producing chains of spores. Severely infected leaves may become chlorotic and distorted before falling. Affected fruits are small in size and malformed.

INTEGRATED DISEASE MANAGEMENT:

- Plants should be grown in well drained soils. Affected plants should be carefully pulled out and destroyed. Replanting should not be done in a pit where the disease has once appeared. The seedling for transplanting should be selected from the nurseries where damping off disease has not occurred because seedlings from contaminated soils act as source of inoculating the field soil with the disease organisms.
- Disinfection of the soil of the pits with 2.5-3% formalin prior to transplanting is quite effective. Two application of formalin in March and May is effective in controlling the disease. The disease can be checked by drenching the soil with Bordeaux mixture or 2% captan. Soil drenching with Calaxin (0.1%) and Topsin-M (0.1%) at bimonthly intervals is very effective for the management of foot rot of papaya.
- Rouging is helpful in reducing disease incidence of leaf curl of papaya. Insect vectors should be kept under control by spraying suitable insecticides. Soil application of carbofuran (1Kg a.l/ha) at the time of sowing and 4-5 sprays of Dimethoate (0.05%) or metasystox or Nuvacron (0.05%) at an interval of 10 days effectively controls the white fly population.
- The affected fruits should be plucked from the plants so that further spread of anthracnose of papaya disease may be avoided. The fruits should be harvested as soon as they nature chemical. Spraying with copper or chloride (3g/litre of water) or carbendazim (1g/litre of water) at 15 days interval effectively controls the disease. Fruits for export should be subjected to hot water treatment or fungicidal wax treatment.
- The powdery mildew of papaya disease is effectively controlled by spraying wettable sulphur (0.3%) at 10 days intervals.

References

Anahosur, K.H. and Patil, S.H. (1983). Effective spray schedule of metalaxyl for sorghum downy mildew therapy. *Indian Phytopathology*.**36**:465-468

Bandyopadhyay, R., Frederickson, D.E., Mclaren, N.W. and Odvody, G.N. (1996). Ergot-a global threat. *International Sorghum and Millets News*. **37**: 1-32.

Brahma, R. N., Asir, R. and Saikia, A. (1991). Efficacy of tilt (Propiconazole) on different wheat cultivars. *Indian Phytopathology*. **44**: 116-118.

Kalappanavar, K., Patidar, R.K. and Kulkarni, S. (2008). Management Strategies of Leaf Rust of Wheat Caused by *Puccinia recondita* f. sp. *tritici* Rob. ex. Desm. *Karnataka Journal of Agricultural Science*. **21 (1)** :61-64

Mani. M.T. and Hepziba, S. J. (2009). Integrated disease management of pearl millet downy mildew caused by *Sclerospora graminicola*. *Archives of Phytopathology and Plant Protection*. **42(2)**:136 - 141.

Pande, S., Mughogho, L K. and Karunakar, R. I. (1994a). Etiology of stalk rot and lodging in grain sorghum. *International Journal of tropical Plant Diseases and Review of Tropical Plant Pathology*. **12**: 219-234.

Pande, S., Mughogho,L.K. and Karunakar,R.I.(1992). Incidence of charcoal rot in sorghum cultivars as affected by sowing date and plant density. *Indian Journal of Plant Protection.* **20**: 162-172.

Patil, V.S. (2000). Epidemiology and management of leaf blight of wheat caused by *Exserohilum hawaiienesis* (Bugnicourt). *Ph.D. thesis,* University of Agricultural Sciences, Dharwad, India.

Ramachandra, C. G. (2000). Studies on leaf blight of *Dicoccum* wheat caused by *Exserohilum hawaiiensis* (Bugnicourt). *M.Sc. (Agri.) thesis,* University of Agricultural Sciences, Dharwad, India.

Shivana, H. and Anahosur, K.H. (1990). Breeding downy mildew resistant varieties. *Indian Phytopathology.***43**:372-374.

Singh, S. D. and Shetty, H.S. (1990). Efficacy of systemic fungicides metalaxyl for the control of downy mildew (*Sclerospora graminicola*) of pearlmillet *(Pennisetum glaucum). Indian Journal Agricultural Sciences.* **60**: 575-581.

Singh, S. D., King, S. B. and Werder, J. (1993a). Downy mildew disease of pearlmillet. Information Bull. No.37. (In En. Summaries in FR, ES) *International Crops Research Institute for the Semi-Arid Tropics,* Patancheru, A.P. 502324.

Singh, S.N. and Gupta, A.K.(2000). Bioassay of fungicides against *Drechslera sativum* causing foliar blight of wheat. *52nd Annual Meeting and National Symposium on Role of Resistance in Intensive Agriculture,* Directorate of Wheat Research, Karnal, p. 25.

Sundaram, N. V. (1962). Studies on parasites of rust. *Indian Journal of Agricultural Sciences.* **32**: 266-271.

Thakur, R .P., Subbarao. K. B., Williams, R. J., Gupta. S.C. ,Thakur D.P. and Guthrie J.E.(1986). Identification of stable resistance to smut in pearl millet. *Plant Disease* **70**:38-41

Thakur, R.P., Rao, V.P. and Williams. R.J. (1988). Registration of four population of pearl millet germplasm with multiple disease resistance. *Crop Sciences.* **28**:381-392.

CHAPTER 25

Agricultural Production Enhancement Strategies for Rainfed Farming Systems of North-west Himalayas In Changing Climate

R. D. Gupta[1], Vikas Sharma[2] and Brij Nandan[3]

[1]*Ex-Associate Dean, Faculty of Agriculture-cum Chief Scientist KVK, Sher-e-Kashmir University of Agricultural Sciences and Technology of Jammu.*

[2]*Junior Scientist, Regional Agricultural research Station, SKUAST-Jammu, Rajouri.*

[3]*Junior Scientist, Pulses Research sub-station, SKUAST-Jammu, Samba*

Introduction

As Agricultural Production is a biological phenomenon, it is bound to suffer from any fluctuations which occur in the *Environment*. *Environment* reflects both *Physical (Land, Air, Water)* and *Biological (Plants and Animals) environments* and are also known as abiotic and biotic environments. A dynamic system where both abiotic and biotic environments are interacting each other to bring forth in its structural and functional changes is called *Ecosystem*. Both kinds of environments are also called biophysical factors. Apart from these factors, socioeconomic conditions also have much influence on "Agricultural Production".

There is no doubt that both kinds of *Environments (Abiotic and Biotic)* have a profound effect on the production and productivity of various agricultural crops directly or indirectly. However, these days whatever happens to the environment in respect of change of climatic elements like temperature, rainfall, hailstorms, snowfalls, cyclones etc., is attributed to *Global Warming*. *Global Warming* takes place when atmospheric gases called *Green House Gases* absorb more heat than normal averages. It refers to an average increase in Earth's temperature over a period of time which in turn causes change in climate. A warmer climate of the Earth may lead to change in rainfall pattern, melting of glaciers and a rise in sea level water. This will eventually result in a wide range of imprints on the life of humans, animals and plants - agricultural crops, including fruit plants and vegetables, and forest tree species, especially in *Dryland* and *Rainfed Farming systems*. Difference between the two is that the former experiences annual rainfall less than 800 mm and that the latter more than 800 mm with semiarid and subhumid climate (Venkateswarlu, 1985).

Indian agriculture relies greatly on dryland farming for its growth and sustenance. Out of India's total land area of 329 million hectares, nearly 143 million ha are under cultivation, and of the latter, 92 million ha come under *Dryland Farming* and *Rainfed Farming*. In the present chapter, agricultural production and future strategies for Rainfed Agroecosystem of North West Himalayas, have been detailed under changed climate caused through *Global Warming*. North West Himalayas comprise of three main states *viz.*, Uttranchal (Uttrakhand), Himachal Pradesh and Jammu and Kashmir, and sub-montane parts of Punjab and Haryana, where about 60-70 per cent of the total arable land depends upon rains (Monsoon and Western Disturbances).

Climate change and its implications on crop productivity

The history shows that for the food crops production, warming is considered better than cooling. However, too much warming is quite injurious to the functioning of normal growth and development of crops.

A joint study conducted by the Research Institute of United Kingdom and India showed that in the years to come yield of various crops would be hit by climate change caused by *Global Warming* due to emission of green house gases - CO_2, CH_4, CFC (Anonymous, 2005; Gupta, 2005a). The concentration of CO_2 is now a bit more than 380 ppm compared to a range of about 200-300 ppm during the past 8,00,000 years. The current concentration of CH_4 is 1800 ppb compared to a range of 400-700 ppb during that time (Anonymous, 2008a). Oxides of nitrogen *viz.*, NO, N_2O, NO_2 are the other green house gases which are emitted into the atmosphere from various industrial and combustion processes, automobile exhaust and agricultural fertilizers, especially nitrogenous ones. Oxides of nitrogen move into the stratosphere and contribute depletion of O_3 layer also. Rise in 2^0 Celsius would reduce the grain yield potential. Warmer areas will show more crop loss. In eastern region of the country, rice yield will decline, whereas in north India including north-western Himalayan states, reduction in the yield of various crops would be offset by abnormal climate and weather.

Based on average values, the world has already warmed about 0.7^0 Celsius since 1800s (Anonymous, 2008b). And if the current rate of emission continues, this may further create a green house effect, resulting in warming of the atmosphere by 1.2 to 3.9^0 Celsius or 1.5 to 4.5^0 Celsius by the years 2030 to 2080. These higher temperatures would cut yield of rice grown in tropics. Rise in such temperatures, will warm-up the oceans, glaciers/polar ice caps may melt causing rising sea levels and thereby threatening the arable lands in countries having vast coastlines including India.

The potential risks to agriculture in India consist of shifts in agroecological zones, decrease in productivity of crops and grazing lands, increase in frequency of droughts and flood (Singh and Sharma, 1999). Change in climate will also affect many animals and plant species both on land and sea which may not be able to adapt to frequent environmental changes. Increased levels of CFCs, N_2O, NO in the atmosphere are believed to have caused severe damage to ozone layer which results in increased UV-rays on the earth, endangering thereby human, animal and plant life. Rapid change in climate will affect ecosystems and as a consequence some plants and wildlife may benefit while others will be unable to adapt fast enough and face extinction (Wittwer, 1998; Anonymous, 2002).

Observance of climate change

Although the *Global Warming* effect in north-western Himalayan states was observed about 25-30 years ago yet its impact was more pronounced during the last decade of the 20th century. The data compiled by the National Oceanic and Atmospheric Administration of the US indicated July, 1998 as the hottest month in the last 148 years. This summer season was also one of the hottest in India when a number of human deaths occurred due to sun strokes.

Conspicuous first impact of Global Warming in Kashmir valley was observed during 1998-1999 when whole of the Kashmir valley suffered from a severe drought of its recorded history. Detailed information has been compiled by Gupta (2005a). During the year 2004, acute shortage of water was found not only in Srinagar but in Lolab valley where a number of areas were affected. During January-February (2003), when the people of Jammu region were shivering due to cold wave, the inhabitants of Kashmir were basking in sunshine as there was no snowfall. There was also no snowfall and rainfall during October-November of 2002-2003 in the Pir Panjal peaks. This also created a drought-like condition in *Kandi* areas of Jammu, Himachal Pradesh, Punjab, Haryana and Uttrakhand and, as a result, farmers could not sow *Rabi* crops. During the winter of 2002-2003, frost damage was quite serious in fruit plants like mango, litchi, kinnow and *ber* grown in various areas of Una (Himachal Pradesh). Similar damage was also noticed in Solan area of Himachal Pradesh (Samra, 2003).

Unexpected heavy snowfall on 30th April, 2004 caused devastation. Thousands of trees were broken and felled. Pir Panjal and Dhaula Dhar range (Himachal Pradesh) had 30 to 40 cm snowfall which affected the production of almond, walnut and apple. Due to untimely rainfall, there was heavy loss to paddy and maize in many parts of Jammu region.

During the failure of monsoon in midway during 2004, the farmers of Jammu asked for contingent plan for sowing their crops to meet any eventuality that might arise due to continuous drought situation. It was because they had already suffered from drought for four consecutive years (2001 to 2004). The orchardists/ farmers of Kashmir have not still forgotten how their crops fruits/vegetables and saffron were affected during the spell of drought.

The *Tsunami* that took place on 26th December, 2004 in South East Asia covering India, Sri Lanka, Indonesia, Thailand and a number of islands had its effect also seen in Jammu and Kashmir, Himachal and parts of Uttrakhand by way of massive snowfall during February, 2005. This was called *Snow Tsunami Effect*. This massive snowfall caused a heavy loss to fruit crops both in Kashmir and Jammu regions.

The people of Kashmir associated with agriculture/horticulture were not happy with the variation in the temperature during May, 2005. The hailstorms caused extensive damage in the apple production. During December, 2006 and January, 2007 the night temperature fell as low as 1^0 Celsius in Jammu which have never been experienced.

Incessant rains in Cold Arid Zone of Ladakh and Lahaul Spiti and Kinnaur during August, 2006 worried the residents of these areas (Sharma, 2006). Heavy rains accompanied with cloud bursts triggered floods in Ladakh and caused widespread damage to crops and buildings.

The valleyites of Kashmir showed their concern about the climatic changes (Koul, 2007) as in the present times rainfall/snowfall no longer had a fixed season but that entails unpredictability. A significant increase of 0.05 to 0.08^0 Celsius in temperature every year in Jammu and Kashmir has been revealed in a study carried out by the Indian Meteorological Department (Koul, 2007).

Joshi pointed out that due to non-occurrence of westerly disturbances upto January 2, 2008 resulted in the minimum temperature to extent of 7.4^0 Celsius. Dry weather due to cold wave has badly affected the production of vegetables and other crops besides fruit trees. Dry weather has also resulted in considerable drop in the water level of the river Jehlum and the Sindh, and their tributaries.

Future Strategies

Growing of pulses

Pulses are known to release nitric oxide (NO) but no one had determined it so far. Now researchers from Bhubaneshwar based Institute of Mineral and Materials Technology have measured NO content emitted from certain pulse farms in eastern India (Indian-specific emission factor). And it was found that Indian emission factors of NO from pulse cultivation are lower than the Intergovernmental Panel on Climate Change Emission Factors (Bhatta, 2008). It is attributed to less use of nitrogenous fertilizer in pulse production in India.

The study indicated that average emission of NO from cultivation of green gram, gram and black gram in two tribal villages of Orissa was in the range of 11 to 20 microgramme (μg) m^2 day^{-1} against 80 to 90 μg m^2 day^{-1} emitted from cereals in Britain. As NO emission under different environmental conditions are different so, it becomes imperative to study the emission values under various agro-ecological conditions.

It is worthwhile to state that hitherto many areas of *Kandi* belt, especially of Jammu, Akhnoor and Samba tehsils were great producer of gram, *kulth* (*Dolichos biflorus*), *moth* (*Phaseolus aconitifolius*) *vis-à-vis* other pulses like green gram, black gram and lentil (Gupta, 2006). Hence, due consideration must be given for growing of pulses in the *Kandi* belts. The fodder crops like jowar and *bajra* should always be sown with pulses *i.e. jowar + moth* and *bajra* + cowpea. After the harvest of these crops, *gobi sarson + toria* should be resorted by the end of 3rd/4th week of September. Sowing of pulses will not only help in ameliorating soil and atmospheric environments but will also assist in vanishing away of the malnutrition particularly of children and women, as pulses are rich source of proteins.

Growing of C4 plants

C_4 plants such as pearl millet, maize, sugarcane and sorghum should be preferred to grow. It is because of more assimilation of CO_2 which is attributed to presence of two types of storage cells in these plants *i.e.* Mesophyll and Bundle sheath cells. Moreover, in C_4 plants, Phosphophenol Pyruvic Acid Carboxylase Enzyme is present, whereas in C_3 plants like Rice and Wheat, Ribulose-1-5-Biphosphate Carboxylase Enzyme is present.

It is pertinent to mention that in *Kandi* areas of Jammu, Punjab and Himachal, maize, pearl millet and sorghum used to be grown on a larges scale before 1947, but now the cultivation of pearl millet and sorghum is done on a small scale. It might have happened due to change in food habits.

Growing of C_4 plants will help to protect the atmosphere by regulating fluxes of CO_2, the amount of which is increasing year after year. Like pulses, these plants require less amount of nitrogenous fertilizers and as such, these will control the flux of N_2O, NO, etc.

Adoption of agroforestry

In the wake of an acute shortage of water in the *Kandi* belts of Jammu, Himachal Pradesh, Punjab, Haryana, Uttranchal and Bundelkhand region adjoining to Uttranchal's *Kandi* belt, agriculture sector has become poor in shape. The monsoons have brought in much less water in the last five-six years (2002-2003 to 2006-2007) for a largely rain-dependent cultivation. Also, there have not been much adequate efforts to manage and conserve the remaining water resources. Only a limited number of tanks are being relied upon for irrigation in most of the areas except for those where there are irrigation canal network under Command Area Development of Jammu and other parts of *Kandi* areas. As a result, the yields of the crops is so low that even the peasants do not obtain the cost of the inputs, which they have invested.

Keeping in view the facts stated above, the *Kandi* belt and Bundelkhand regions calls for adoption of Agroforestry Systems.. Earlier *Kandi* belt and Bundelkhand regions had one of the finest forests in India and all the demands of local people in respect of fuelwood, fodder, food, timber used were met from these local resources. The situation has now steadily deteriorated due to deforestation and mismanagement of natural resources. Due to deforestation, there is not only emission of CO_2 in the atmosphere but there is also an acceleration of soil erosion, loss of soil fertility and low crop productivity. To mitigate such effects, the benefits provided by trees are best sustained by integrating them into agriculturally productive landscape, a practice known as *Agroforestry*.

Various multipurpose tree species can be grown under various Agroforestry Systems in *Kandi* belt and Bundelkhand region. The systems can be designed based on the farmers requirements - fodder, fuelwood, timber and fruit. While growing trees, local species should be preferred to grow (Rai, 2007).

Silviculture *i.e.* planting of fodder trees and introduced grasses has been observed to be the only option for regenerating and improving the Himalayan grasslands (Misri *et al.*, 2004). Tree species like *Albizia lebbeck, Bauhinia variegata, Morus alba, Grewia optiva* have been found most suitable in this system of agroforestry. *Cenchrus ciliaris, Congo signal, Guinea* were the main grass species and *Siratro, Stylosanthes* and white clover the main legumes introduced.

Promotion of horticulture

Due to change of climate and lack of water in the rainfed areas, fruit plants should be preferred to grow depending upon the area. As entire *Kandi* range is almost bare of vegetation and is subjected to all types of damages due to heavy soil erosion, landslides and floods, so fruit plants such as mango, guava, orange, sweet lime, lemon, litchi etc. should be planted during mid-July to September. Strawberry is also being grown. In the temperate areas, apple, pear (*Bhagu Gosha*), cherry, almond should be grown. Walnut, commonly known as *akhrot* is grown in many parts of the northwest Himalayas between elevation of 1200 and 2100 m. But its best quality is found in valley of Kashmir. It is grown for edible fruit and the fine grained wood used for furniture. Walnut oil is used by painter as drying oil. Mulberry is often planted for edible fruit and leaves for silkworm rearing (Gupta, 2007). Its leaves are also used for treating fever, cold, cough and high blood pressure. Quince is most hardy for temperate area. These days, Himachal has come to be known as the pioneer of hill area development. It has earned a name as the "Apple State" of India and has become "Fruit Bowl" of the country, especially in apple cultivation. Apple fruits have a number of medicinal values. It helps in lowering cholesterol level, reduce the risk of breast, colon and lung cancers, and also aid in digestion. Apple fruit also prevent neurodegenerative diseases like Parkinson's and Alzheimers (Gupta and Kher, 2009).

The Ladakh region is famous for good quality apricot. The dry apricot called *"Chuli"* is not only popular in the region but in other parts of the country also. Seabuckthorn is common along Indus river, roadsides and wastelands both in Ladakh and Lahaul Spiti. Its fruits are edible and rich source of Vitamin C and has a number of medicinal values. Fruits are also used for making jam, juice, sauce and concentrate drink.

In the changed climate of *Kandi* belt, indigenous drought hardy fruit plants are required to be grown. *Amla (Phyllanthus emblica), lassora* (*Cordia dichotoma*), *ber* (*Zizyphus mauritiana*) and another species of *ber* (*Zizyphus nummularia*) are some such fruits which have proved beneficial for the *Kandi* belts of Jammu, Punjab and Himachal Pradesh. Apart from these fruits, there are also some other fruit trees which can be successfully grown in this area. These are: *jamun* (*Syzigium cumini*), *bael* (*Aegle marmelos*), *phalsa* (*Grewia asiatica*), *caronda* (*Carissa carandes*) and *girna* (*Carissa spinarum*). Cultivation of *phalsa* is being much popularized in the Bari Badhori area and other parts of Raya Suchani. Papaya (*Carca papaya*), guava (*Psidium guajava*), *kinnow*, orange, *galgal* are the other fruits which proved useful in agroforestry as tested under Bari Badhori Watershed Management Project (Jamwal and Gupta, 2007).

The agroclimatic conditions in Uttranchal permit profitable production of a wide variety of temperate and subtropical fruits. They are apples, apricots, peaches, plums, walnuts, lemons, limes, litchi etc. Improved technology for production of most of these fruits is now available with the Horticulture Research Station, Chaubatia in Almora district. Besides, vegetables can be grown during off season. Potatoes, cauliflowers, peas, beans, tomatoes, cowpea etc., can be grown in Uttranchal hills successfully, especially at a time when these are not available in plains.

All efforts should now be made for setting up new orchards mainly in the community lands. Marketing facilities should be undertaken through cooperatives so that middle men are eliminated. This will bring forth improvement in the existing system of marketing of fruits and vegetables, which is far too inadequate and inefficient to provide any incentive to the orchardists/vegetable growers.

It has emerged from the foregoing facts that this is the high time for all of us to cultivate fruit plants wherever it is possible as they not only act as source of food but have medicinal values also. Moreover, the fruit plants like other trees purify the air using CO_2 and produce O_2, and, thereby, maintain ecological balance and control the emission of green house gases.

Promotion of floriculture/herbal plants

Flowers have been grown in India since pre-historic times and constitute an integral part of our heritage and culture. Although there is a vast variety of winter annuals which add beauty and colour to the gardens during autumn/winter in north India yet ten most popular and widely grown annuals are : (i) Marigold, (ii) Asters, (iii) Calendula, (iv) Candytuft, (v) Antirrhinum, (vi) Lurkspur, (vii) Mesembryanthemum, (viii) Nasturtium, (ix) Flox and (x) Salvia. Chrysanthemum, Dahlia, Petunia, Jasmine, Bougainvillea, etc. are the other flower plants which can be tried. Out of these annuals, marigold, asters, antirrhinum and nasturtium can be grown in the north west Himalayan conditions.

Cultivation of marigold has a lot of potential not only in plains of Jammu region but also in the hills of Jammu, Udhampur and Doda districts. In Chenani area of Udhampur district, it is already being cultivated. Its cultivation is also being done in many parts of Kangra and Solan districts of Himachal.

Jammu region of Jammu and Kashmir state possesses a lot of scope to grow Gladiolus. It is a herbaceous plant growing from a solid fibrous coated bulb (or corm) with long narrow, plated leaves and terminal one-sided spike of bright coloured irregular flowers. There are number of species of Gladiolus. Its cultivation is already being done in Jammu plains (Nazir, 2008) and can be grown in hills also, and should be tried to grow there. If it becomes successful, then cultivation can play a vital role in turning the economy of the floriculture growers.

Hills of Uttranchal also offer a conducive climate for cultivation of flowers. Roses, gladiolus, carnation, chrysanthemum and orchids can be grown in these hills with considerable benefits. Indeed, floriculture in hills can contribute a lot to the economy of the region.

Floriculture has, infact, emerged as a potential foreign exchange earner in many parts of India. Once upon a time, floriculture was mere a gardener's activity but now it has become a bread and butter supplier. It has emerged as a lucrative profession with a much higher potential for returns than most of the field and horticultural crops. Hence, it should be popularized among the orchardists/ farmers.

Like Jammu and Kashmir and Himachal Pradesh, Uttranchal hills are also well known for growing of herbal plants which are of great medicinal value.

Shifting from fossil fuel based agriculture to biodiverse organic farming :

Recently released review on the economics of "Climate Change" issued by the United Kingdom Government warns that unless radical action is taken, climate change could seriously threaten the society. Reducing emissions through shifting from fossil fuel based industrial agricultural system to biodiverse organic systems can contribute significantly to ameliorate the climate chaos (Shiva, 2007). Biodiverse organic farming and localized food systems can solve all aspects of the ecological crises caused by industrial chemicalised agriculture.

Why biodiverse organic farming? It took about three decades for the International Rice Research Institute (IRRI) to discover that pesticides used indiscriminately during "Green Revolution" are unnecessary (Partap Singh, 2003). But by then the use of pesticides had already polluted the fertile soils and contaminated the ground water which have finally affected the public health (Singh, 2006). Similarly, indiscriminate use of chemical fertilizers has not only created sickness in soils but has also posed health hazards both among the humans, animals and plants (Gupta and Singh, 2006). Health hazards in humans are due to increased toxic residues of pesticides and/or fertilizers in various food stuffs (Gupta *et al.*, 2005).

Accumulation of heavy metals in soils, incidence of fluoride, eutrophication of waterbodies and depletion of ozone layer are the other effects caused by the imbalanced use of fertilizers. Thus, to maintain balance and harmony with the Nature and also among socio-economical and ecological dimensions of farming operations, "Organic Farming" in diversified agriculture is the right answer to tackle the issues raised by the environmentalists (Dubey and Kaistha, 2005; Singh and Saxena, 2005).

In view of the facts as stated above and emission of green house gases from the nitrogenous fertilizers like urea, ammonium sulphate, calcium ammonium nitrate during nitrification and denitrification, there is an increasing awareness of "Organic Farming" practices the world over.

"Organic Farming", thus, means farming in spirit of organic relationship where the farmers mostly render the use of organics (organic manures, biofertilizers, vermicompost etc.) instead of inorganics (chemical fertilizers, pesticides). Organic farming is very much native to India. Any farmer keen in organic farming would, therefore, first look at the ancient Indian agricultural practices which are more than 4000 years old. This practice has maintained the soil fertility status over a long period of time. For the last five years *i.e.* since 2005, the farmers of Khakrola village in Himachal Pradesh have been working tirelessly to usher another green revolution. This is through organic farming. These farmers, most of them farm women, have been shunning the use of pesticides and chemical fertilizers to grow organic foodgrains, fruits and vegetables with the help of herbal sprays and vermicompost. As a result of these ecofriendly organic practices, the farmers of the area (150 in number) have not only improved the yield of the crops/fruits/ vegetales and enhanced its marketability, but have also become eligible for Asian Certification on the quality of the organic produce (Lalway, 2006).

The experiments done on organic farming at the Khakrola village are part of a Rs. 15 million model project initiated during 2003 by the M.R. Morarka Foundation. This is Rajasthan based Non-Government Organisation that is working in the Agriculture Department of Himachal Pradesh and funding the initiative to switch whole of the state over to organic farming. The farmers of Khakrola are, in fact, among a group of 5,657 peasants who have been registered under Morarka Foundation's Organic Project. This project is being implemented in nine blocks of Shimla district. These blocks are : Narkanda, Jubbal, Rohru, Chirgaon, Chaupal, Mashobra, of which Khakrola is a part of Vasantpur, Theog and Rampur.

Integrated use of organic and inorganics

Use of fertilizers will remain a key element of strategy to increase the production of crops in view of the increased population of humans which is presently 1180 millions. But the use of fertilizers has a number of harmful effects. So the best possibility is to use both organics (FYM, compost, vermicompost, biofertilizers, green manuring) and inorganics (chemical fertilizers). However, the NPK chemical fertilizers should be applied in the ratio of 4:2:1. It would be much better if the fertilizers are applied on the basis of soil test. Thus, the combined use of organics and inorganics is called Integrated Nutrient Management. Use of organics and inorganics has been found useful in many Indian situations.

When FYM is applied with N, P, K fertilizers there is an improvement in the yield of maize in the Kandi belt. The maximum yield was realized with recommended dose of NPK applied. Half through chemical fertilizers and half through FYM. Similarly in rice, application of half of the recommended dose of NPK through fertilizers and half through green manuring has been proved useful (Gupta et al., 2005). According to Acharya et al. (2001), the contribution of N, P, K, NPK and FYM was 19, 37, 13, 64 and 31 per cent in case of maize and 82, 0, 1, 83 and 17 per cent in wheat under Himachal soil conditions. Further, it was reported that FYM had positive effective on tuber yield of potato and its application @ 20 t/ha not only reduced optimum dose of N from 181 to 163 kg N/ha but also increased the yield level from 23.5 to 31.2 t/ha. FYM @ 10 and 20 t/ha also improved the fertilizer use efficiency by 10 and 35 per cent, respectively. Role of FYM in rice-wheat rotations in improving soil fertility and crop yields has also been tested in research station of Central Himalayas (Singh et al., 2001). It has been found that OC, total N, available P and K increased consistently with increasing levels of FYM. Wheat crop recorded higher yields in FYM treated plots.

Adoption of watershed management approach

In spite of 1000 to 2500 mm of rainfall in north west Himalayas, there is an acute shortage of water in the hills, especially during summer and winter seasons. It is because most of rainfall water runs of the steep slopes carrying soil with it and causes soil erosion and landslides/landslips. The remedy is intensive soil and water conservation measures like bench terracing, construction of water shortage structures (big or small), renovation of *Baulies* (springs), tanks and wells through watershed management.

Watershed management work in the World Bank's *Kandi* Project in Himachal Pradesh, Punjab and Haryana has shown that successful water harvesting structures are possible especially in the lower and foot hills (Kapta, 2004). Such structures have also proved beneficial in the *Kandi* belt of Jammu (Jamwal and Gupta, 2007) in development of agriculture and allied disciplines. Thus, watershed management is a holistic approach in which all major activities like forest management, adoption of soil an water conservation measures including water harvesting structures, management of crops, fruit trees/vegetables, grassland, animals are involved. All these approaches and innovations can be simultaneously addressed in an integrated way, on an area planning basis within watersheds having area, ranging from 100 to 500 ha. As it covers all agricultural and rural developmental programmes, so these will not harm much to the environment and ecology of the areas. Moreover, management of forests *vis-à-vis* other plants grown will take care of green houses gases also.

Change in certain package practices

It has been found that if developing countries, including India, can make some changes in the planting time of the crops, use irrigation more wisely and judiciously use manures and fertilizers through Integrated Nutrient Management, then production of some cereals (rice, wheat, maize) would resist the effect of climate and yield of crops would enhance.

Such type of technologies should also be practised among the fields of the farmers dwelling in various parts of Uttrakhand and Himachal Pradesh after conducting research by the Scientists of Agricultural/Horticultural Universities. This requires more attention for the farmers of Rudraprayag (Uttrakhand), especially those living in Hilaungad Watershed, where many of the farmers, like farmers of Kashmir, are shifting from agricultural to non-agricultural activities. A recent study conducted by Sharma (2008) indicated a remarkable shift from farm based to non-farm based activities in the said area. Now only 14 per cent of the people are practising agriculture but this figure was about 51 per cent about 10 years ago (upto 1998). The major cause of this drift was change of climate, resulting into poor natural resources like lack of water and low soil fertility due to heavy soil erosion.

New initiatives

With the new initiatives like National Food Security Mission, Rashtriya Krishi Vikas Yojna and National Agriculture Extension System by the Government of India and concerted efforts of Indian peasants *vis-à-vis* scientists of ICAR Institutes and State Agricultural Universities, the foodgrain production in 2007-2008 has touched a record high of 227.32 million tonnes. Role of favourable climate in such foodgrain production cannot be ruled out. The foodgrain production in this context is as follows :

Crops	Year [Foodgrain Production (in million tonnes)]	
	2003-2004	**2007-2008**
Foodgrains	213.19	227.32
Rice	88.53	95.68
Wheat	75.15	76.78
Maize	14.98	18.54
Cotton	13.73	23.19

Source : Department of Agriculture and Cooperation, MoA, GoI

Suggestions for future

Evolution of tougher/hardier varieties of crops :

(i) Due to change in climate, there are certainly lot of challenges to the farmers in the near future. Change in temperature is one of them, and scarcity of water is another. In warmer or drier areas, especially *Kandi* belt beset by greater extremes of droughts and floods during rainy season, farmers will have to change crop management practices. They will have to grow hardier/ tougher varieties and should be prepared for the constant change in the way they operate.

For this purpose, the plant breeders are required to come forward in this line *i.e.* they must evolve suitable varieties of maize, pearl millet and other crops. Similarly, they have to think from this angle for the farmers of *Karewas* of Kashmir and *Changar* areas of Himachal Pradesh.

(ii) It is believed that with increased levels of CO_2 (Anonymous, 2002), the seed production of rice and soybean increased by 42 and 20 per cent, respectively, whereas per cent increase in corn and wheat seeds were 5 and 15 per cent. Research from this angle is required to be done under north western Himalayan soil conditions.

Curbing of deforestation and encouraging afforestation

The depletion of forests, that is deforestation should be ruthlessly curbed and afforestation should be encouraged at all levels. While afforesting the deforested areas and/or wastelands, growing of local species should be preferred.

Introduction of good public transportation

(i) The real breakthrough in solving vehicular pollution problem can be achieved if all the cars, three-wheelers and local buses are run on compressed natural gas (CNG). CNG has already brought about a revolution in the transport sector of metro cities like New Delhi and Bangalore, and can be an option before the north west Himalayan states.

(ii) The use of catalytic converters should be made mandatory to suppress carbon monoxide and oxides of nitrogen.

(iii) Use of biodiesel prepared from *Jatropha* tree species is the need of the day.

Social forestry scheme

A number of social forestry schemes have been formulated by the Central and State Governments and the same must be introduced. This will help to protect the green gold of the north west Himalayas and in turn will decrease the concentration of green house gases.

References

Acharya, C.L., Sood, M.C. and Sharma, R.C. 2001. Indigenous nutrient management practices in Himachal Pradesh. In Indigenous Nutrient Management Practices – Wisdom Alive in India (Edited by Acharya et al.). Indian Institute of Soil Science, Nabi Bagh, Bhopal, MP, India.

Anonymous. 2002. Higher CO2 levels are mixed blessings for agriculture. The Kashmir Times. 59 (283) : 13 [October 10].

Anonymous. 2005. Mercurial weather in valley has residents confused. The Kashmir Times. 62 (136) : 6 [May 17, Tuesday]

Anonymous. 2008a. Greenhouse gases levels highest : Analysis. The Tribune. 128 (139) : 13 [May 13, Tuesday].

Anonymous. 2008b. Global warming : Farmers to face heat. The Tribune. 125 (124) : 13 [May 2, Monday].

Bhatta, A. 2008. Environment science lentil report - pulses in India contribute less to global warming. Science and Environment Fortnightly, Down To Earth. 17 (11): 46.

Dubey, Y.P. and Kaistha, B.P. 2005. Components of organic farming system. Farmers Forum. 5 (9) : 22-23.

Gupta, R.D. 2005a. Environmental Degradation of Jammu and Kashmir Himalayas and Their Control. Associated Publishing Company, Karol Bagh, New Delhi, India.

Gupta, R.D. 2005b. Organic farming - need, concept and merits : A review. Farmers Forum. 5 (9) : 6-9.

Gupta, R.D. 2006. Agriculture of Kandi belt. The Daily Excelsior. 42 (298) : 1 [Sunday Magazine, October 29].

Gupta, R.D. 2007. Mulberry : A tree of multipurpose uses. The Daily Excelsior. 43 (222) : 3 [Sunday Magazine, August 12].

Gupta, R.D. and Abrol, V. 2006. Agroforestry, a sound land use system for Himalayan Rainfed Ecosystem. Farmer's Forum. 6 (3) : 19-22.

Gupta, R.D. and Singh, H. 2006. Indiscriminate use of fertilizer poses health hazards. Farmer's Forum. 6 (9) : 20-24.

Gupta, R.D. and Kher, R. 2009. Apple a day keeps the doctor away. DAV's Ayurveda. Vol. 1, Issue 12, February, 2009 : pp. 16-17.

Gupta, R.D., Kher, D. and Jalali, V.K. 2005. Organic farming : Concept and its prospective in Jammu and Kashmir. Journal of Research, SKUAST-J. 4 (1) : 25-37.

Jamwal, G.S. and Gupta, R.D. 2007. Watershed management project : A case study of Bari-Badhori (J&K). In : Natural Resource Management for Sustainable Hill Agriculture (Edited by Arora et al.). Soil Conservation Society of India, Jammu Chapter, SKUAST-J, Chatha, Jammu, J&K State, India.

Kapta, S.K. 2004. Promising participation watershed development in hills. Watershed News. 19 (3) : 27-39.

Koul, A.A. 2007. Global climatic changes robbing Kashmir valley of abundant snowfall. The Kashmir Times. 64 (340) : 1 & 4 [December 7, Thursday].

Lalway, N. 2006. Back to farming basics. The Kashmir Times. 63 (257) : 6.

Misri, B., Khola, O.P.S., Dev, I., Sareen, S. and Radotra, S. 2004. Natural resource enhancement and conservation through silviculture in the mid hills of Himachal pradesh. National Conference on Resource Conserving Technologies for Social Upliftment (December 7-9). NASC, New Delhi, Extended Abstracts, pp 132-134 (Central Association of Soil and Water Conservation) Research and Training Institute, Dehradun, Uttranchal, India.

Nazir, M. 2008. Jammu pride : Colour revolution in the making. The Daily Excelsior. 44 (285) : 3 [Sunday Magazine, Oct].

Partap Singh, S. 2003. Is the modern science the real pest? The Daily Excelsior. 39 (159): 7.

Rai, A.K. 2007. Agroforestry : Possible alternatives for carbon sequestration. Development Alternatives. 17 (9) : 12.

Samra, J.S. 2003. Impact of climate and weather on Indian agriculture. Journal of Indian Society Soil Science. 51 (4) : 418-430.

Sharma, N. 2008. Hilaungad watershed in Rudraprayag. Study reveals shift from agriculture activities. The Tribune. 128 (155) : 17 [June 5, Thursday].

Sharma, S.P. 2006. Cold desert and pouring rain. The Glimpses of Future. 21 (228) : 4 [August 19].

Shiva, V. 2007. Why we need to go organic : Responding to the agrarian crisis and climate chaos. Farmers Forum. 7 (1) : 4-5.

Singh, B.V. and Saxena, S. 2005. Sustainable organic farming. Farmers Forum. 5 (9) : 17-21.

Singh, G.P.I. 2006. Green revolution and after : The public health perspective. Farmers Forum. 6 (6) : 16-21.

Singh, M. and Sharma, T.R. 1999. Global warming : Its causes and implications on crop productivity. Farmers and Parliament. 36 (11) : 8-9.

Singh, R.D., Rao, K.S. and Bisht, K.K.S. 2001. Indigenous nutrient management technology in Central Himalaya. In Indigenous Nutrient Management Practices - Wisdom Alive in India (Edited by Acharya et al.). Indian Institute of Soil Science, Nabi Bagh, Bhopal, MP, India.

Venkateswarlu, J. 1985. Dryland farming. Central Research Institute for Dryland Agriculture (ICAR), Hyderabad, India.

Wittwer, S.H. 1998. The changing global environment and world crop production. Crop Sciences Recent Advances. The Haworth Press Inc., pp 291-299.

CHAPTER

26 Post Harvest Technology and Value Addition for Dryland Crops

Monika Sood and Raj Kumari Kaul

Division of Post Harvest Technology, Faculty of Agriculture, Sher-e-Kashmir University of Agricultural Sciences and Technology of Jammu, Udheywalla, Jammu.

In India, dryland agriculture is an important source of livelihood. It accounts for more than 70% of the cultivated area of the country, but contributes only 42% of the national food basket. The yields in dry tracts are low and there is always the risk of a total crop failure as a result of erratic rainfall. Thus dryland agriculture is faced with the twin problems of climatic instability and low productivity (Parvathi *et al.* 2000). One important aspect of dryland agriculture is the fact that production is seasonal, which means that grains must be stored for long periods by traders, procurement agencies and consumers.

Post harvest technology studies in respect of food grains, orchard crops, vegetables and other food materials aim at determining the gains derived through value addition activities and preventing losses to make more of food materials available for human consumption, animal production as well as to the advantage of the society at large (Ojha, 2005).

Post harvest losses

In the food grains production and utilization system, the post harvest losses occur at all stages of operations. There is a sizeable quantitative and qualitative loss of crops during different post harvest operations like threshing, winnowing, transportation, processing and storage. But these losses can be avoided by adopting improved post harvest practices evolved through researches. The reduction of losses, especially those caused by insects, rodents, microorganisms, birds and animals can increase available food supplies, particularly in developing countries and more so in backward pockets where the losses are very high and the need is the greatest.

Losses vary by crop, variety, year, pest and pest combination, length of storage, methods of threshing, drying, handling, storage, processing, transportation and distribution rate of consumption and according to both the climate and the culture in which food is produced and consumed.

Table 1. Estimated post-harvest losses of paddy at producers' level.

Operations	Losses (percent to total production)
1. Transport from field to threshing floor	0.79
2. Threshing	0.89
3. Winnowing	0.48
4. Transport from threshing floor to storage	0.16
5. Storage	0.40
Total	2.71

Anonymous, 2007

Post harvest operations

Post harvest operations are those activities which are carried out from the stage immediately after harvesting to the final stage of utilization of the products. In general they include all the operations which take place after harvesting and which are required to make an agricultural produce suitable for immediate consumption and/or storage. These commonly include threshing, winnowing, cleaning, drying, storage, processing, by-product utilization and so on. In simple words, these activities are termed as "value addition" activities so as to improve the quality, storability and utilization efficiencies. Thus the product is transformed into different grades, shapes and sizes which provide opportunities to the producers to earn more income than what he gets by selling the raw products without value addition. In dryland agriculture, an important aspect of postharvest operations is the need to ensure that the produce is kept free from rot, pests and rodents. Given the low productivity of the system as a whole, it is important to see that the produce which is ultimately obtained is edible and palatable. Numerous post harvest technologies including improved material and better equipment, have been introduced (Table 2) to make the process faster, easier and more profitable (Jain, 1985). Let us discuss products from cereals, horticulture crops, oilseeds and legumes to understand the utility of post harvest operations which are carried out to convert the raw material into value added products suitable for human consumption.

Table 2: Post harvest operations

S.No.	Activities	Traditional methods	Improved methods
1	Threshing	Manual beating or Animal treading	Mechanical threshing with important designs of these machines
2	Cleaning	Manual action with ordinary baskets	Mechanical cleaning with manual or mechanical power`
3	Final cleaning	Manually operated SUPA simple devices	Manual cleaner cum graders or power operated machines
4	Drying	Open yard sun drying	By means of solar dryers or heated air dryers using mechanical power

5	Storage	Earthern storage bins or bag storage	Metallic bins, stone bins or brick bins of improved designs
6	Milling	Traditional methods using hand pounding devices or stone wheels or oil *ghannis*	Modern rice or flour mills of different capacities
7	Utilization of by-products	Direct feeding to animals	Using solvent extraction for separation oil from cakes or making other food products
8	Marketing	Selling the raw material to middleman at low prices	Selling the graded products or processed materials to super bazaar with good profits

Ojha, 2005

Cereals

Paddy

Value addition activities for converting paddy into rice involves paddy shelling by rubber roll shellers and rice polishers or metallic hullns or central fugal shellers. The flow chart for processing of paddy/rice is given in Fig 1. Processing operations aim at improving the raw materials by 'primary processing' and transforming into the new products by 'secondary and tertiary processing' activities.

Rice is a staple food and used by many ways as under (Chand and Kumar, 2002):

Staple food : Rice is used as a staple food by more than 60 percent of World population. Cooking of rice is a most popular way of eating. There are many ways of domestic use like *khichadi, pulav, kheer, zeerarice, idli, dosa* etc.

Starch : Rice starch is used in making ice-cream, custard powder, puddings, gel, distillation of potable alcohol, etc.

Rice bran : It is used in confectionery products like bread, snacks, cookies and biscuits. The defatted bran is also used as cattle feed, organic fertilizer (compost), medicinal purpose and in wax making.

Rice bran oil : Rice bran oil is used as edible oil, in soap and fatty acids manufacturing. It is also used in cosmetics, synthetic fibers, plasticisers, detergents and emulsifiers. At present, about 6 lakhs tonnes of rice bran oil is produced from 35 lakhs tonnes rice bran per annum in the country. It is nutritionally superior and provides better protection to heart.

Flaked rice : It is made from parboiled rice and used in many preparations.

Puffed rice : It is made from paddy and used as whole for eating.

Parched rice : It is made from parboiled rice and is easily digestible. In India, about 4-5 percent of total supplies of rice is used as parched rice.

Rice husk: It is used as a fúel, in board and paper manufacturing, packing and building materials and as an insulator. It is also used for compost making and chemical derivatives.

Rice broken : It is used for making food item like breakfast cereals, baby foods, rice flour, noodles, rice cakes, *Idli* and *dosa* etc. and also used as a poultry feed.

Rice straw : Mainly used as animal feed, fuel, mushroom bed, for mulching in horticultural crops and in preparation of paper and compost.

Paddy as a seed: The paddy is used as seed. The proportion utilized for seed purpose varies from 2 to 6 percent of total production

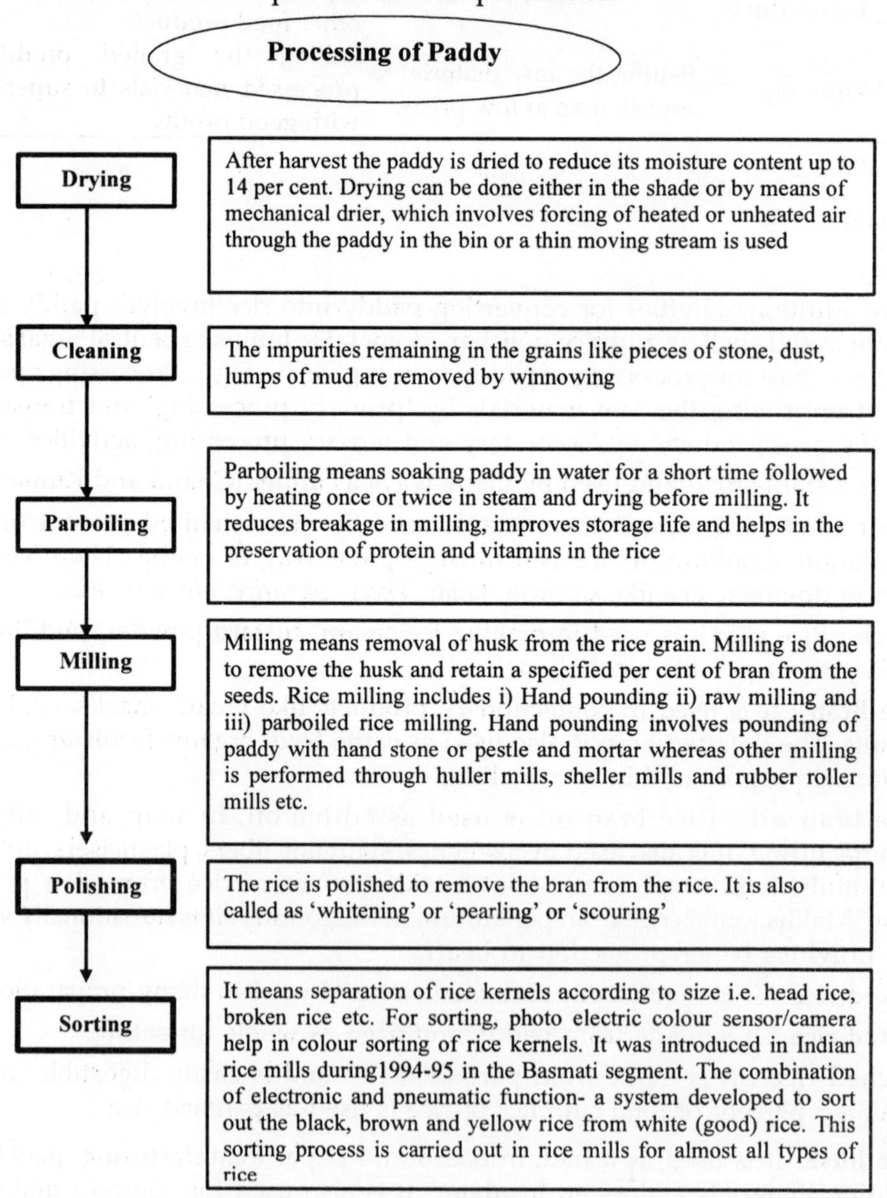

Fig 1: Flow chart for processing of Paddy

Barley

In the case of traditional barley, the tough fibrous hull is literally cemented to the kernel and must be removed using a milling process commonly referred to as pearling. The degree of pearling (typically stated in %) refers to the amount of kernel removed during processing. The first stage or mildest form is termed "blocking" which only removes the hull and a small amount of the outer bran layer. The second pearling stage removes an additional 10% of the kernel, mostly pericarp and the germ fraction, and the resulting product is referred to as "pot barley". The third and final stage removes another 10% of the kernel and produces "pearled barley". The remainder of the seed coat, aleurone and a portion of the subaleurone layers are removed in this final stage. Pearled barley typically represents 60-70% of the original barley kernel. Note pearled barley, from any one of the three pearling stages, can be used as the starring material to produce barley grits, meal, flakes, flour. Pearling of covered barley has a similar function to impact dehulling of standard oats, i.e. hull removal. However, in the case of barley additional tissue layers are removed especially during the second and third pearling stages (Marconi *et al.* 2000).

Other milling techniques, such as roller milling commonly used on other cereal grains can be applied to both dehulled and hulless barley. Although the use of alternative milling procedures is not currently a common practice in the barley industry, these techniques could be used to produce a wide variety of specialty barley products highly enriched in beta-glucan (Andersson *et al.* 2000). While not a typical barley product, bran is produced in both pearling and roller milling. Bhatty (1997) reported that 30% pearling removed all of the pericarp, testa, aleurone and subaleurone in hulless barley.

Another largest use of barley grain is for malt. Globally, 30% of the world barley production is used for malting purpose and 70% for feed use. In addition to barley, wheat and rye are also malted but barley grain has been preferred to other grains. The reasons why barley is commonly used for malt are its husk protecting the coleoptiles during germination process and filtering, firm texture of barley grains and tradition. About, 90% of malted barley is utilized for malting beer and the remaining for food substitutes. Barley malt can be substituted in to a lot of food stuffs such as biscuits, bread, cakes, desserts, etc. (Table 3).

Table 3: Food uses of Malt as by-product

Food Stuff	Colour	Enzyme	Flavour	Sweetness	Nutrition
Biscuits and crackers	X	X	X	X	X
Bread	X	X	X	X	X
Breakfast cereal			X	X	X
Cakes	X		X	X	X

Dessert	X		X		
Ice cream	X		X		
Malted food drinks		X	X	X	X
Meat products	X				
Sauces	X		X	X	
Soft drinks	X		X	X	X
Type of malt products used	Soluble Extract (SE)	SE or flour	SE or flour, flake	SE	SE flour, flake

Bramforth and Barclay, 1993

Wheat

In the semi-arid unirrigated tracts of India the productivity of wheat is very low. The farmer's risk does not end when the crop matures, grain may be lost during harvest because of shattering and spillage or birds, rodents and insects may consume/damage it in the field or in storage. Whereas, early harvesting results in the grains with higher moisture content, which in turn may attract mould infestation resulting into development of aflatoxin, the postharvest losses can be reduced to half with the use of available technology viz., timely harvest, use of proper harvesting and threshing equipments, safe storage, prophylactic and curative measures to check infestation. However, farmers are not fully aware of the postharvest losses during harvesting and storing (Chakravarthy, 1995).

Milling of Wheat – In our country, there are about 2,60,000 small flour mills engaged in primary milling and 820 (1999) large flour mills using about 10.5 million tonnes of wheat.

Traditional Stone Grinder – Whole wheat is ground with bran and germ.

Modern Flour Mill –The object is to obtain maximum amount of flour from endosperm without any bran or germ content. Generally, yield of white flour is about 70 per cent and mill feed (Bran – 12, Germ – 3 and shorts 15 per cent) 30 per cent by weight.

The following steps are involved in wheat milling -

Wheat	Receiving, drying and storage.
Cleaning	To remove impurities like sticks, stones, sand, dirt, other food grains, defective food grains etc. Lastly, the wheat is washed.
Conditioning	The temperature of wheat should not be raised above 47^0 C so that gluten quality is not affected. Hydrothermal treatment is used for conditioning because both moistening and heating are carried out simultaneously

Grinding	Is carried out by roller mills - break roll, reduction roll systems and scratch system generally. In break rolls the bran is cracked and kernel is broken. The endosperm adhering to bran is milled.
Packaging and Storage	End products are packed in waterproof bags and stored in cold dry conditions.
Blending	Increasing health consciousness amongst the consumers, some blending of flours is being carried out e.g. soybean flour, fortification of flour by adding Calcium carbonate, vitamins A and D, Thiamine, Riboflavin.

Initial processing like dehusking, shelling, drying and cleaning add value to wheat and reduce handling and storage costs. The value addition in bread industry is only 12 per cent against 92 per cent in USA. Before consumption, wheat is parboiled, wheat flour (*atta*), refined flour (*maida*) and grits (*suji, dalia*) are utilized for various end uses. Wheat flour is obtained by 5-10 hp burr mills, whereas *maida* and *suji* are produced in roller mills with 13 per cent bran and 3 per cent germ as by products. Wheat contains about 2 to 3 per cent germ, which is generally mixed with bran and sold as cattle feed. Wheat germ is a rich source of protein (25-30 per cent and vitamin E) and can be used in Biscuits, breakfast food and high protein drinks. Besides, the wheat gluten dried powder can be mixed with flour to produce slimming diets, crisp breads, breakfast foods, breads, etc. to increase their texture and nutritional value. In so far as utilization of different varieties of wheat is concerned, *T.aestivum / vulgare*, the common bread wheat is preferentially used for making *Chapaties*, bread, biscuits, whereas *T.durum*, known as Macaroni wheat is preferred for making *suji*, macaroni, spaghetti, vermicelli, etc. *Triticum dicoccum*, known as emmer wheat (commercially known as ***Khapli***) is commonly used in South India for the preparation of breakfast food called ***Upuma*** (Faruqi and Singh, 1981). Wheat is used for different purposes and hence quality requirements vary on its end use. Hard wheat (*T.aestivum*) with strong gluten and > 12.0 per cent protein has been found suitable for bread making, for biscuits soft wheat with less gluten and < 11.0 protein is best suited. Whereas, for pasta products (*T.durum*), hard wheat with gluten < 12.5; protein < 10.0 and 7.0 FPM carotene content is required.

Sorghum and millets

Sorghum and millets can have excellent quality for processing but to obtain that inherent quality a value added, chain securing, identity preserved grain for processing into profitable upscale urban products is necessary. Processing technology is not the major obstacle to successful production of products. Consistent supply of a modest quality grain is the major constraint.

Major constraints to sorghum and millet utilization

- Lack of consistent, uniform quality grain supplies
- Logistics/markets
- Subsidized imported cereals
- Extension of existing processing technology unavailable
- Few shelf-stable convenience foods
- Governmental policies - VAT on sorghum-RSA
- Subsidized Maize, rice or wheat-based food systems
- Poor image of sorghum and millet
- Nutritional myths - tannins, poor digestibility
- Grain moulds

The milling quality of sorghum and millet is determined mainly by kernel shape, density, hardness, structure, and presence of a pigmented testa, pericarp thickness and color. Kernels with a high proportion of hard endosperm, white, thick pericarp without a pigmented testa have outstanding dehulling properties. Soft floury kernels disintegrate during dehulling and cannot be milled efficiently. For hand dehulling, a thick starchy mesocarp reduces labour 50% or more. Long, slender pearl millet kernels have very poor dehulling properties, while spherical kernels have the highest yields of decorticated grain (Fig 2). The white tan food sorghums have significantly improved yields of light colored flour and decorticated kernels.

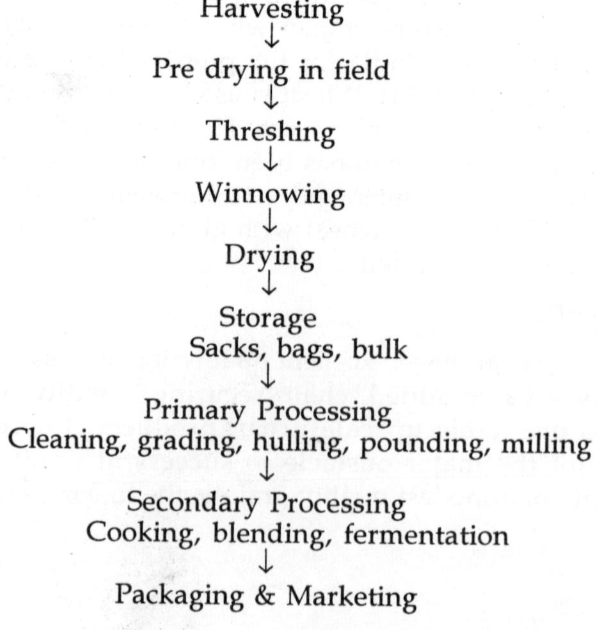

Fig 2: Total post harvest system for sorghum and millets

The proper sorghum and millet cultivars can be processed into a wide variety of very acceptable commercial food products. These grains can be extruded to produce a great array of snacks, ready to eat breakfast foods, instant porridges and other products. The flakes of a waxy sorghum obtained by dry heat processing can be used to produce granola products with excellent texture and taste. Tortillas and tortilla chips have been produced from sorghum and pearl millet alone or with maize blends. The sorghum products have a bland flavour while pearl millet products have a unique strong flavour and colour. The critical limitation is again cost efficient, reliable supplies of grain.

Neither sorghum nor millet have gluten proteins, so to produce yeast-leavened breads, they are usually substituted for part of the wheat flour in the formula. The level of substitution varies depending upon the quality of the wheat flour, the baking procedure, the quality of the sorghum or millet flour and the type of product desired. In biscuits, (cookies) up to 100% sorghum or millet flour can be used. The non-wheat flour gives a drier more sandy texture so the formula must be modified. White sorghum has a definite advantage over maize and millet in composite flours because of its bland flavour and light colour (Goel *et al.*2007).

Horticulture crops

Mango

The freshly harvested raw or ripe mangoes are subjected to primary and secondary processing activities to transform the raw materials into value added useful finished products for human and animal consumption. Unripe mangoes are graded, washed and then used for making pickles or powder (*amchur*). The operations like washing, boiling, drying, size reductions, packaging etc. are performed to convert raw mangoes into useful products. Similarly, ripe mangoes are subjected to washing, grading, deskining, crushing, filtration, boiling, drying or dehydration to make jam, squash, fruit bar, nectar etc. Also, the mango stone is crushed and used for preparing animal feed or extraction of mango fat (Ojha, 2005).

Ber

Another very important fruit of arid region is ber which has been accepted by the farmers on larger scale. The harvested fruits are sorted to discard the damaged, overripe, unripe and misshapen fruits. Then the fruits are graded. Precooling of the fruits at 10°C immediately after harvest increases the shelf life by about 3 days when subsequently stored at room temperature. The storage life can be prolonged to 30-40 days by storing at 3°C and 85-90% humidity. Preharvest spray of 1% calcium nitrate and dipping of the fruits before storage in 500ppm captaf also improves their shelf life (Khurdiya 2001a). Ber can be processed to prepare *murabba*, candy, dehydrated ber, pulp, jam and ready-to-serve beverage.

Karonda

Karonda requires 2-3 pickings. On an average, each plant provides 3-5kg fruits. Both unripe and ripe fruits are harvested. There is no standard practice for grading and packing of fruits. Fruits after harvesting are kept in shade. Undesirable or blemished fruits are sorted out. Good fruits packed in baskets are marketed. Storage life of fruits depends upon the stage of harvest. Processing of fruits into value added products like ready-to-serve beverage, pickles and candy is mostly done (All and Rab, 2000).

Aonla

Aonla fruits are widely used in the form of fresh fruit or powder in various preparations. A number of products including sauce, candy, dried chips, powder, chutney and *murabba* can be prepared from aonla fruit. Fresh fruits are highly acidic and unsuitable for direct consumption. Therefore fruits are processed into products such as preserves, pickles and powder. A substantial quantity of the ascorbic acid of aonla may be lost in an ordinary method of pickling, but much of it may be retained by boiling the fruits in water for few minutes and then putting them in a heavy salt solution (Goyal *et al.* 2008).

Pomegranate

Pomegranate fruits of good quality fetch fairly good price in the market and are not much used for the processing purposes. However, low quality fruits including cracked, damaged, spotted or small sized fruits may be utilized for the processing. Fruits are processed into quality juice and syrup. The delicious juice of the pomegranate blends well with other fruit juices also. Fruit juice easily ferments and may be used for making wine. Jelly may also be prepared from the juice. The dried seeds, marketed commercially as *anardana,* are widely used as condiment prepared from the sour type of the pomegranate (Khurdiya, 2001b).

Phalsa

A fully grown phalsa shrub gives 5-6kg fruits. Just after picking, the fruits are packed in well aerated bamboo baskets and sent to markets. Phalsa fruits are very perishable. They cannot be stored for more than 2-3 days under ambient conditions. They can be graded according to size and colour. Fruits should be packed in bamboo baskets cushioned with gunny cloth or paper cuttings for distant markets (Khurdiya, 2001a). Storage life of fruits depends on the stage of harvest. Fruits harvested at turning stage can be stored for 2-3 days at room temperature and about 7 days in cold storage. A ready-to-serve beverage can be formulated from phalsa fruit juice. Syrup can also be prepared from phalsa in which clear juice is mixed with an equal amount of sugar and preserved with sodium benzoate (Khurdiya, 2001b).

Custard apple

The custard apple is almost entirely consumed as a desert, its pulp may be mixed with milk to form a delicious drink. Extraction of the pulp is a major constraint in processing. The problems encountered during the processing of custard apple are the development of bitterness, unpleasant repulsive off-flavour in the pulp on heating beyond 65°C. The treatment of the pulp with 0.2% pectinase enzyme and four hours incubation period were found to be the optimum conditions for the extraction of juice. The organoleptic evaluation of the RTS beverage showed that the beverage prepared by using 20% juice was the most acceptable. The organoleptic properties of custard apple wine were comparable with those of the grape wine properties (Kotecha *et al.* 1995).

Jamun

Jamun has never been exploited commercially for the preparation of different products. The importance of jamun lies in its refreshing and curative properties. Jamun fruits can be processed into excellent quality fermented beverages such as vinegar or cider and non-fermented Ready-to-Serve beverages and squashes. A good quality jelly can also be prepared from its fruits. The seeds can be processed into powder, which is very useful to cure diabetes (Khurdiya, 2001b). Fully ripe fruits are harvested daily by hand picking or shaking the branches and collecting the fruits on a polythene sheet. Jamun trees need a number of pickings, since all fruits do not ripen at a time. Jamun fruits are highly perishable. They can be stored only up to 2 days at ambient temperature. Pre-cooled fruits, packed in perforated polythene bags, can be stored for 3 weeks at 8-10°C and 85-90% humidity (Khurdiya, 2001a). Fruits are normally packed in bamboo baskets and transported to local markets.

Jackfruit

Jackfruit production is seasonal. The jackfruit yields once a year. The majority of fruits are marketed locally by farmers. Some of them are brought to the roadside or local markets. Jackfruits which are not of good quality are locally utilized for chip making. In addition to chips, other products prepared from jackfruit are *papads* and *halwa*. These products have good keeping qualities. Jackfruit processing was standardised at CFTRI (Berry and Kalra, 1998) to produce a variety of products like canned jackfruits bulbs in syrup, squash, raw jack pickle, roasted jack seeds, jack seed flour and candied jackfruit.

Tamarind

The separation of pulp from the tamarind seed is rather difficult due to its low water content. The industrial process for the manufacture of tamarind juice concentrate in India is also based on the extraction of the pulp with boiling water (Khurdiya 2001b). Tamarind pulp is the fruit base in the preparations of the jams, jellies and icecreams. A process for the clarification of tamarind juice by treatment of 0.12 to 0.15% gelatin was also developed.

Oilseeds

In India, about 20 million hectare area is under oil seed crops. Most of these crops are grown under unirrigated and on marginal lands with low productivity status. Nutritional oils are the richest source of energy (9cal/g) and provide essential fatty acids and fat soluble vitamins. It is essential not only to increase the production of oil seed but also to reduce the losses during harvesting and post harvest operations of oil crops. The losses which can be minimized by adopting improved technology already available in the market are presented in Table 4.

Table 4: Improved technology for minimising losses

Operation	Losses (%)
Harvesting	5-15
Threshing	2-5
Cleaning	2-3
Handling and storage	10-15

Ojha, 2005

Oilseed storage

After harvesting and drying the oilseed crops, the seeds can be stored in improved storage structure depending on the quantity and environmental conditions. In case of small quantity of such materials, the metal bins, pusa bin and stone bins can be successfully used. However, for the large quantities of such stocks, the water proof godowns or warehouses are preferred to stock bags full of oilseeds. Storage of oilseeds under high humid conditions would not be desirable and must be avoided. Arid regions are best suited for storage of food materials without much of precautions except fumigation and moisture entry in the godown for structure (Ojha, 2005).

Oil extraction

Traditional oil extraction equipment of India is the bullock-drawn '*ghani*' which leaves 10-15 per cent oil with cake. The mechanical power *ghanis* have now been introduced but they also leave 8-12% oil with cake. As far as possible, the total oil content of the seed should be extracted but it is not possible with mechanical devices. Hence, the solvent extraction plants have been installed in the country (Ojha, 2005). These plants have ability to extract nearly total quantity of oil leaving less than 0.5 per cent with the cake. The deoiled cakes are excellent for use as animal feeds.

Legume milling technology

Dal milling is the third largest processing industry in India after rice milling and wheat milling. Milling of pulses means removal of outer husk and splitting the grain into two equal halves. Generally, the husk is much more tightly held by the kernel of some pulses than most cereals. Therefore, dehusking of some pulses poses a problem. The method of alternate wetting and drying is used to facilitate dehusking and splitting of pulses (Fig 3). Dal milling is generally done in two steps:

1. Loosening of husk by wet or dry methods.
2. Dehusking and splitting into the cotyledons using appropriate machines.

In India the dehusked split pulses are produced by traditional methods of milling. At the family level, the *chakki* (sheller) is used to split pulses. In traditional pulses milling methods, the loosening of husk by conditioning is insufficient. Therefore, a large amount of abrasive force is applied for the complete dehusking of the grains, which results in high losses in the form of brokens and powder. Consequently, the yield of split pulses in traditional mills is only 65 to 70 per cent in comparison to 82 to 85 per cent potential yield (Chakravarthy, 1995). The amount of oil mixed with pulses vary from place to place from 50g/q. of grain to 400g/q. Similarly, addition of water also varies from 4 to 20kg/q. of grain. For loosening of husk and its complete removal, 3to7 passes through the roller in case of pigeon pea are given to the conditioned grain (Ojha, 2005).

Measures for minimizing post harvest losses

Following preventive measures should be adopted to avoid post harvest losses:

- Harvest at proper stage of maturity to reduce losses.
- Use proper method of harvesting.
- Adopt modern mechanical methods, to avoid the losses in threshing and winnowing.
- Use improved technique for processing.
- Adopt cleaning and grading for remunerative prices to avoid financial loss.
- Use good packaging materials for storage and in transport like HDPE bags.
- Adopt proper technique in storage.
- Apply pest control measures during storage.
- Proper handling i.e loading and unloading of crops with good transportation facilities at farm and market level reduces losses.

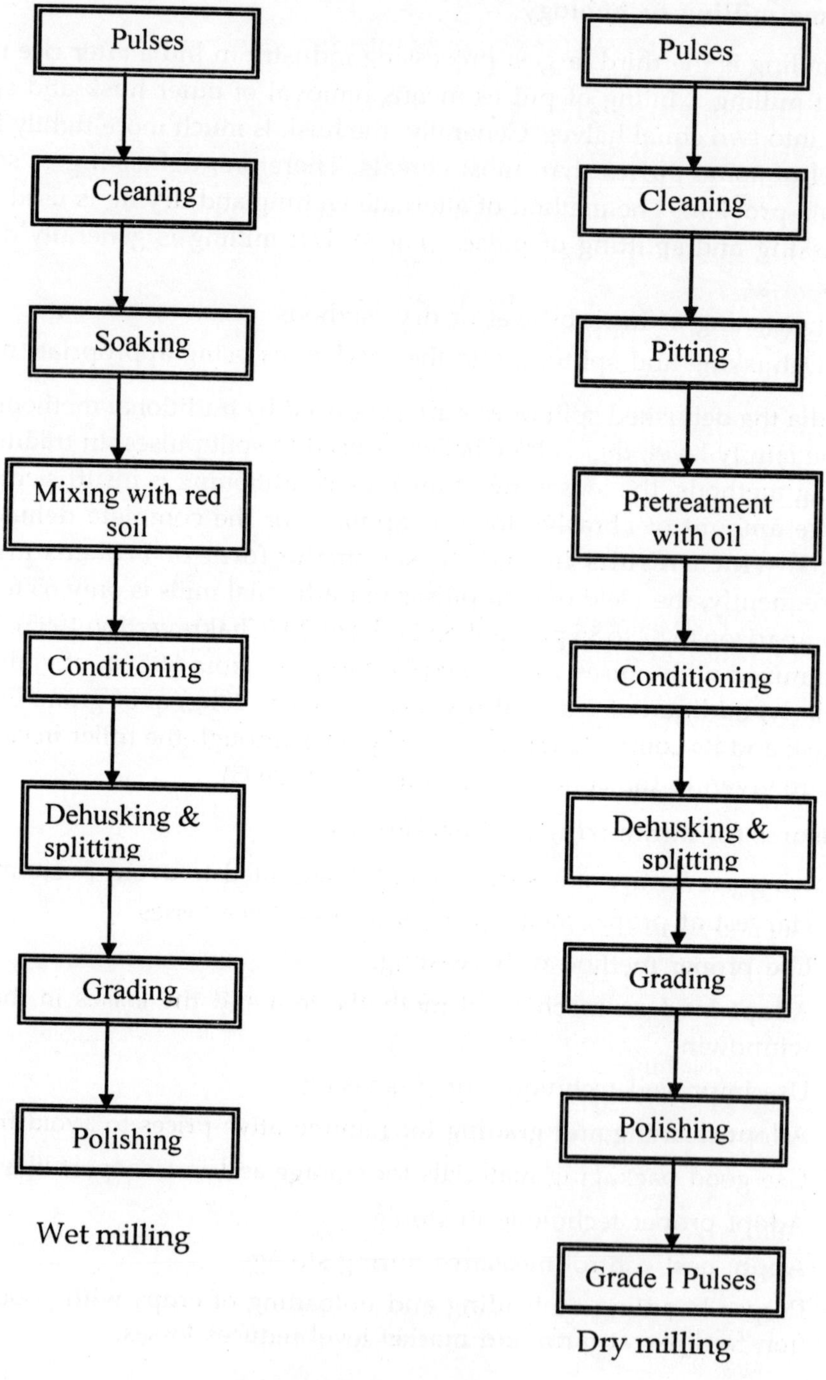

Pulses → Cleaning → Soaking → Mixing with red soil → Conditioning → Dehusking & splitting → Grading → Polishing

Wet milling

Pulses → Cleaning → Pitting → Pretreatment with oil → Conditioning → Dehusking & splitting → Grading → Polishing → Grade I Pulses

Dry milling

Fig: 3 Milling of pulses

Strategy for value-added products

The best strategy for developing convenient, shelf-stable products is to use identity preserved crops to produce high-value products that can be priced slightly lower than imported products. The target should be middle class where sufficient prices can be obtained to provide profits for all. There is no need to develop low cost, inferior quality foods that do not provide significant profits. Following strategies should be adopted for maximization of profits.

- Identify upscale products
- Supermarkets
- Use low input, appropriate technologies
- Use identity preserved food grain and other crops
- Specify variety and hybrids
- Educate farmers and producers
- Economics-share value-added processing profits with members of the supply chain.

References

All, R. and Rab, F. 2000. Research needs and new product development from underutilized tropical fruits. Acta Hort. ISHS. 5:18:241-248

Andersson, A. A., Andersson. R. & Aman, P. (2000). Air classification of barley flours. Cereal Chemistry. 77.463-467.

Anonymous, 2007. Report on the proceedings of the National Post Harvest conference, Ministry of Agriculture (MoA). Gaborone, Botswana.

Berry, S.K. and Kalra, C.K. 1998. Chemistry and technology of jackfruit (*Artocarpus heterophyllus*)- a review. Indian Food Packer. 3:62-76.

Bhatty, R. S. (1997). Milling of regular and waxy starch hull-less barleys for the production of bran and flour. Cereal Chemistry. 74:93-699.

Bramforth, C.W. and Barclay, A.H.P. 1993. Malting technology: the uses of Malt Barley: Chemistry and Technology. A.A.C.C. Inc. St. Paul, Minnesota, USA. pp298.

Chakravarthy, A. 1995. Post harvest technology of cereals, pulses and oilseeds. Oxford and IBH publishing Co. pvt ltd, New Delhi.

Chand, R. and Kumar, P.2002. Long term changes in course cereal consumption in India. Indian Journal of Agricultural Economics. 57(3): 13.

Faruqi, N.Y.Z and Singh H.P. 1981. Chapati Making Properties of Indian Wheats- A review. Agricultural Marketing Vol.XXIV (1), pp 105.

Goel,A.K., Kumar, R. And Mann, S.S. 2007. Post harvest management and value addition, Daya publishing house, Delhi. 74-97, 272-291.

Goyal, R. K., Patil, R. T., Kingsly, A. R. P., Walia, H. and Kumar, P. 2008. Status of post harvest technology of aonla in India-Review. American Journal of Food Technology, 3 (1): 13-23.

Jain, H.K.1985. Indian Agriculture in 2000 A.D. New Delhi. Indian National Science Academy. pp1-6.

Khurdiya, D.S., 2001a. Post harvest management of underutilized fruits for fresh marketing. Winter school on exploitation of underutilized fruits. Pp 266-274.

Khurdiya, D.S. 2001b. Post harvest management of underutilised fruits for processed products. Winter school on exploitation of underutilised fruits. Pp 291-298.

Kotecha, P.M., Adsule, R.N. and Kadam, S.S. 1995. Processing of custard apple: preparation of ready-to-serve beverage and wine. Indian Food Packer 49(4):5-11.

Marconi, E., Graziano, M. & Cubadda, R. 2000. Composition and utilization of barley pearling by-products for making functional pastas rich in dietary fiber and beta-glucans. Cereal Chemistry. 77:133-139.

Ojha, T.P. 2005. Post Harvest technology for dryland crops. In: Sustainable development of dryland agriculture in India (Ed. Singh, R.P.). Scientific publishers.pp 461-472.

Parvathi, S., Chandrakandan, K. and Karthikeyan, C. 2000. Women in dryland practices in Tamil Nadu, India. Indigenous knowledge and development monitor. 1-6

CHAPTER

27

Agricultural Marketing System and Status of Indian Agriculture

Sudhakar Dwivedi and Manish Sharma

Division of Ag. Economics and Statistics, Faculty of Agriculture, Sher-e-Kashmir University of Agricultural Sciences and Technology of Jammu, Chatha, Jammu-180009.

Introduction

Marketing of agricultural produce is considered as an integral part of agriculture, since farmers are encouraged to make more investment and to increase production, if they are paid fair price of their produce and vice-versa. In India, agriculture was done formerly on a subsistence basis; the villages were self-sufficient, people exchanged their surplus goods and services within the villages on a barter basis. With the development of infrastructural facilities, agriculture has become commercial in character, the farmers opt those enterprises of agriculture that fetch better price and returns.

Agricultural marketing involves in its simplest form the buying and selling of agricultural produce. In olden days, this definition of agricultural marketing may be accepted, when the village economy was more or less self-sufficient, the marketing of agricultural produce presented no difficulty, as the farmers sold his produce direct to the consumers on a cash or barter basis. But now a days marketing of agricultural produce has to undergo a series of transfers or exchanges from one hand to another before it finally reaches to the consumer.

The Indian council of Agricultural Research defined it as, "Agricultural marketing involves three important functions i.e. (i) assembling (concentration), (ii) preparation for consumption (processing), and (iii) distribution (dispersion).

The National Commission on Agriculture defined agricultural marketing as "Agricultural marketing is a process which starts with a decision to produce a saleable farm commodity and it involves all aspects of a market structure of system, both functional and institutional based on technical and economic considerations and includes pre and postharvest operations, assembling, grading, storage, transportation and distribution.

Importance

It is necessary to improve the marketing system to aid the process of agricultural development for two reasons.

- If the surplus produce does not move to the market to bring additional revenue to farmers, it may work as disincentive to increase production.
- If the system does not supply foodgrains such as oil, fruits, vegetables, fish, meat, eggs etc. at reasonable prices to the consumers at the time and place needed by them, increased production has no meaning in a welfare society.

  These are two aspects of marketing of agricultural produce. These two aspects are also known as agricultural business system. The following are the basic objectives of an efficient marketing system:

- To enable the primary producers / farmers to reap the best possible benefits.
- To provide facilities for lifting all produce, the farmers are willing to sell at an incentive price.
- To reduce the price spread between the basic producer and ultimate consumer.
- To make available all product of farm origin to consumers at reasonable price without impairing the quality of the produce.

Nature and features of Agricultural produce

Agricultural produces are different in nature and contents from industrial goods in the following respects:

- Agricultural produces tend to be bulky and their weight and volume are great for their value in comparison with many industrial goods.
- The demand on storage and transport facilities is heavier and more specialised in case of agricultural products than in case of industrial products.
- Mostly agricultural commodities are highly perishable. Although some crops such as rice retain their quality for longtime, but most of farm produce are perishable and cannot retain long on way to the final consumer without suffering loss and deterioration in quality.
- Moreover agricultural produces are available only in their season but such conditions of seasonal availability is not found in industrial products.
- Agricultural produces are to be found scattered over a vast area and its collection posses a big problem, while in case of industrial goods this is not.
- There are many varieties and kinds in farm produce and so it is difficult to grade them.
- Most of the farmers of our country have small size of holding, thus they have small surpluses and further they have to sell their produce just after harvesting of the crop and get low price.

- Lastly, both demands and supply of agricultural products are inelastic. A bumper crop, without any minimum guaranteed support price from the govt. may spell disaster for the farmer. Similarly, the farmer may not really be in a position to take advantage of shortages or deficit crop.

Requirement for efficient marketing system

Indian farmers still suffer from lack of prosperity and that is mainly due to the defects in the marketing of agricultural produce. At the present moment India lacks organised market for the sale of agricultural produce. For efficient and prosperous agriculture, well organised marketing is essential. There is the excessive number of middlemen who intervene between the farmer and the final consumer. The absence of knowledge of market conditions and price fluctuations in such markets, due to the illiteracy of the farmers places the farmers at the mercy of moneylenders and middlemen. It is high time that marketing facilities for agricultural goods be improved. Let us now discuss how an efficient marketing system can stimulate production and increase welfare.

- An efficient marketing can ensure a fair price for the marketable surplus which in itself is an incentive to the producer to grow more.
- The efficient marketing system can promote capital formation in agricultural sector and thus, may help in raising the productivity.
- It can support agricultural credit and credit institutions. Thus, financing of agriculture at reasonable rate of interest may expand it in terms of volume and coverage.
- An efficient system of marketing can ensure price stability and this is an important factor in planning for agricultural production.
- Marketing agencies can ensure a regular and timely supply of agricultural inputs at controlled prices.
- An efficient marketing system ensure an adequate marketable surplus. To maintain the tempo of economic development, it is necessary to meet the increasing demand for food, raw materials and other agricultural products by the non-farming population. This can be done by increasing the marketable surplus. It means that the presence of proper and effective distribution machinery, the availability of transport, communications, storage and warehousing facilities, and the prevalence of fair marketing practices may induce farmers especially small farmers to bring their surplus produce to the market.
- The efficient marketing system ensures proper price spread between producers and consumers. There is a large army of intermediaries between the producer and the actual consumer. This army get fat at the expense of both the producers and consumers, by indulging in many fraudulent practices. Thus, efficient marketing systems are must for ensuring incentive prices to the farmers and reduce the price spread between the farmer and the consumer.

- It ensures proper availability of agricultural inputs and consumer products to the farmer at reasonable prices. The middle man handling the supplies of seed, fertilizers, pesticides and other inputs tends to manipulate the marketing system to his advantage, causing a loss to the farmer and the aggregate production in the economy. So, an efficient system of marketing must be required to ensure the availability of agricultural products at fair prices to the farmers.

Agencies through which agricultural produce is marketed

The transfer of produce or commodities takes place through a chain of middlemen or agencies. In the primary market, the main functionaries are the farmer, the village itinerary merchant, pre-harvest contractors, commission agents etc. In the secondary market, the processing and manufacturing agencies are the additional functionaries. Financing agents like sarraffs, banks and cooperatives may also take part. In the export or terminal market, the commercial analyst and shipping agents also get involved in the transfer of commodities.

These functionaries have their own setup. They may be individual, partners or cooperatives who may buy and sell on ready and future basis, a price determined by forces of supply and demand. Every functionary renders some services in the process of marketing and also earns varying margin of profit for himself and at the same time bears risks involved in the process. This procedure makes marketing rather complicated and inflates the price of the produce at which it is ultimately sold to the consumer.

Recent Scenario of Indian Agriculture

Statistics of Land Utilization

Land utilization statistics refers to that statistics which relate to utilization of the entire tract of land and which comes within the geographical limits of a country. Generally, it is not possible to utilize every inch of land that a country possesses, as some part of it may be taken up by forests, some areas may not be easily accessible and there might be considerable difficulties in the utilization of the others. To have a full picture of land utilization it is necessary to collect a large variety of statistics relating to total area, area under forests, area not available for cultivation, area of other uncultivated land and current fallows, net area sown, area devoted to different crops and the area and crop irrigated and unirrigated. Statistics of land utilization are available in India since the year 1884. However, over a period of time the geographical coverage and scope has been gradually expanding from year to year which render them incomparable.

Table 1.Details of land utilization in India during 2004-05 and 2005-06 (In Million Hectares)

S. No.	Classification		2004-05	2005-06
I	**Total Geographical area**		328.73	328.73
II	**Reporting area for Land Utilization Statistics (1 to 5)**		305.31	305.27
	1.	Forest	69.72(22.80)	69.79(22.90)
	2.	Not available for cultivation (A+B)	42.34	42.51
	(A)	Area under Non-Agril. Uses	24.76(8.10)	25.03(5.80)
	(B)	Barren & Un-culturable Land	17.58(5.80)	17.48(5.70)
	3.	**Other uncultivated Land excluding Fallow Land (A+B+C)**		
	(A)	Permanent Pastures & Other Grazing Lands	10.42(3.40)	10.42(3.40)
	(B)	Land under Misc. Tree Crops & Groves not Included in net area sown	03.38(1.10)	03.38(1.10)
	(C)	Culturable Waste Land	13.16(4.30)	13.12(4.30)
	4.	Fallow Lands (A+B)		
	(A)	Fallow lands other than current fallows	10.71(3.50)	10.50(3.40)
	(B)	Current Fallow	14.43(4.70)	13.67(4.50)
	5.	Net area sown (6-7)	141.14(46.20)	141.89(46.50)
	6.	Total cropped are	190.42	192.80
	7.	Area sown more than once	49.28	50.90
	8.	Cropping intensity	134.90	135.90
III	Net Irrigated area		58.87	60.20
IV	Gross irrigated area		80.00	82.63

(Anonymous. 2008). Note: Figures in parentheses indicate the percent to total.

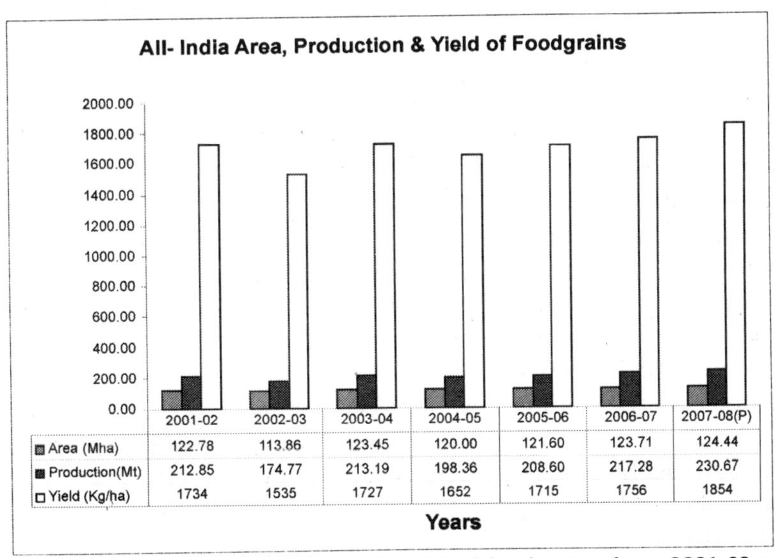

Figure 1: All India area, production & yield of foodgrains from 2001-02 to 2007-08.

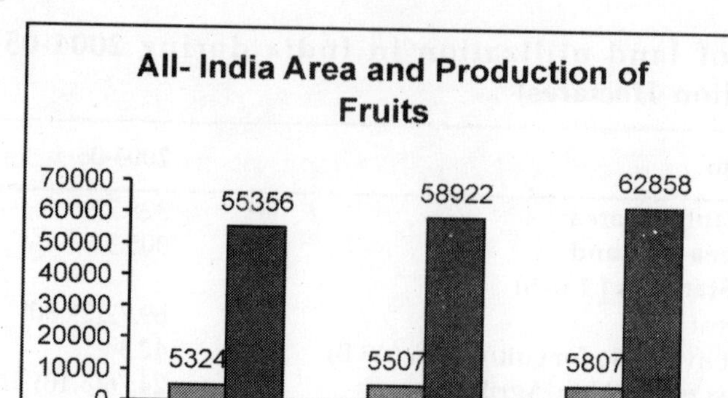

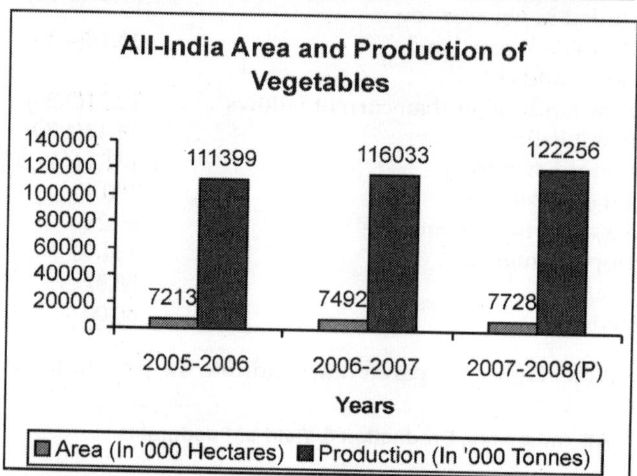

Figure 2: All India area and production of fruits & vegetables from 2005-06 to 2007-08. **(Anonymous. 2008)**

From the figures 1 and 2 it is depicted that Foodgrains, Fruits & Vegetables in India are at increasing trend but at slow rate. Per capita production of foodgrains had increased from 183 kg during early 1970s to 207 kg by mid1990s, even though the country's population increased by more than 50%. However after mid 1990s, foodgrains production could not keep pace with the population growth. Per capita production of cereals declined by 17 kg and pulses by 3 kg during the past decade. Total demand for cereals would grow to 218.9 mt by 2011-12 and would reach 261.5 mt by 2020-21. Demand for pulses during this period would grow to 16.1 mt and 19.1 mt, respectively. Domestic demand for foodgrains is projected to reach 235.0 mt by 2011-12 and 280.6 mt by 2020-21. It is important to mention that these projections do not include export demand.

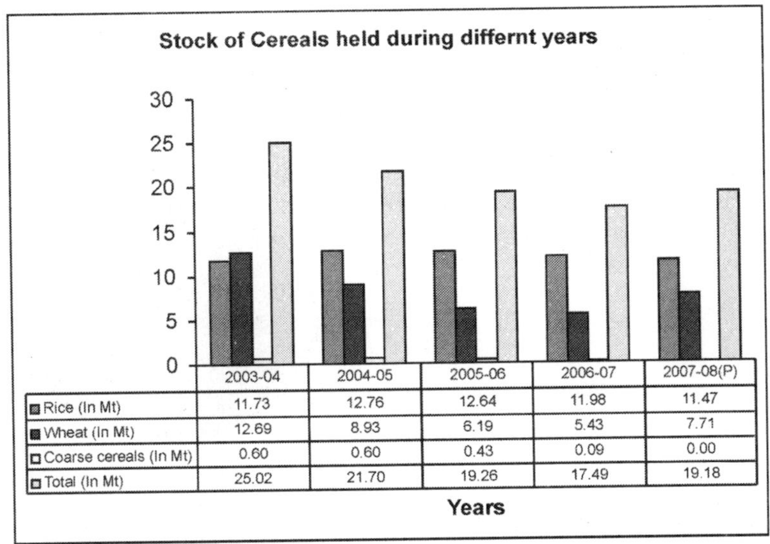

	2003-04	2004-05	2005-06	2006-07	2007-08(P)
Rice (In Mt)	11.73	12.76	12.64	11.98	11.47
Wheat (In Mt)	12.69	8.93	6.19	5.43	7.71
Coarse cereals (In Mt)	0.60	0.60	0.43	0.09	0.00
Total (In Mt)	25.02	21.70	19.26	17.49	19.18

Figure 3: Stock of Cereals held by Central and State Agencies (as on I[st] January) (Anonymous. 2008)

The figure 3 depicting stock of cereals indicate that the stock available during 2006-07 and 2007-08 is low as compared to the stock available during 2005-06 (due to some natural calamities) whereas the population of India is increasing at the rate of 1.57% annually which is an alarming indicator apprehending food crisis in future. The maximum stock of cereals was available during 2003-04 which is 25.02 mt. It was high due to better yield during that agriculture year.

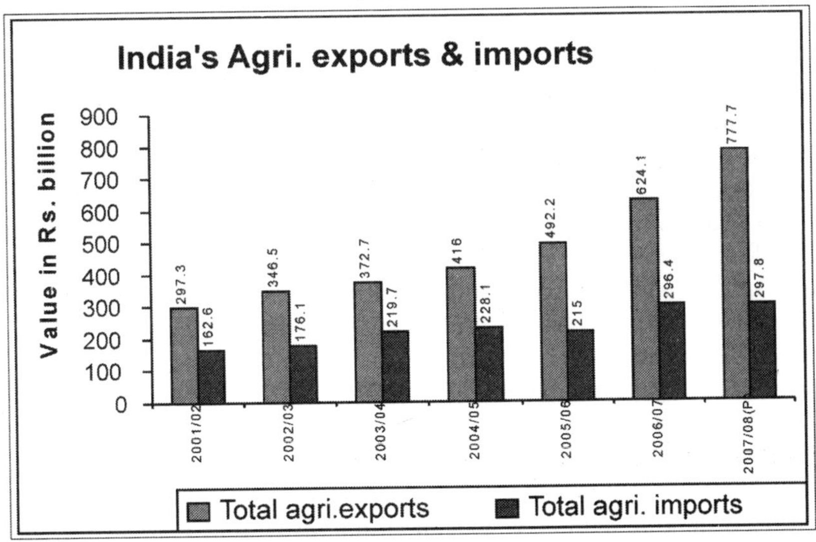

Figure 4: India's changing Agricultural exports & imports during 2001-02 to 2007-08

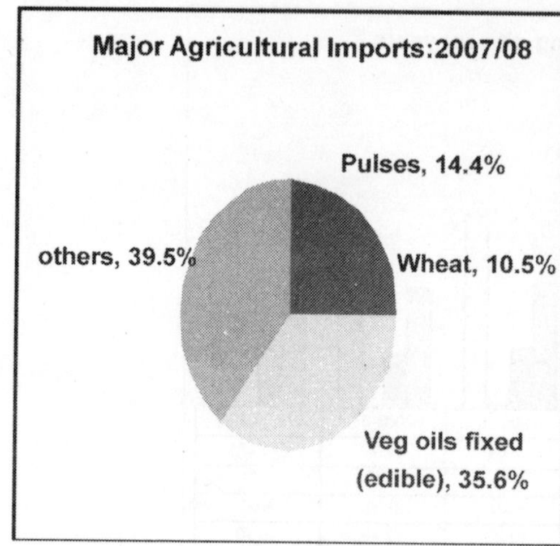

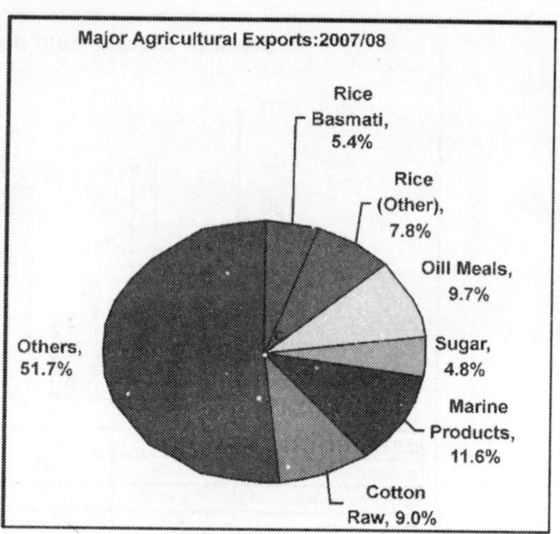

Figure 5: Major agricultural imports & exports during 2007-08(TE) (% distribution). [(Anonymous. 2008) Note: Data for 2007-08 are provisional.]

From the figures 4 and 5 related to India's agricultural exports and imports system it is revealed that during year 2007-08 India exported more than 60% as compared to the base year 2001-02. This directly increases the Indian economy. On the other hand the India importing low agriculture products during the year 2007-08 as compared to the base year 2001-02. The maximum quantity importing is edible oil which is 35.6%, Pulses (14.4%), Wheat (10.5%) and others (39.5%). The TE 2007-08: value of the agril. imports is $6.3 bn and value of agri. exports is $14.7 bn. On the other hand, these facts indicate that now India is a self sufficient country to meet its future requirements in agriculture products, which will play a vital role in the progress of Indian economy.

References

Anonymous. 2008. Agriculture Statistics at a Glance 2008, Directorate of economics and Statistics, Ministry of Agriculture, Govt. of India.

Agricultural Research Data Book (2008). Published by Director, Indian Agricultural Statistical Research Institute, New Delhi. INDIA.

B.P.Tyagi(2005). Agricultural Economics and Rural Devolpment. Daya Press, Saket, Meerut, U.P.

Compendium on Agricultural Marketing and rural Agri-business by Directorate of Extension Education, SKUAST-Jammu, 07-11 August, 2007.

http://www.dacnet.nic.in

http://www.economywatch.com/database/

http://www.indiancommodity.com/statistics/

R. N. Soni (2005). Leading issues in agricultural economics. Vishal Publishing Co. Jalandhar, Punjab.

S. Subba Reddy, P. Raghu Ram, T.V. Neelakantha Sastry and I. Bhawani Devi(2005). Agricultural Economics. Oxford & IBH Publishing Co. Pvt. Ltd., New Delhi.

CHAPTER

28 Information and Communication Technologies - A Viable Means for Technology Dissemination

P.S. Slathia and S.K. Kher

Division of Ag. Extension, Faculty of Agriculture, Sher-e-Kashmir University of Agricultural Sciences and Technology of Jammu, Chatha, Jammu-180009.

Indian agriculture sector is leveraging the information and communication technologies (ICT) to disseminate right information at right time. As, it is said that information is essential for survival. In order to be really instrumental, information must be transformed into knowledge. In the process of development, ICTs can be instrumental by providing flow of information and knowledge to the people concerned in such a way that they can choose to make best use of it. The development and proliferation of electronically communicated information has accelerated economic and social change across all areas of human activity worldwide and it continues to do so at rapid pace. ICT diffusion in developing countries is also in progress. The concept of ICT was coined by Stevenson in 1997, it includes range of technologies like Radio, TV, Telephone, Internet and Satellite based communication system.

ICT is an integration of technologies and the process to distribute and communicate the desired information to the target audience and making the target audience more participative.

ICT refers to various technologies which interlink information technology such as, personal computer with communication technology as telephone and telecommunication networks (Chapman and Slaymaker, 2002).

According to Michiels and Van Crowder (2001), ICT's are a range of electronic technologies which when covered in new configuration are flexible, adoptable, enabling and capable of transforming organization and redefining social relations. ICT's are therefore expanding assembly of technologies that can be used to collect, store and share information between people using multiple devices and multimedia. Basic assumptions of ICT's are:

Equality - ICT can achieve equality in the provisions of communication reaching particularly, with reference to geographical location.

Quality - ICT's help to bring an improvement in quality of subject matter to be transferred.

Quantity - IT can lead to the quantity of population to whom the messages must reach. Mass media helps in a great way in this direction.

Economics – IT can help to reduce cost of communication.

Table: ICT Profile of India

Total population	1.05 billions
Literacy level (natural languages)	52 %
Literacy level (English)	5 %
Computer ownership per 1000 inhabitants	0.6
Telephone density	7
People connected to Internet	0.42 %
Number of websites in national languages	20,000
Number of websites in English	1,30,000

Source: Asian-Pacific Development Information Programme, UNDP, Bangkok.

Most commonly IT tools available for transfer of technology in Indian agriculture are radio, television, video, telephone, computer and internet.

Radio

Radio is one of the oldest IT tool started broadcasting in India in 1927, rural broadcasting started in the year 1935. Full fledged farm unit was established in 1966 for broadcasting programmes for farmers and farm women regularly. All India Radio has more than 200 radio stations, the network of 332 with 149 medium wave and 128 FM transmitters provides Radio coverage to a population of 98.82% spread over 89.51% area of the country. The stations of All India Radio broadcast various farm and home programmes directed to rural audience. There are more rural radios than in urban areas. Radio programmes are helpful in

- Creating awareness among the people.
- Help in changing their attitude.
- Motivating farmers by highlighting the achievements of other farmers.
- Enhances credibility among the people.
- Reinforce learning.

Community radio has in recent times, however, started finding roots with several initiatives such as VOICES, AID, MANA Radio, Deccan Development Society and the Kutch Mahila Vikas Sangathan (Kutch Women's Progress Cooperative). Most of these initiatives are partnerships with the State's All India Radio.

Television

Television is one of the powerful media close to the rural masses. In transfer of technology process TV plays an important role especially at awareness and interest stages of adoption. Message through it is helpful in motivating, stimulating, inducing and also in changing attitude of farmers. First ever, telecast originated from the studio in Akashwani Bhawan, New Delhi on 15th of Sep, 1959. A regular service with news bulletin started in 1965. TV programme especially for agriculture extension started in year 1975-76.

TV Viewers by Regions in India (1998-99)(Million)

States/UTs	Home Viewers			Other Viewers	Total Viewers
	Urban	Rural	Total		
Delhi	9.8	-	9.8	1.4	11.2
Uttar Pradesh	15.3	21	36.3	15.7	52
Rajasthan	6.6	7.7	14.3	5.5	19.8
Punjab	4.2	9.1	13.3	3.4	16.7
Haryana	2.8	7.4	10.2	2.2	12.4
Himachal Pradesh	0.8	2.9	3.7	1.6	5.3
Jammu & Kashmir	1.7	7.4	3.1	1.2	4.3
North	41.2	49.5	90.7	31	121.7
Maharashtra	28.2	16.7	45	15.8	60.8
Gujarat	13.1	9.3	22.4	6	28.4
Madhya Pradesh	10.9	12.5	23.4	11.6	35
West	52.3	38.5	90.8	33.4	124.2
Tamil Nadu	20.1	11.9	32	5.4	37.4
Andhra Pradesh	19.5	14.2	33.7	9.9	43.6
Karnataka	14.3	9.6	23.9	4.9	28.8
Kerala	6.3	9.7	16	4.5	20.5
South	60.2	45.4	105.6	24.7	130.3
West Bengal	21.1	11.3	32.4	7.4	39.8
Orissa	4.3	6.3	10.7	5.5	16.1
Bihar	7.9	12.3	20.2	11.1	31.3
North East Zone	4.4	7.4	11.8	4.3	161.1
East Zone & North East Zone	37.7	37.3	75	28.3	103.3
India	191.4	170.7	362.1	117.4	

Source : http\\www.indiastat.com

Telephone

Telephone has also been a popular IT tool for the farmers and farm women. This powerful electronic machine that was a farmers dream earlier has become reality, as the farmers can immediately make use of it to address their farm related problems. Call centres for the farmers have been established known as Kissan

Call Centres. These call centres are operational in 08 selected locations covering all the states. Farmers can call anytime on the toll free number 1551 to get technical guidance from the agriculture experts.

According to the Telecom Regulatory Authority of India (TRAI), at the end of April 2009, the total number of telephone connections reached 441.47 million. With this growth, the overall tele-density reached 37.94 at the end of April 2009.

According to Business Monitor International, India is currently adding 8-10 million mobile subscribers every month. It is estimated that by mid 2012, around half the country's population will own a mobile phone. This would translate into 612 million mobile subscribers, accounting for a tele-density of around 51 per cent by 2012.

The Internet

The Internet is a global system of interconnected computer networks that interchange data by packet switching using the standardized Internet Protocol Suite (TCP/IP). It is a "network of networks" that consists of millions of private and public, academic, business, and government networks of local to global scope that are linked by copper wires, fiber-optic cables, wireless connections, and other technologies. The Internet carries various information resources and services, such as electronic mail, online chat, file transfer and file sharing, online gaming, and the inter-linked hypertext documents and other resources of the World Wide Web (WWW).

The increase in internet users during past few years can be assessed from table below. The table reveals that from 1998 to 2006, the percentage increase has reached from 0.1% in 1998 to 3.7% in 2007. The percentage has been taken between users and population i.e. what percentage of population uses the internet.

Table: Internet Usage and Population Statistics

YEAR	Users	Population	% Pen.	Usage Source
1998	1,400,000	1,094,870,677	0.1 %	ITU
1999	2,800,000	1,094,870,677	0.3 %	ITU
2000	5,500,000	1,094,870,677	0.5 %	ITU
2001	7,000,000	1,094,870,677	0.7 %	ITU
2002	16,500,000	1,094,870,677	1.6 %	ITU
2003	22,500,000	1,094,870,677	2.1 %	ITU
2004	39,200,000	1,094,870,677	3.6 %	**C.I. Almanac**
2005	50,600,000	1,112,225,812	4.5 %	C.I. Almanac
2006	40,000,000	1,112,225,812	3.6 %	IAMAI
2007	42,000,000	1,129,667,528	3.7 %	IWS

CI-computer industry almanac, IWS- information warfare site, ITU-International Telecom Union, IAMAI-Internet and Mobile Association of India

As per India online internet usage survey 2007, only 37% of Indian internet users come from top 10 cities like Delhi, Mumbai, Kolkata, Hyderabad etc. which signifies that not only urban people use internet but rural people are equally interested in internet and are being benefitted by it. Some of the facts and figures about internet facility and its usage can be imagined from the table below.

Number of internet subscribers in India, end 2005	6.13 million [ID; 2006]
Internet subscriber to user ratio	1:10
No. of broadband connections	2 million (Dec. 2006)
Number of people using internet	70 million (As per estimate of NASSCOM) 39 million (As per Google India director)
Internet penetration in Mar 2006 [internet users not necessarily subscribers]	4.5% of Indian population

Source: NASSCOM-National Association of Software and Service Companies

Video-Conference

Video is also one of the important IT tool for transfer of technology. Experiments with small format videotape have aroused the interest of farmers in developing countries. Video tapes are seen as an ideal medium to promote motivation, attitude change, behavior reinforcement, community participation and entertainment. Videotapes have tremendous utility for imparting training interventions.

Similarly, videoconference (also known as a *videoteleconference*) is a set of interactive telecommunication technologies which allow two or more persons at different locations to interact via two-way video and audio transmissions simultaneously. It has also been called visual collaboration and is a type of groupware. This can be as simple as a conversation between two people at different sites. Video-conferencing is playing a vital role in transfer of technology to farm sector.

Videoconferencing provides students with the opportunity to learn by participating in a 2-way communication platform. Furthermore, teachers and lecturers from all over the world can be brought to classes in remote or otherwise isolated places. Students from diverse communities and backgrounds can come together to learn about one another. Students are able to explore, communicate, analyze and share information and ideas with one another. Through videoconferencing students can visit another part of the world to speak with others, visit a zoo, a museum and so on, to learn. In a nutshell, videoconferencing facilitates:

- Motivation
- Attitude change and behavior reinforcement
- Timeliness information and feedback

In agriculture, the process of making available the fruits of research is the function of information disseminators with other equal partners. These are information stakeholder (research organizations), information providers and finally information seekers. The role and relevance of these partners from information generation to dissemination to the concerned through super highway (ICT).

Information stakeholders: First partner of the system is the stakeholder of information. Concerned ministries and individuals who have information to be disseminated for growth, development and governance of people. For example, agriculture research organizations, ministry of agriculture and cooperation, state governments, NGO's, Directorate of Extension etc. The information generated through research, government departments, information related with government acts and regulations, technologies created from time to time from proper governance and well being of people need to be disseminated. These stakeholders need that this information may reach to the concerned people without any loss of time so that, it could be implemented and adopted.

Information infrastructure providers: Next partner of the information is the information provider. They are generally the ministry of information and broadcasting through unbound transmission media like Radio, TV transmitters, Ministry of telecommunication through transmission media like Fiber optics technology, Telephone, Satellite network and internet connectivity and ministry of power. This is a highway of disseminating the information globally. Their role is to develop infrastructure in such a way so that information generated digitized and stored by stakeholders can be disseminated easily even in the remote corners of the country.

Information processors and disseminators: Next partner of the system is the information processor and disseminator. These are the IT professionals, professionals of extension system, KVK, ATMA centres and service providers. Their role is not only to process and digitize the information available but also to assess the need of information seekers.

Information seekers: The final partner of the system is information seekers and users. They may be farmers, agencies, organizations etc. It has been observed that, availability of such infrastructure is mainly concentrated in urban areas but not in remote areas. The information needs of these people are varying depending upon their economic conditions, occupation, topography, education, social and political participation. Though, these facilities are available upto block level yet they are helpless because they live in such areas where these infrastructures are not available.

ICT enhancements to agricultural programs – the four "C's"

***Connectivity* and access**

- Providing or enhancing rural connectivity through private or public telecenters and ISP (internet service provider) promotion
- Developing inexpensive access alternatives to traditional PCs, such as cell phones, pagers, and PDAs (Personal digital Assistant)
- Reducing the overall cost of information access in rural areas
- Ensuring access to key intermediaries and stakeholders (e.g.,women, extension agents, community centers, etc.)

***Capacity building* through social networks and training**

- Developing ICT training activities for key players such as agricultural policy makers, planners, researchers, extension agents, or other intermediaries such as community radio stations and NGOs
- Using ICTs to enable or strengthen distance education and/or teacher support for agricultural education and rural communities
- Creating virtual communities for both non-specialists and specialists
- Linking productive value chains through web portals and ICTs

***Content* and application development**

- Creating and stocking online digital libraries with new or repackaged digital material targeted at small farmers and community intermediaries
- Producing new or adapted agriculture decision support tools targeted at agricultural users, particularly women, small farmers, and local intermediaries

***Conducive* governance and policy**

- Policy advocacy, especially in the telecommunications sector, but also in commerce, agriculture, culture, and privacy law
- Creating investment funds and incubators for rural and agriculturally oriented ICT ventures

Some of the potential applications of ICTs in agricultural extensions are as follows:

1. Quick access to farm technological information with cost effectiveness.
2. Access to farm technologies, which are available on millions of computers spread all over the world.
3. Quick and cost effective communication of information all over the world.
4. Effective decision making related to farm production by the farmers with the help of timely and up to date information.
5. Better prices and profit for farm communities through electronic commerce.
6. Providing information to help the marginalized farmers and farm women.

7. Some of the specific applications of ICTs in agricultural extension in the present scenario will be:
   a) Interactive information network catering to extension worker, students, rural youth, entrepreneurs and farmers.
   b) Virtual academy for extension workers and farmers to provide education on different aspects of agriculture and rural development.
   c) Distance learning programmes and networks for extension personnel, students, farm youths and farmers.
   d) Action network on important areas to solve specific problems.
   e) Provision of information on weather, pests, diseases, price and market situation.
   f) Network to develop small agro-based enterprises.
   g) Indigenous knowledge and agro-ecosystem network.

The use of ICTs in extension provides several key strengths in relation to traditional media. However, ICT also comes with some range of weaknesses. A brokering role for agriculture extension could help balance strengths and weaknesses. Potential strengths and weaknesses are:

Key strengths

- a new range of additional media that can be part of the communication for development "mix" of traditional and/or appropriate media;
- where accessible, these new media have features that enable bottom-up articulation and sharing of information on needs and local knowledge;
- can increase efficiency in use of development resources because information is more widely accessible;
- can result in less duplication of activities because information is more widely accessible
- they tend to reduce communication costs (often dramatically) in comparison to other available communication choices;
- they provide global access to information and human resources; and
- rapid speed of communication - locally, nationally and globally.

Key weaknesses

- can lead to technological dependence;
- capital cost of technologies, and the cost of on-going access and support can be high;
- there is an inherent need for capacity building;
- lack of accessible telecommunication infrastructure in many rural and remote areas

- ICT projects sometimes lack to integrate with existing media, and local communication methods and traditions.

Information and communication technology has the potential and opportunities to help rural people to leap frog some of the traditional barriers to development by improving access to information, expanding their market base and enhancing employment opportunities. For the development in agriculture sector lot of public and private sector initiatives in ICT network has been taken.

Public Sector Initiatives

National informatics centre (NIC) under ministry of communication and information technology is acting as nodal service to look after information development and networking in government, corporate and cooperative sectors for decision support. Major initiatives taken for informatics network are:

- **AGRISNET (Agriculture Resource Information System Network):** Maintains a portal as a prime resource for information on various agricultural markets across India.
- **ICT based village knowledge centre:** Being taken as a mission in each of the villages in the country. At present, most of the state governments have set up IT departments and IT parks.
- **AGMARKNET (Agriculture Marketing Information Network):** Facilitating generation and transmission of prices and arrival of information from agricultural produce markets. About 7000 wholesale and 35000 rural markets and the web based dissemination of this information.
- **DACNET (Department of Agriculture Network):** Is an e-governance project executed by NIC to facilitate dissemination and exchange of agricultural information.
- **CIC (Community Information Centre):** A joint effort by department of IT, NIC and state governments for establishments of ICT facilities in all the blocks of the nine states in particular namely, Arunachal Pradesh, Assam, Manipur, Nagaland, Mizoram, Meghalaya, Sikkim, Tripura and J&K. The beneficiaries are the people of local communities at block level. CICs enables access to database containing wide range of information, employment opportunities, agriculture marketing. They provide a forum for rural youths operating from far flung areas to come to gain skill in computer education and to exchange knowledge and information.
- In the state of Jammu & Kashmir, the CICs were established in the year 2004. These CICs enable the access to database containing information on agriculture marketing, weather information, employment opportunities etc. A study was conducted to access the role of CICs in agriculture and rural development in Jammu region of J&K. It was found that CICs helps to provide agri related information to the farmers i.e. market access to rural people prefer-

ably to potential farmers to the tune of 46 MPS, promotion of entrepreneurship development in agri based ventures 25 MPS from a pooled sample of respondents. Besides, these centers create IT awareness for rural youths, provide e-employment notification to rural youths.

- **ARISNET (Agriculture Research Information Systems Network):** For the purpose of planning and management of agricultural research and scientific communication by ICAR.

Private Sector Initiatives

Information Village Research Project

The project was developed by Prof. M.S. Swaminathan research foundation as a part of its programme of taking the benefits of emerging and frontier technologies to the rural poor. Modern communication and information technologies were found to have great potential to contribute in this respect. An international, interdisciplinary dialogue organized by the foundation 1992, conducted an analysis of a range of issues involved. The dialogue participants concluded that ICTs would have a major role in promoting sustainable agriculture and rural development in developing world. The foundations approach to dissemination of new technologies in rural areas is premised on the statement of its founder, Prof. M.S. Swaminathan -"Whatever a poor family can get benefit from, the rich can also get benefit; the reverse does not happen". Thus involvement of ultra poor in rural areas in managing the use of ICTs was considered essential for success of this project. Other critical issue was the need to involve the rural women. Project was started in south India in 1998 with the objectives:

- Setting up of village information shops that enabled rural families access of basket of modern information and communication technologies, training educated youth especially women in rural areas in operating information shops.
- Training the rural youth in the organization and maintenance of the system.
- Maintenance, updating and dissemination of information on entitlements in rural families using an appropriate blend of modern and existing channels of communication.

ITC E-Chopal

ITC limited is a company whose business depends upon agriculture raw material. About, 5200 e-chopal kiosks cover about 31000 villages in 6 states of India viz. M.P., U.P., Maharashtra, Rajasthan, Karnataka and A.P. - reaching more than 3 million farmers. ITC's vision is to cover 1,00,000 villages by 2010 serving more than 10 million farmers. The philosophy of ITC is "sustainable shareholder value through serving society". ITC e-choupal introduced the farmers to World Wide Web (WWW) and available information sources on internet. The site and data uplink are maintained by ITC infotech, India, Ltd. The e-choupal site provide links to important information and knowledge resources related to farming and other developmental opportunities like weather information, based on which the farmers could decide agriculture operations. E-choupals are village internet kiosks managed by Sanchalaks (Coordinators). It has facilitated information decision making among the farmers by providing real-time information in local language and customized knowledge.

DRISHTI: A multiple rural service provider through information kiosks owned by local villagers operating in the states of Assam, Tamil Nadu, Chattisgarh, UP, Bihar, Orrisa and Haryana.

TATA Kisan Kendra: Operating in the states of UP, Punjab and Haryana for agriculture services free of cost.

Parry's Corners Portal: covering more than 150 villages for advisory services to sugarcane growers.

Gyandoot: The Gyandoot project began in Nov. 1999 and is serving multi-lakhs of people. Important information of the population such as land ownership and records, income, domicile and loan record is computerized.

Samalkya Agritech: In Andhra Pradesh, a private sector initiative using cooperative structure with financial visibility. Farmers get technical know-how regarding the recent advances in agriculture.

SARI: Project through telephone and internet in the villages of Madurai in Tamil Nadu.

N-Louge: Having project sites in the states of Tamil Nadu, Andhra Pradesh, Karnataka, Madhya Pradesh, Gujarat and Maharashtra.

Honey Bee network by SHRISHTI: Society for research initiatives for sustainable technologies and institutions, a project from people to people networking.

Sewa Gujarat: Self employed women association, Gujarat has initiated for development and strengthening self-help groups in Gujarat.

Lok Mitra: This project in Himachal Pradesh envisages setting up of district-wise INTRANET connecting information booths in rural areas.

Computers on wheel: Pingali Rajeshwari, the entrepreneur in AP provides internet on motor bikes and helps the resource poor in remote areas in harnessing the benefits of ICT for information related to market, weather, plant protection, animal health, etc.

Jai Kisan: An NGO in Uttranchal has set up kisan suchana Kendra across the state and promotes IT technology in rural areas.

Initiatives in other development countries.

- Electronic delivery of agricultural information to rural communities in Uganda.
- Potato Extension and Training Information System (PETIS) in Mauritius.
- Farmers Information Technology Services (FITS) in Phillipines.
- Information system pilot projects Cajamarca, Peru.
- Upper west commerce associates, Ghana.
- Coffee growers collectives, Guatemala.
- Grameen Telephone project in Bangladesh.
- Tele-centre project in Mexico.
- The Mango Information Network (MIN) of Phillipines to link researchers, extension workers and farmers.

The rural livelihoods approach emphasizes analysis of gender roles in the maintenance and enhancement of family livelihoods. For example, at an international scale, women are twice as likely as men involved in agricultural activities. Women have principal roles in small holding subsistence farming, agri-business, and food processing. In most developing countries, women's income is significant to the rural household, and women are often the heads of rural households, especially where remittance economies are strong.

With regard to ICTs, there is a clear gender bias. Women significantly lag behind men in their access to ICTs, (O'Farrell, 2003). ICTs and extension reflect the development interventions to incorporate gender:

- training that uses local resource people, such as women entrepreneurs who are interested in, and use, ICTs;
- opportunities to develop leadership capacity within farmers' organizations and rural women and youth groups by engaging in ICT policy debates;
- local development and testing of content oriented to women and their needs;
- training of agricultural workers government and non-government organizations in gender issues and their application to the use of ICTs in agricultural extension and rural development;
- networking with other communities, countries and regions to share information and develop strategic alliances for advocacy and action.

The effective use of ICT requires capacity building. IT plays a crucial role in operationalization and spreading extension reforms. IT strengthening of public system can help enabling the extensionists to gather, store, retrieve and disseminate a broad range of information needed by the farmers thus, transferring it from extension workers to knowledge workers to realize much talked about bottom-up, demand-driven technology generation, assessment, refinement and transfer. ICT can help in the extension system in such agricultural developments.

References

Anonymous (2003). Activities of USAID peers and other organizations, Future direction in agriculture and ICTs at USAID.

Bhal, R. (2005). Dissemination of agriculture information through communication super highway.

Chapmen, R and Slaymaker, T. (2002). ICTs, and rural development. Renew of literature, current intervention and opportunities for action, working paper 192, ODI, London.

http:// www. ibef.org/industry/telecommunications. aspx

K, Vijay Ragavan (2005). ICTs and agriculture Extension; opportunities and challenges. National seminar on green to evergreen; challenges to extension education, IARI, New Delhi. 15-17, 2005, pp 132-39

Lead paper presented in superhighway in National seminar on green to evergreen; challenges to extension education, IARI, New Delhi. 15-17, 2005, pp 121-31.

Michiels', S.I, and Crowder Van, L. (2001). Discovering the magic box; Local appropriate of information and communication technologies (ICTs). SDRE, FAO, Rome.

Munyua, H (2000). Information and communication technologies for rural development and food security, lesson from field experiences in developing countries, SD Dimensions, Nov. 2000, FAO, Rome.

O' Farrell, C. (2003). Gender and Agriculture in the Information Society. Presentation to CTA's 6th Consultative Expert Meeting of its Observatory on ICTs. CTA, Netherlands. http:// www. cta. int/observatory 2003/ ppt_presentations/Gender.pdf

Richardson, D. 2005. ICTs Transforming Agricultural Extension - Summary Report of CTA's 6th Consultative Expert Meeting of its Observatory on ICTs. CTA, Netherlands.

Sarkar, J. (2004). Information and Communication Technology; Internet paper presented in National workshop on communication support for sustaining extension services, BHU, Varanasi

Singh Baldeo, Padaria, R.N. (2006). Revitalizing extension system and ICT utilization in Technology transfer. Lead paper presented in National seminar of Indian Society of Extension Education on ICT; opportunities and challenges for revitalizing extension system at Navsari, Gujarat. 27-29, Dec. 2006, pp 11-22.

Slathia, P.S., Bhagat, G.R, Kher, S.K. (2006). Community Information Centres: Paradigm for rural development in J&k State. Paper presented in National seminar of Indian Society of Extension Education on ICT; opportunities and challenges for revitalizing extension system at Navsari, Gujarat. 27-29, Dec. 2006, pp -96.

Slathia, P.S., Paul, Narinder, Bhagat, G.R. (2006). Opinion of information technology professionals towards community information centres in J & K. Rajasthan Journal of Extension Education, Udaipur, No. 14,

Telecom Regulatory Authority of India, Information note to the Press (Press Release No. 61 / 2007), 20 Jun 2007.

Veerabhadraiah, V. (2006). Information and communication technology for revitalizing extension system. Lead paper presented in National seminar of Indian Society of Extension Education on ICT; opportunities and challenges for revitalizing extension system at Navsari, Gujarat. 27-29, Dec. 2006.

Index